Metal-Oxo and Metal-Peroxo Species in Catalytic Oxidations

Volume Editor: B. Meunier

With contributions by
W. Adam, F. Banse, J. Bernadou, A.G. Blackman, R.M. Burger, R.H. Crabtree, H. Fujii, J.-J. Girerd, W.A. Herrmann, F.E. Kühn, B. Meunier, C.M. Mitchell, C.R. Saha-Möller, D. Schröder, H. Schwarz, S. Shaik, P.E.M. Siegbahn, A.J. Simaan, W.B. Tolman, J.S. Valentine, Y. Watanabe, O. Weichhold, D.L. Wertz

Springer

The series *Structure and Bonding* publishes critical reviews on topics of research concerned with chemical structure and bonding. The scope of the series spans the entire Periodic Table. It focuses attention on new and developing areas of modern structural and theoretical chemistry such as nanostructures, molecular electronics, designed molecular solids, surfaces, metal clusters and supramolecular structures. Physical and spectroscopic techniques used to determine, examine and model structures fall within the purview of Structure and Bonding to the extent that the focus is on the scientific results obtained and not on specialist information concerning the techniques themselves. Issues associated with the development of bonding models and generalizations that illuminate the reactivity pathways and rates of chemical processes are also relevant.

As a rule, contributions are specially commissioned. The editors and publishers will, however, always be pleased to receive suggestions and supplementary information. Papers are accepted for *Structure and Bonding* in English.

In references *Structure and Bonding* is abbreviated *Struct.Bond.*
and is cited as a journal.

Springer WWW home page: http://www.springer.de

ISSN 0081-5993
ISBN 3-540-66943-4
Springer-Verlag Berlin Heidelberg New York

CIP Data applied for

Typesetting: Scientific Publishing Services (P) Ltd, Madras
Production editor: Christiane Messerschmidt, Rheinau
Cover: Medio V. Leins, Berlin
Printed on acid-free paper SPIN: 10702329 02/3020 – 5 4 3 2 1 0

97
Structure and Bonding

Springer
Berlin
Heidelberg
New York
Barcelona
Hong Kong
London
Milan
Paris
Singapore
Tokyo

Volume Editor

Professor Dr. Bernard Meunier

Laboratoire de Chimie de Coordination du CNRS
205 route de Narbonne
31077 Toulouse Cedex 4, France
E-mail: bmeunier@lcc-toulouse.fr

Editorial Board

Preface

Free radical oxidations of hydrocarbons in the liquid phase are still the only large-scale method for obtaining most of the oxygen-containing chemical raw materials. These ≪autooxidation reactions≫ involve the reaction of the triplet state of dioxygen with a carbon-centered radical as elementary step. For the last twenty years, many research groups have worked in the field of transition-metal catalyzed oxidations having in mind two main challenges: (i) the understanding of the mechanism of oxidation reactions catalyzed by mono- and di-oxygenases at the molecular level (how to hydroxylate an aliphatic C—H bond at room temperature with retention of configuration?) and (ii) the use of the knowledge accumulated on oxygenases with biomimetic models to create new metal-catalyzed oxidation methods as selective as possible, without free radical intermediates.

Probably more than 2000 articles have been published on metal-catalyzed oxidations at ambient temperature between 1980 and 1999. The present volume is not an attempt to produce exhaustive reviews on all these catalytic oxidations performed in mild conditions (the so-called biomimetic or bioinspired oxidations), but is a selection of different contributions to describe the current knowledge on the different metal-oxo and metal-peroxo species which are involved in catalytic oxidations. These highly reactivity metal-oxygen species are the key intermediates in heme- or non-heme oxygenases, as well as in metal-catalyzed oxidations. The accumulated knowledge on these reactive entities will facilitate a better understanding of the different intermediates involved in the catalytic cycles of enzymes like cytochromes P450, non-heme oxygenases, copper oxygenases, in the oxidation of water by the manganese-center of Phyotosystem II, in the chemistry of dioxiranes and also in DNA cleavage performed by activated bleomycin. The description of these metal-oxo and metal-peroxo species is based on physico-chemical data, reactivity results and also on theoretical calculations.

The titles of the ten different contributions are as follows:
1. Active iron-oxo and iron-peroxo species in cytochrome P-450 and peroxidases; oxo-hydroxo tautomerism with water-soluble metalloporphyrins (Bernard Meunier and Jean Bernadou), **2.** Nucleophilicity of iron-peroxo porphyrin complexes. (Joan S. Valentine and Diana Wertz), **3.** Characterization of high-valent oxo-metalloporphyrins. (Yoshihito Watanabe and Hiroshi Fujii), **4.** Characterization, orbital description, and reactivity patterns of transition-metal oxo species in the gas phase (Detfel Schröder, Helmut

Schwarz and Sason Shaik), **5.** Theoritical aspects on high-valent metal-oxo or -peroxo species in methane monooxygenases and the manganese-center in Photosystem II. (P.E.M. Siegbahn and R.C. Crabtree), **6.** Characterization and properties of non-heme iron-peroxo complexes (Jean-Jacques Girerd, Frédéric Banse and Ariane J. Simaan), **7.** Copper-dioxygen and copper-oxo species relevant to copper oxygenases and oxidases (William B. Tolman and Allan Blackman), **8.** Rhenium-oxo and rhenium-peroxo complexes in catalytic oxidations (Fritz E. Kühn and Wolfgang A. Herrmann), **9.** Structure, reactivity and selectivity of metal-peroxo complexes versus dioxiranes (Waldemar Adam, Catherine M. Mitchell, Chantu R. Saha-Moeller and Olivier Weichold) and **10.** Nature of Activated Bleomycin (Richard M. Burger).

We hope that these different contributions on metal-oxo and metal-peroxo species will help the readers in their efforts to better understand the mechanistic aspects of metal-catalyzed oxidations. This volume illustrates new directions for chemists engaged in the discovery of new efficient and selective oxidation catalysts, molecular biologists interested by enzyme-catalyzed oxidations and pharmacologists working on the development of potential antitumoral or antiviral drugs able to degrade the nuclei acids of tumor cells or viruses.

As Editor of this volume, I am deeply grateful to the authors of the different contributions. They all agreed with enthusiasm to participate in the preparation of the present Volume despite the pressure which is on the shoulders of talented researchers.

Toulouse, March 2000 Bernard Meunier

Contents

Contents of Volume 93

Bonding and Charge Distribution in Polyoxometalates: A Bond Valence Approach

Volume Editor: D.M.P. Mingos

Active Iron-Oxo and Iron-Peroxo Species in Cytochromes P450 and Peroxidases; Oxo-Hydroxo Tautomerism with Water-Soluble Metalloporphyrins

Bernard Meunier[1], Jean Bernadou[2]

Laboratoire de Chimie de Coordination du CNRS, 205 route de Narbonne, 31077 Toulouse cedex 4, France

[1] *E-mail: bmeunier@lcc-toulouse.fr*

[2] *E-mail: bernadou@lcc-toulouse.fr*

Heme-containing monooxygenases are able to catalyze two different classes of oxidation reactions. The first class includes oxygenation reactions (hydroxylation, epoxidation, *N*- or *S*-oxide formation, etc.) which are mediated by an electrophilic oxidative species. The second class is represented by the oxidative deformylation of aldhehydes and involves a nucleophilic oxidant as active intermediate. The reductive activation of molecular oxygen by cytochromes P450 generates a nucleophilic iron(III)-peroxo species which produces by protonation an electrophilic high-valent iron-oxo [formally an iron(V)oxo] responsible for electrophilic oxygen atom transfers. The nucleophilic properties of the iron(III)-peroxo intermediate in cytochrome P450 are due to the porphyrin ring acting as electron reservoir and also to the negative charge accumulated on the proximal cysteine during the initial reduction step of the catalytic cycle. The nature of the high-valent iron-oxo species generated in the catalytic cycle of heme-peroxidases will be also discussed. Among the different methods for studying the oxygenation reactions mediated by high-valent metal-oxo porphyrin complexes, the recent discovery of the "oxo-hydroxo tautomerism" provides a useful tool to investigate the mechanism of O-atom transfer reactions in aqueous media.

Keywords: Metal-oxo, Metal-peroxo, Electrophilic, Nucleophilic, Cytochrome, Peroxidase, Oxo-hydroxo tautomerism, Metalloporphyrin

List of Abbreviations

CBZ	carbamazepine
Cpd I or II	Compound I or II of a heme-peroxidase, an iron(IV)-oxo radical-cation species, or an iron(IV)-oxo species, respectively
CPO	chloroperoxidase
DFT	density functional theory
ee	enantiomeric excess
$F_{20}TPP$	dianion of the *meso*-tetrakis(pentafluorophenyl)porphyrin ligand
KIE	kinetic isotope effect
HRP	horseradish peroxidase
L	a neutral ligand (e.g., pyridine) providing two electrons to a metal center via a donor-acceptor bond
m-CPBA	*meta*-chloroperobenzoic acid

5-MF	5-methylene-2-furanone
NADH	(NADPH) reduced form of nicotinamide adenine dinucleotide (phosphate)
PPIX	protoporphyrin-IX
TDCPPS	dianion of the *meso*-tetrakis(2,6-dichloro-3-sulfonatophenyl)porphyrin ligand
TF_4TMAP	dianion of *meso*-tetrakis(2,3,5,6-tetrafluoro-4-*N*,*N*,*N*-trimethylaniliniumyl)porphyrin ligand
TMP	the dianion of *meso*-tetramesitylporphyrin ligand
TMPyP	the dianion of *meso*-tetrakis(4-methylpyridiniumyl)porphyrin ligand
X	an anionic ligand (halide, deprotonated cysteine RS^-, superoxide anion $O_2^{\bullet -}$, monoanion of hydrogen peroxide HOO^-, . . . etc.)

1 Introduction

The cytochromes P450 constitute a large family of cysteinato-heme enzymes (over 500 members) present in all forms of life (plants, bacteria, mammals) and they play a key role in the oxidative transformation of endogeneous and exogenous molecules [1–3]. These enzymes are monooxygenases able to catalyze the insertion of one oxygen atom of dioxygen within many different substrates, the second oxygen atom of O_2 being reduced to a water molecule. The two electrons of this reductive activation of molecular oxygen are provided by NAD(P)H via a reductase. These enzymes are located in the membrane of the endoplasmic reticulum and are able to perform various difficult oxygenation reactions. Cytochromes P450 catalyze the hydroxylation of saturated carbon-hydrogen bonds (Eq. 1), the epoxidation of double bonds, the oxidation of heteroatoms, dealkylation reactions, oxidations of aromatics, ... etc.:

$$RH + O_2 + 2e^- + 2H^+ \xrightarrow{\text{Cytochrome P450}} ROH + H_2O \quad (1)$$

Being a triplet (two unpaired electrons in the ground state), molecular oxygen is unreactive towards organic molecules at low temperatures. The reaction of dioxygen with the single state of organic substrates is spin-forbidden [4]. Consequently, the oxygenation of organic molecules at physiological temperatures must involve the modification of the electronic structure of one of the partners. Living systems mainly use enzymes to modify the electronic structure of dioxygen to a form which is adapted for the desired oxidation reaction. This modification can be performed by metal-dependent oxygenases, like cytochromes P450 or non-heme metalloenzymes (e.g., methane monooxygenase), or by flavin-containing enzymes which do not possess metal-based prosthetic groups. Forty years after the isolation and the characterization of cytochromes P450 [5], the exact nature of the active species responsible for the oxygen insertion step is still a matter of debate. After an "iron-oxenoid" period during the 1970s [5] came the "high-valent iron-oxo" period, mainly based on elegant works using single oxygen atom donors with cytochrome P450 itself and with chemical models employing synthetic metalloporphyrins [3, 6–10]. More recently, the hypothesis of an electrophilic "iron(III)-hydroperoxo" appeared in the literature [11], partly inspired by oxidations catalyzed by non-heme iron complexes and DNA sugar oxidation performed by activated iron-bleomycin within tumor cells.

In Sect. 2 we will discuss the nature of the different species generated during the catalytic cycle of cytochromes P450 and heme-peroxidases taking into consideration the knowledge which has been accumulated on these enzymes themselves and from biomimetic oxidations over the past two decades [12]. In Sect. 3 we will then present a survey of recent observations of an "oxo-hydroxo tautomerism" that has been collected on oxygenation reactions catalyzed by water-soluble metalloporphyrin complexes [13]. This oxo-hydroxo tautomerism is a useful tool which can be employed in mechanistic studies of oxygen atom transfer performed by high-valent metal-oxo species in aqueous solutions.

2 Iron-Oxo and Iron-Peroxo Species in Cytochromes P450 and Peroxidases

Before entering into the details of the catalytic cycle of cytochromes P450, we will first report the essential structural features of this fascinating class of monooxygenases. Subsequently, the main steps of the catalytic cycle of heme-peroxidases will be presented.

2.1 Structural Aspects of Cytochromes P450 and Substrate Binding

The name cytochrome P450 arises from the fact that the reduced protein efficiently binds carbon monoxide with a strong absorption band (the Soret π–π^* band) of the PPIX-Fe(II)-CO chromophore at 450 nm. This spectroscopic property has been very useful since the early studies on cytochromes P450 to monitor the presence of this enzyme in the microsomal fraction of liver tissues [3]. Cytochrome P450 is no longer a black box. In 1986, Poulos et al. provided the first tri-dimensional structure of the cytochrome $P450_{cam}$ of *Pseudomonas putida* (Scheme 1) which catalyzes the stereospecific hydroxylation of the exo C5-H bond of camphor (Scheme 2) [14, 15].

This microorganism has the ability to use this terpene as its only source of carbon and energy, camphor being hydroxylated and then metabolized to isobutyrate and acetate.

The electrons which are necessary at different steps of the catalytic cycle are provided by NADH-coupled flavoprotein (putidaredoxin reductase) via an iron-sulfur protein (putidaredoxin). $P450_{cam}$ is a 45,000 Dalton polypeptide chain containing a single ferric protoporphyrin-IX and a cysteine, Cys-357, as axial ligand (a proximal cysteinato ligand is also present in chloroperoxidase).

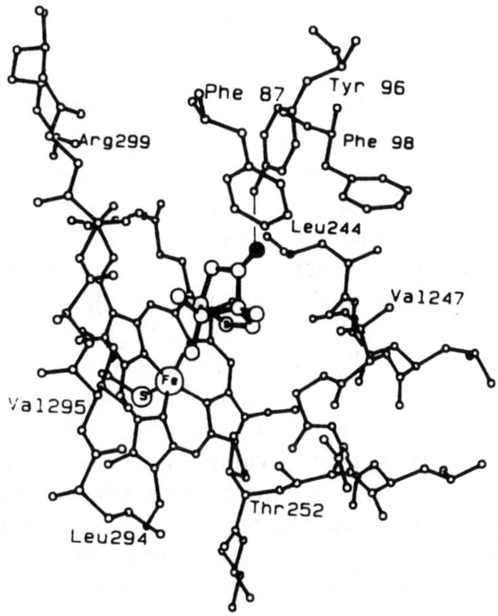

Scheme 1. Structure of the cytochrome $P450_{cam}$ of *Pseudomonas putida*. Reproduced with permission from [14]

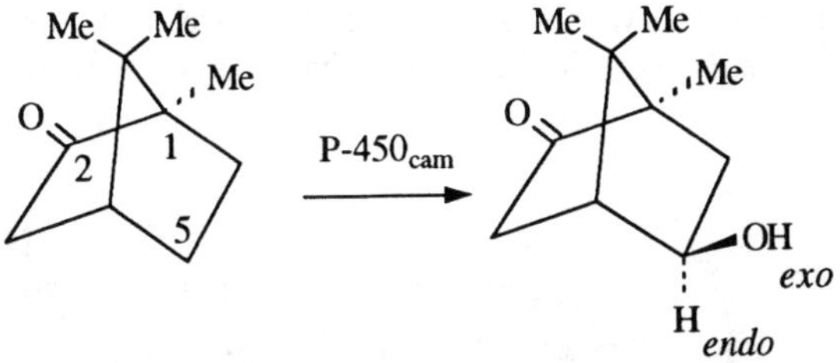

Scheme 2. Stereospecific hydroxylation of the exo C—H bond at position 5 of camphor by cytochrome $P450_{cam}$

As in many other heme proteins, the Fe^{III} state of the resting enzyme equilibrates between the low spin (S = 1/2) and the high spin state (S = 5/2), but the low spin state is favored in the absence of substrate and the sixth position of the octahedron is occupied by a water molecule (an aqua ligand). In the presence of camphor, the spin equilibrium of the iron(III) center is shifted toward the high spin form and the axial water molecule is removed, as well as the other water molecules located within the rather hydrophobic pocket of the active site of this monooxygenase. This modification of the coordination sphere of the metal center induces a change in the redox potential of the iron center, shifting from −300 to −170 mV after substrate binding, which facilitates the reduction of the pentacoordinated ferric center by the reductase in the next step of the catalytic cycle.

The kinetic and thermodynamic parameters of the reversible binding of camphor in cytochrome $P450_{cam}$ have been well studied by Griffin and Peterson [16]. The association of camphor is a second order reaction with a rate constant of 4.1×10^6 M^{-1} s^{-1}, while the dissociation is a slow first order reaction with a rate constant of 6.0 s^{-1}. The camphor binding corresponds to an entropy change (ΔS) of 26 cal mol^{-1} K^{-1} and a free-energy change (ΔG) of −7.7 kcal mol^{-1} at 21 °C. The substrate binding is an entropy-driven process, as the favorable entropy change associated with the removal of water molecules from the hydrophobic active site by camphor binding represents the major component of the overall binding energy [16].

The catalytic cycle of cytochrome P450 is triggered by the entry of the substrate into the active site displacing the axial water molecule (Scheme 3). Consequently the iron is more displaced from the plane of the porphyrin ring: 0.44 Å compared to 0.30 Å in the resting state of the enzyme. In addition, Tyr-96 which binds with the keto group of camphor acts as a probe for the polarity changes within the active site of the enzyme when camphor binds. The sensitivity of Tyr-96 to environmental polarity favors the access of bulk water molecules in the product-enzyme complex formed with 5-*exo*-hydroxycamphor, thus facilitating the product release from the enzyme [17].

X-ray diffraction studies on $P450_{cam}$ crystallized with and without the camphor substrate show the existence of a hydrogen bond interaction between the 2-keto group of camphor and Tyr-96 in addition to weak enzyme hydrophobic interactions with Phe-87, Leu-244, Val-247, Thr-252, and Val-295. All these different low-energy interactions are sufficient to dictate the exact fitting of the substrate inside the active site, namely putting the 5-exo

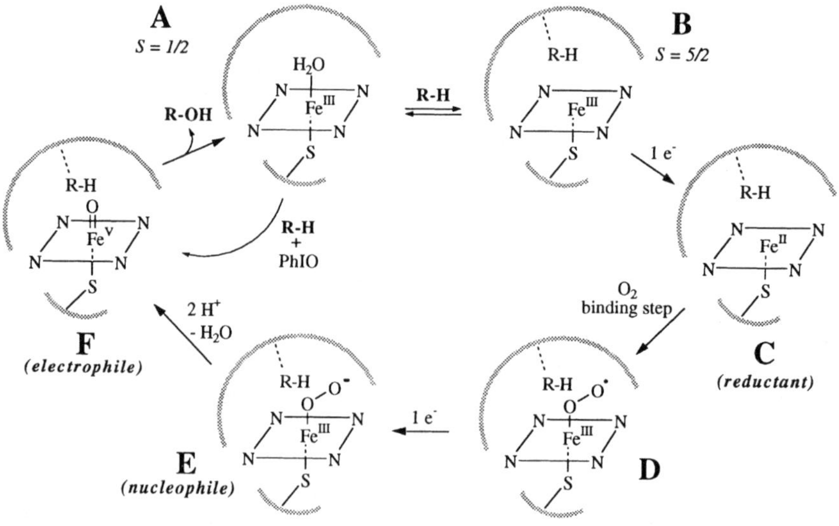

Scheme 3. Schematic representation of the different possible intermediates in the catalytic cycle of cytochrome P450

hydrogen atom close to the putative active iron(V)-oxo species, then making possible the specific hydroxylation at the 5-exo position. This specificity is modified when using analogs of camphor [18].

The crystal structures of $P450_{terp}$ (an α-terpineol monooxygenase), $P450_{eryF}$ (a monooxygenase involved in the biosynthesis of erythromycin), and $P450_{BM3}$ (a fatty acid hydroxylase) have been determined [18, 19]. In all these bacterial enzymes, a threonine residue is present in the active site to deliver protons to the iron-dioxygen intermediate during catalysis. Recently, a structure of $P450_{cam}$ modified with two electroactive ferrocenyl groups has been solved [20].

2.2 Catalytic Cycle of Cytochromes P450

2.2.1 *First Reduction Step*

After the substrate binding step discussed above (**A** → **B** in Scheme 3), the next step is the reduction of the iron(III) center to a ferrous state. A large positively charged area on the outside of P450s is involved in the interaction of the oxygenase with the reductase, suggesting that the contact between both partners is primarily electrostatic [19]. The rate of the first reduction step is relatively slow ($k = 35\ s^{-1}$). The reduced form of cytochrome P450 (intermediate **C** in Scheme 3) is an extremely efficient reducing agent. For example, polyhalogenated hydrocarbons (halothane, carbon tetrachloride, etc.) produce a stable iron(II)-carbene species via a reductive dehalogenation [3].

The nature of the sulfur-iron bond is probably modified in the ferrous intermediate **C** compared to the starting ferric derivative **A**. As a matter of fact, the sulfur-iron bond distance changes from 2.20 Å in the ferric $P450_{cam}$ (**A**, Scheme 3) to 2.35–2.41 Å in the CO-adduct of intermediate **C** with the ferrous ion only displaced by 0.02 Å from the porphyrin plane [21]. This important structural change has been discussed in terms of coordination chemistry and the possible modification of the nature of the sulfur-iron bond. Despite the high resolution of the X-ray structure, the authors preferred to attribute this distance modification to experimental errors rather than to a possible modification of the nature of the iron-sulfur bond. The reduction of **B**, a L_2X_3 ferric complex, would formally lead to a negatively charged L_2X_3 ferrous complex or to a L_3X_2 ferrous complex with a neutral cysteine ligand (RSH, i.e., an L ligand) instead of a cysteinato ligand (RS-, an X ligand). The protonation of the proximal cysteine would give rise to a cytochrome having a Soret band at 420 nm, excluding the hypothesis of a full protonation of the sulfur atom, but theoretical calculations [22] and data on models systems [23] demonstrate that a negative charge exists on intermediate **C** (this charge is delocalized on the whole prosthetic group, including the sulfur atom, and is certainly playing a key role in the heterolytic cleavage of the O—O bond of the ferric-hydroperoxo complex leading to the high-valent iron-oxo species. The cysteinyl sulfur atom is hydrogen bonded to the amide proton of three amino acids in $P450_{cam}$ – see [18, 24]). This charge is not explicitly indicated on intermediates **C** to **E** in Scheme 2, like in any other text-book on cytochrome P450, for simplification, but plays a key role in the electron density of the iron-dioxygen intermediate. The representation of this negative charge will be also discussed later (see Scheme 11).

The reduced ferrous cytochrome $P450_{cam}$ has a d^6 high spin state. The next step is the binding of dioxygen to the ferrous state of this metalloenzyme.

2.2.2
Oxygen Binding Step

Triplet dioxygen reacts with ferrous cytochrome $P450_{cam}$ with a second order rate constant of $1.7 \times 10^6\ M^{-1}\ s^{-1}$ to produce a stable dioxygen adduct ($K_{aff} = 10^6\ M^{-1}$) [2]. One electron from the iron center and one from triplet oxygen are pairing to create an iron(III)-oxygen bond. This oxygen-iron complex (compound **D** in Scheme 3) is relatively stable, but can dissociate to an iron(III) and superoxide anion with a rate constant of $0.01\ s^{-1}$ at room temperature. Within the enzyme, the release of superoxide is followed by its disproportionation and generates hydrogen peroxide, a source of harmful hydroxyl radicals (such a step is called a "decoupling reaction" in the vocabulary of P450 chemistry). So, the intermediate **D** can be regarded as an η^1-superoxide ion coordinated to an iron(III) center with an unpaired electron on the terminal oxygen atom. An O—O stretching vibration has been observed at $1141\ cm^{-1}$ in the resonance Raman spectrum of $P450_{cam}$ under catalytic conditions [25]. The Fe-O—O bending mode has been observed at $401\ cm^{-1}$ by resonance Raman spectroscopy with an angle of approximately 125–130° [26].

The electronic structure and the chemical properties of this ferrous-dioxygen intermediate of P450 enzymes are different from the analogue stable ferrous-dioxygen states of molecular oxygen carriers such as hemoglobin and myoglobin. In these heme-dioxygen carriers, the reduced ferrous state is neutral with the proximal position being occupied by a histidine nitrogen atom. The basicity of the proximal ligand increases the stability of the dioxygen-iron adduct [27]. In this respect, a proximal cysteine is not as good as a histidine ligand to stabilize a Fe^{II}-heme-dioxygen adduct.

2.2.3
Transfer of the Second Electron (Formation of a Nucleophilic Iron-Peroxo Species)

The second reduction step is the rate determining step in all different cytochrome P450s. This relatively slow step ($k = 17\ s^{-1}$ in cytochrome $P450_{cam}$) generates a negatively charged iron(III)-peroxo complex (intermediate **E** in Scheme 3) which is probably quickly protonated at this stage. This intermediate with a red-shifted band at 350–450 nm with a split Soret band has been observed by UV-visible spectroscopy in a D251N $P450_{cam}$ mutant (the aspartic residue at 251 being replaced by an asparagine) [28, 29]. The aspartic residue plays a key role in the kinetics of the proton transfer to this intermediate **E**. In the mutant Asp251Asn, the rate-determining step is not the reduction step, but the proton transfer to generate an Fe—OOH entity (this protonated form of complex **E** is not depicted in Scheme 3). The turnover rate for camphor hydroxylation of this mutant is 1–2 orders of magnitude slower than native $P450_{cam}$ [29].

This non-protonated iron(III)peroxo complex **E** is an obvious candidate to explain some particular reactions catalyzed by cytochrome P450s which obviously involve a nucleophilic oxidant intermediate. This is the case with both the oxidative decarbonylation of aldehydes and the final step of the aromatization of the cycle A of androstenedione in the biosynthesis of estrone. The intermediate **E** is certainly a stronger nucleophile than the corresponding protonated form Fe^{III}—OOH. However, both entities can be involved in these nucleophilic oxidations depending on the relative rates of the nucleophilic addition of the non-protonated iron-peroxo and of its protonation. If the proton transfer is faster than the nucleophilic attack, then only Fe^{III}—OOH is responsible for the nucleophilic oxidations observed with some substrates.

Aldehydes are usually oxidized to carboxylic acids by cytochrome P450s, but a second reaction pathway has been identified with some particular substrates. Cyclohexene is formed during the oxidation of cyclohexanecarboxaldehyde catalyzed by $P450_{2B4}$, a liver P450 induced by phenobarbital (Scheme 4) [30]. This reaction was not observed when iodosylbenzene was used as single oxygen atom source, indicating clearly that the high-valent iron-oxo intermediate was not involved in this reaction. The same oxidative deformylation has been observed with other aldehydes including citronellal [31]. The carbon atom of the aldehyde function is eliminated as formic acid. The initial nucleophilic attack of the iron-peroxo entity on the carbonyl function is probably followed by a concerted reaction involving 6-electron like in a Grob fragmentation (see Scheme 4 for a description of the reaction pathways).

A second classical example of the role of a nucleophilic iron-peroxo species in a P450 cycle is the third step of the aromatization reaction of the A-ring of androst-4-ene-3,17-dione to estrone catalyzed by human placental aromatase [32, 33]. This reaction involves three consecutive oxidative steps: two hydroxylations at the 19-methyl group and a final oxidative decarbonylation of the intermediate 19-aldehyde. All three reactions are stereospecific: the first

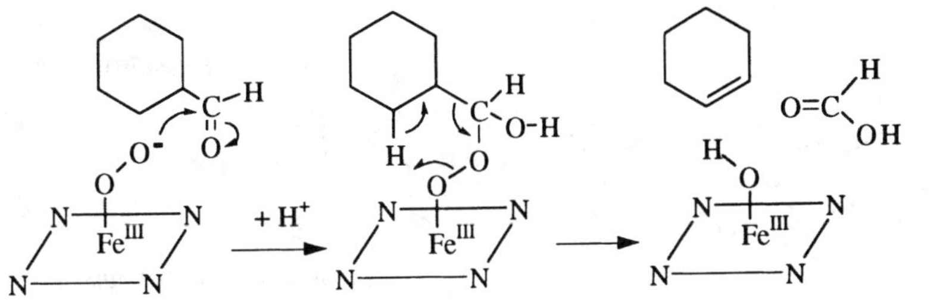

Scheme 4. Oxidative decarbonylation of cyclohexanecarboxaldehyde to cyclohexene and formic acid catalyzed by cytochrome P450

hydroxylation at the 19-methyl occurs with retention of configuration, the second removes the 19-pro-*R* hydrogen yielding a *gem*-diol which produces the 19-aldehyde derivative (see Scheme 5).

The final step is the nucleophilic attack of the iron-peroxo intermediate generated during the third catalytic cycle (the substrate remains effectively within the active site for three cycles). This decarbonylation reaction also involves the stereospecific removal of the 1β and the 2β hydrogens to produce the phenolic A-ring of estrone while the 19-methyl group is eliminated as formic acid.

This type of nucleophilic oxidation of an iron-peroxo onto a carbonyl group has also been evidenced in the final step of the formation of NO during the oxidative degradation of arginine catalyzed by NO synthases, a family of heme-enzymes able to produce nitric oxide in vivo [34]. However, it should be mentioned that the oxidation of heme to biliverdin by heme oxygenase might be due to an electrophilic iron(III)-peroxo species [35].

In general, the main reactions catalyzed by P450 enzymes are the incorporation of an oxygen atom into a substrate: hydroxylation of an aliphatic C—H bond, epoxidation of an olefin, oxidation of an heteroatom, etc. All these oxidations involve an electrophilic metal-oxo species as we will see below.

2.2.4
First Protonation Step (Formation of a Nucleophilic Iron-Hydroperoxo Species)

Threonine-252 in $P450_{cam}$ plays a key role in the protonation of the iron-peroxo intermediate (see Scheme 6). This threonine residue is highly conserved among cytochrome P450s.

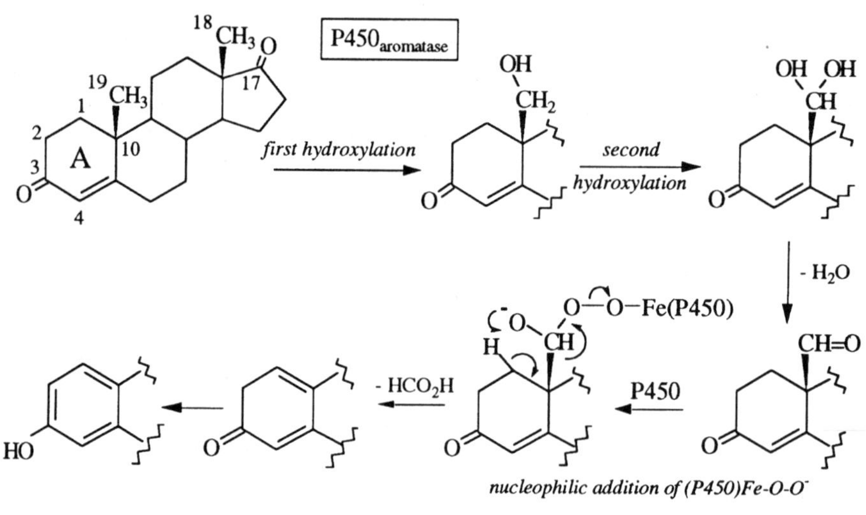

Scheme 5. Mechanism of the aromatization of the A-ring of androst-4-ene-3,17-dione to estrone catalyzed by human placental aromatase

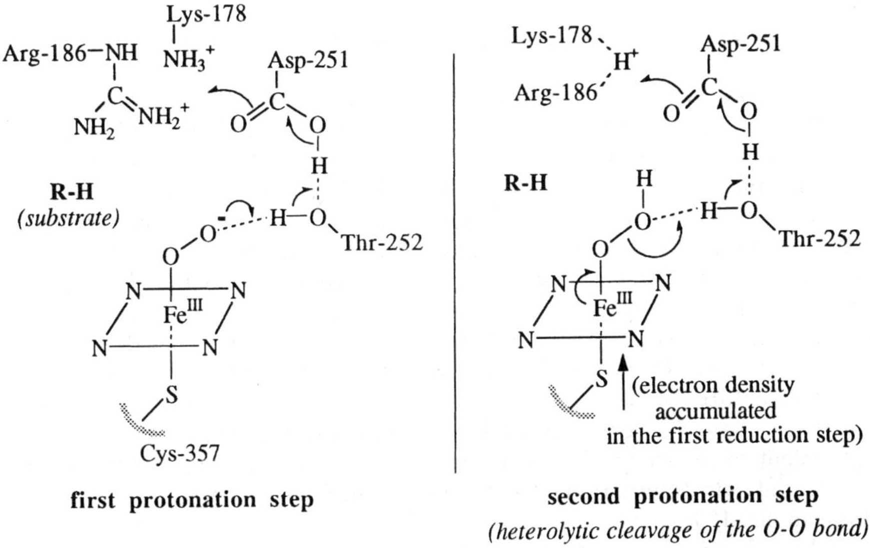

Scheme 6. Essential role of Thr-252 and Asp-251 residues in proton delivery to the iron-dioxygen species in cytochrome P450s

The first protonation produces an Fe^{III}—OOH intermediate which can behave as a nucleophile (see above for the discussion on the reduced nucleophilicity of this protonated form compared to the corresponding non-protonated species, intermediate **E** in Scheme 3). The close interaction of Thr-252 with Asp-251 allows fast proton transfers from protonated forms of Lys-178 and Arg-186. When Thr-252 is replaced in $P450_{cam}$ by an alanine, the mutant enzyme is producing mainly hydrogen peroxide and water molecules in large excess compared to the expected camphor hydroxylation (the so-called uncoupled reactions) [36, 37].

The essential role of the Asp-251 in the proton transfer to the Fe^{III}—O—O^- intermediate (**E** in Scheme 3) has been studied in cytochrome $P450_{cam}$ using site-directed mutagenesis, crystallography, and kinetic solvent isotope effects [29]. The turnover rates of the Asp251Asn mutant in various proton-deuterium mixtures have been determined. The kinetic solvent isotope effect is larger, with a value of 10 compared to 1.8 for the wild-type enzyme. The hydrogen bond network created with the aspartic residue is broken in the mutant and the dioxygen adduct is more open to bulk water molecules.

(Side-commentary: it should be noted that the oxidation states of the iron centers of intermediates 4a and 4b in Fig. 1 of [29] are not correct: the protonation of the terminal oxygen atom is not going to change the oxidation state of the iron center from II to III. In fact, the oxidation state of the iron should be already III in 4a).

2.2.5
Second Protonation Step (Generation of an Electrophilic Iron-Oxo Species)

As for the first protonation step, the proton which is necessary to produce an iron(V)-oxo species and a water molecule after protonation of the terminal oxygen atom of the Fe—OOH entity is also provided by Thr-252, also involving the contribution of Asp-251, Lys-178, and Arg-186. This protonation is probably assisted by the negative charge accumulated on the proximal cysteine in the first reduction step (see Sect. 2.2.1) which is one of the driving forces of the heterolytic cleavage of the O—O bond which generates the electrophilic high-valent Fe^{V}-oxo species. Since the protonation steps are faster than the second reduction step in wild P450s, no life-times can be provided for the Fe—OOH and Fe=O species. Both are very short-life entities. For this reason, the exact nature of the hydroxylating species in cytochrome P450 is still a matter of debate. However, recent data presented in meetings suggest that the high valent iron-oxo species might be fully characterized by crystallography (Sligar et al., 9th International Conference on Biological Inorganic Chemistry, Minneapolis, 1999).

In this second protonation step, the formal oxidation state of the iron center increases from III to V with the formation of an iron-oxygen double bond (Scheme 7).

The electronic structure of the iron(V)-oxo species can be described as a singlet state (structure $\mathbf{F_1}$ in Scheme 7) and is probably the ground state of the active electrophilic species in P450. However, recent elegant calculations by Shaik et al. propose that the reactive species is a triplet state with a single iron-oxygen bond and a strong radical character on the oxygen atom (structure $\mathbf{F_2}$) [38]. The third possible structure $\mathbf{F_3}$ corresponds with a transfer of one electron from the porphyrin ring to the metal center. Such Fe^{IV}-oxo radical-cation species has been fully characterized in heme-peroxidases (the so-called Compound I, see Sect. 2.5 and [39] for a recent book on peroxidases). These three high-valent structures correspond to a perferryl, an iron(IV)-oxyl or a ferryl porphyrin radical-cation state for $\mathbf{F_1}$, $\mathbf{F_2}$, or $\mathbf{F_3}$, respectively. The iron(V)-oxo species $\mathbf{F_1}$ has not been characterized at the enzyme level, but with synthetic metalloporphyrins, in fact only with manganese derivatives up to now (see Sect. 2.4 on manganese(V)-oxo porphyrin complexes).

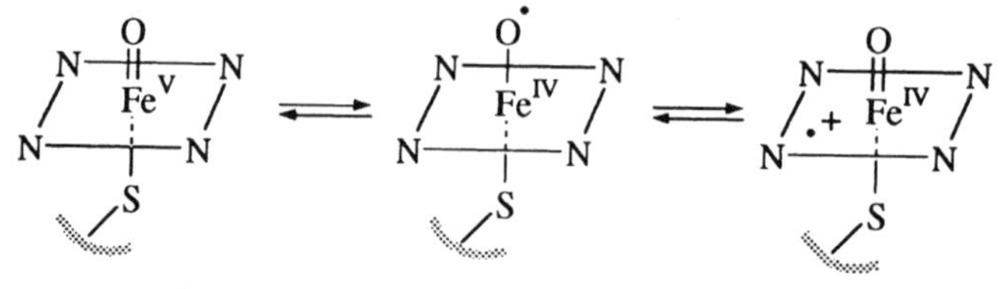

F_1 perferryl (singlet state)   **F_2 iron(IV)-oxyl** (triplet state)   **F_3 ferryl porphyrin radical-cation**

Scheme 7. Representation of the different possible structures of the electrophilic high-valent iron-oxo species in cytochrome P450

It should be noted that an iron(III)-oxene is another possible limit form for the representation of the active electrophilic species of P450. However, such an electronic structure has not been observed when oxidizing iron(III) porphyrins with oxygen atom donors (in all cases the oxidation state of the iron center is at least IV as determined by magnetic data on isolated complexes, see [6–10]). Consequently, there is no longer any reason to use the name "iron-oxene" for the reactive iron-oxo species of P450 enzymes.

The possible formation of an iron-oxo species as the key intermediate in oxygenation reactions catalyzed by cytochrome P450 has been evidenced by using a single oxygen atom donor as oxygen source with the enzyme itself [3] or with synthetic metalloporphyrins used as P450 chemical models [6]. We will discuss now the main aspects of the mechanism of the hydroxylation of aliphatic C—H bond by P450. Two main mechanisms have been characterized depending on the nature of the substrate and on the enzyme itself: a cage-controlled radical mechanism (the "oxygen rebound" mechanism, see [6]) or a concerted mechanism without formation of a radical intermediate based on the use of ultrafast "radical clocks" (see [40, 41] for recent works on this aspect). Both possible reaction pathways are depicted in Scheme 8.

Hydroxylation reactions catalyzed by P450s are not always stereospecific; the loss of stereochemistry has been shown with a tetradeuterated norbornane derivative and with many other substrates [2, 42]. The partial loss of retention of configuration can easily be explained by the oxygen rebound mechanism with a rate of the formation of the carbon-oxygen bond being very close to the

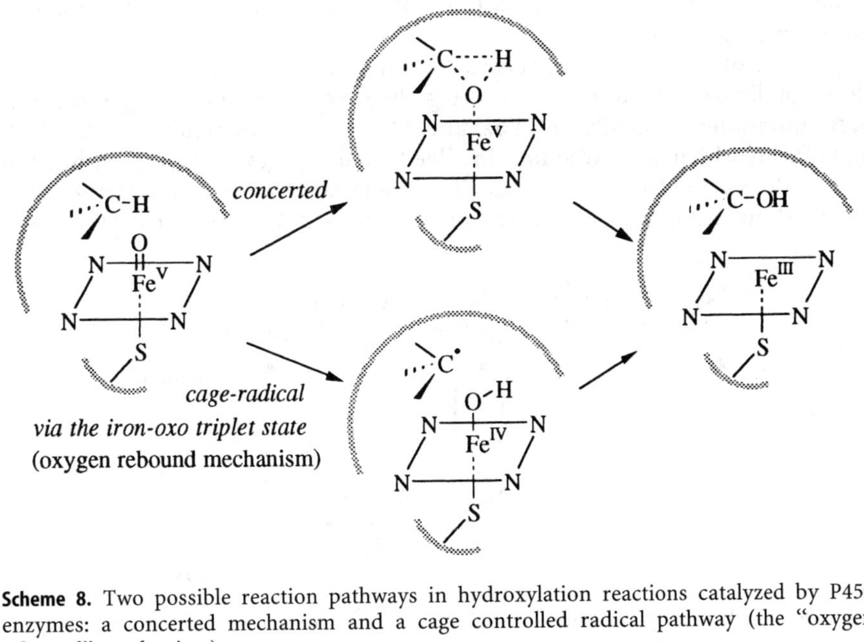

Scheme 8. Two possible reaction pathways in hydroxylation reactions catalyzed by P450 enzymes: a concerted mechanism and a cage controlled radical pathway (the "oxygen rebound" mechanism)

rate of the inversion of configuration of the intermediate radical generated after abstraction of an H-atom by the high-valent iron-oxo species (see Scheme 9).

The abstraction of a hydrogen atom by the active species of P450 is consistent with the high intrinsic isotope effects observed in hydroxylation reactions catalyzed by these heme-monooxygenases [2, 43, 44]. The k_H/k_D values range from 5 to 12, depending on the substrates and the category of P450. In addition, it should be noted that the value of the intrinsic isotope effect of the hydroxylation step can be masked by the other enzymatic steps, reversible substrate binding, product release, etc. In fact, the KIE values determined with P450 and *t*-BuO$^\bullet$ (a pure H-atom abstractor) in a series of different substrates are very similar, suggesting that both reagents have in common an H-atom transfer step [44].

Kinetic data on the hydroxylation reaction catalyzed by P450s have been obtained by using several different radical clocks to probe the rate of the oxygen atom transfer step [40, 41]. The absence of ring opening products in the hydroxylation of *trans*-1-methyl-2-phenylcyclopropane (the rate constant for ring opening is $2 \times 10^9\ s^{-1}$) with two different P450s (from rat or rabbit) indicates that the rate constant for the formation of the C—O bond in the alcohol product should be above $2 \times 10^{11}\ s^{-1}$ [40]. Using related cyclopropane derivatives, Newcomb et al. estimated that the maximum lifetime of a putative C-centered radical intermediate in the hydroxylation by $P450_{2B1}$ should be less than 1 picosecond [41]. These kinetic data strongly indicate that substrate-protein interactions within the active site (the enzyme is "holding" the substrate) probably have a key role in reducing the ring-opening rates of these radical clocks compared to their rates in solution and also contribute to enhance the rates of the C—O bond formation above rate values controlled by diffusion in solution.

The role of an iron(V)-oxo species as being the only species responsible for electrophilic oxygen atom transfers (e.g., hydroxylation and epoxidation) has been questioned recently by Vaz and Coon based on studies with P450 mutants in which the threonine involved in the proton delivery within the active site was replaced by alanine. They obtained two contradictory results. Epoxidation and hydroxylation rates with the T302 A mutant of $P450_{2B4}$ were

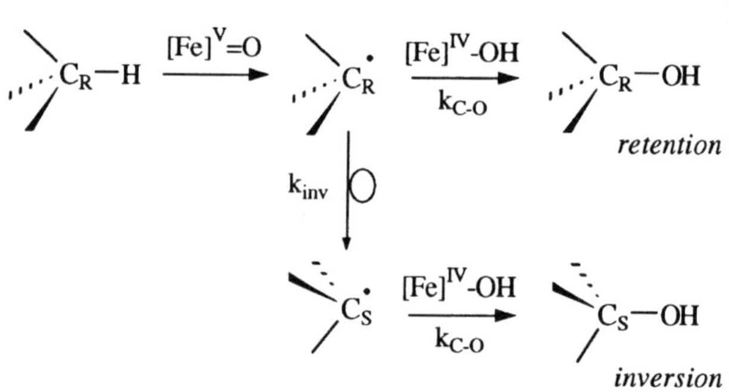

Scheme 9. Possible loss of retention of configuration in hydroxylation reactions involving an oxygen rebound mechanism

reduced whereas epoxidation rates were increased with the T303 A mutant of $P450_{2E1}$ [45, 46]. These data have been interpreted as evidence of the electrophilic properties of the Fe^{III}—OOH intermediate (this species being reported as being able to behave as nucleophile in aldehyde deformylation or as electrophile in epoxidation). However, the observed differences in product formation between the wild and the mutant enzyme can be explained by an enhanced proton delivery by bulk water as observed in the Asp-251 mutant of $P450_{cam}$ [29] and also by a facilitated approach of the olefin to the iron-oxo species. The mechanistic proposals by Newcomb, Coon, and Vaz are summarized in Scheme 10.

The fact that the iron(III)-hydroperoxo intermediate still has a negative charge in its coordination sphere makes it a poor candidate as electrophilic agent. The distribution of charges on the P450 intermediates described in Scheme 10 is rather confusing in terms of the formal classical rules used in coordination chemistry for the description of metal complexes. The presence of two negative charges on the terminal oxygen of the iron-peroxo species is not correct: this terminal oxygen atom being already engaged in a covalent bond with the other oxygen atom, it can carry only one negative charge. In fact, the second negative charge is delocalized on the proximal cysteinato ligand (see Sect. 2.2.1) and should be indicated outside of the complex as described in Scheme 11.

The iron(III)-peroxo species with two negative charges is probably the "super-nucleophile" in P450 intermediates (see above for the mechanism of aldehyde deformylation). Then the first proton, delivered by the threonine or directly by bulk water, is added to the terminal negatively charged oxygen atom of the peroxo which is a more basic site than the proximal sulfur with its delocalized negative charge. The iron(III)-hydroperoxo species obtained after this first protonation step still has one negative charge which is on the side of the proximal cysteine and not on the oxygen atom of the hydroperoxo ligand (see the structure of the nucleophilic iron(III)-hydroperoxo in Scheme 11).

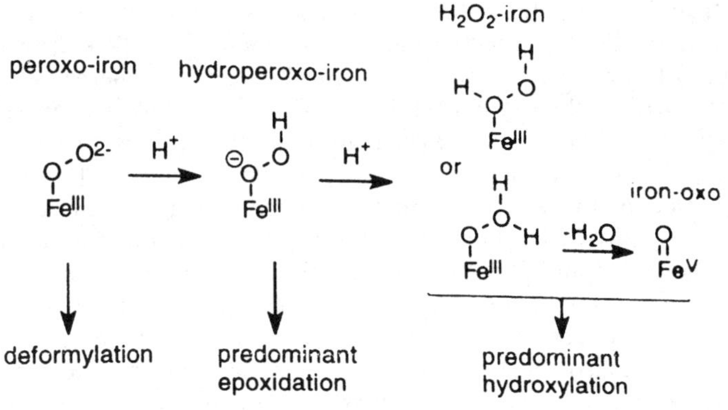

Scheme 10. Different oxidant species generated by P450 enzymes according to Vaz et al. Reproduced with permission from [46]

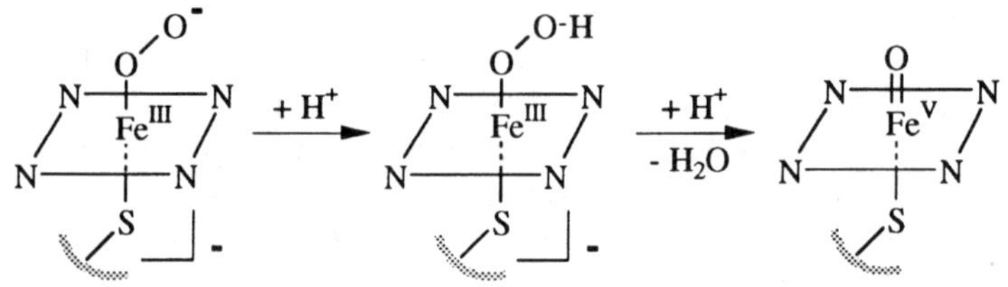

iron(III)-peroxo *"super" nucleophile* **iron(III)-hydroperoxo** *nucleophile* **iron(V)-oxo** *electrophile*

Scheme 11. Formal descriptions of iron(III)-peroxo, iron(III)-hydroperoxo, and iron(V)-oxo species with indication of the negative charges

The second proton delivery by the threonine to the hydroperoxo ligand gives two neutral molecules, the iron(V)-oxo and a water molecule.

The development of DFT calculations, or related methods, will certainly be very helpful to describe in terms of molecular orbitals the electronic structures and the properties of these different reactive P450 intermediates which cannot be easily isolated and characterized by classical spectroscopic methods [47, 48].

We will now report a few recent examples of the reactivity of metal-hydroperoxo and metal-oxo porphyrin complexes. This non-exhaustive presentation will only provide the necessary background related to reactive intermediates of the catalytic cycle of cytochrome P450 enzymes.

2.3 Reactivity of Iron(III)-Peroxo Porphyrin Complexes

The nucleophilicity of iron(III)-peroxo porphyrin complexes has been recently illustrated by Valentine et al. [49, 50]. The high reactivity of negatively charged iron(III)-peroxo species having a side-on structure has been demonstrated by using electron deficient olefins (unsaturated ketones or quinones). The peroxo complex $[Fe^{III}(TMP)O_2]^-$ reacts quickly with menadione (2-methyl-1,4-naphthoquinone) to produce the corresponding 2,3-epoxide whereas no reaction was observed with cyclohexene [49]. In contrast, this latter electron-rich olefin is easily epoxidized by single oxygen atom donors (iodosylbenzene, hypochlorite, monopersulfate) using metalloporphyrin as catalysts [6, 7].

When replacing the electron-donating porphyrin ligand by an electron-withdrawing porphyrin containing perfluorophenyl substituents at the meso positions, the nucleophilicity of the corresponding iron-peroxo is reduced. But the presence of an axial ligand can restore its reactivity by opening the side-on peroxo to a more reactive end-on structure [50]. The peroxo complex $[Fe^{III}(TMP)O_2]^-$ is also able to mimic the deformylation step of aromatase [51].

What do we need to modify the reactivity of an iron-peroxo porphyrin complex, to go from a nucleophile to an electrophile active species able to epoxidize electron-rich olefins? In order to remove electron density from the iron-peroxo, highly electron-deficient porphyrin ligands should be used [52] associated with an electron-withdrawing proximal ligand. These properties are both the opposite of the electronic properties of the prosthetic group of P450; the natural porphyrin PPIX is a very good electron reservoir and the proximal cysteine ligand, associated to a network of amino acids, is also able to accumulate electron density. Both factors are strongly against (in favor of) electrophilic (nucleophilic) behavior of iron-peroxo intermediates in cytochrome P450s.

It should be noted that non-heme metal-peroxo complexes which are able to react with electron-rich olefins (di- or tri-substituted aliphatic olefins) have electron-attracting ligands in their coordination sphere. This is the case for rhenium-oxo-peroxo catalysts (see [53] for a recent article, and the chapter by Herrmann and Kühn in this volume) or of iron-peroxo polypyridine complexes when the iron center is still able to behave as a Lewis acid with respect to the terminal oxygen atom of the peroxo motif (see [54], and the chapter by Girerd, Banse and Simaan in this volume).

2.4 Characterization and Reactivity of High-Valent Metal-Oxo Porphyrin Complexes

The first high-valent iron-oxo porphyrin complex has been isolated by Groves et al. by oxidation at low temperature of Fe(TMP)Cl with *m*-chloroperbenzoic acid [55]. The resulting "green" compound is an iron(IV)-oxo complex with a radical-cation on the porphyrin ring which resembles the Compound I of heme-peroxidases (see [6, 7, 56] for review articles on the preparation and characterization of high-valent iron-oxo porphyrin complexes). Many other Compound I models have been obtained with various differently substituted metalloporphyrins. However, in all these model systems, the exchange between the spin $S = 1$ of the ferryl group with the spin $S' = 1/2$ of the porphyrin radical-cation was found to be strongly ferromagnetic in contrast to the weak ferromagnetic coupling usually observed with Cpd I derivatives [57].

Up to now, nobody has been able to identify an iron(V)-oxo. One possible explanation is probably related to the fact that most of these high-valent iron-oxo-complexes have no proximal anionic ligand or a neutral one as in horseradish peroxidase which has a proximal histidine [58]. The P450 models with a proximal sulfur-containing ligand are more reactive than the corresponding models without this cysteine analogue and they might be the only suitable P450 models to allow the isolation of a true perferryl complex [59]. Oxygen-containing proximal ligands reduce the oxygenase activity of synthetic metalloporphyrins compared to the corresponding complexes with nitrogen-containing axial ligands [60].

The electron reservoir capacity of porphyrin ligands is in fact highly adapted to favor the formation of high-valent oxidation states of a metal center, still keeping a sufficient lability of the metal-oxygen bond in order to obtain a facile oxygen atom transfer. The stability of high-valent iron- or manganese-oxo species is highly dependent on the nature of the ligands. For example, an inert d^2 square-pyramidal manganese(V)-oxo stable at room temperature has been prepared with a diamido ligand [61, 62]. An additional example of lability vs stability of high-valent species is the case of nitrido-manganese(V) porphyrin complexes which are kinetically inert with organic substrates, but are able to transfer the nitrido motif to chromium(III) porphyrin via a two-electron redox process mediated by a heterobimetallic μ-nitrido intermediate [63]. The same nitrido-manganese(V) porphyrin complexes can be transformed into nitrogen atom-transfer agents after acylation of the nitrido ligand to generate a labile acylimido-manganese(V) porphyrin complex [64].

Both Mn^{IV}-oxo and Mn^{V}-oxo porphyrin complexes have recently been characterized (see [65, 66] for early attempts). A (Por)Mn^{IV}=O complex has been characterized by X-ray absorption spectroscopy [67]. The Mn—O bond distance is 1.69 Å and the complex has a $S = 3/2$ spin state corresponding to a high spin d^3 configuration. More recently, it has been possible to characterize a manganese(V)-oxo porphyrin complex by stopped-flow spectrophotometry [68]. Produced with hypochlorite or monopersulfate, (TMPyP)(X)Mn^{V}=O exhibits a Soret band at 443 nm between that of (TMPyP)Mn^{IV}=O (428 nm) and Mn^{III}(TMPyP) (462 nm). The half-lifetime of this Mn^{V}-oxo is 25 ms at room temperature at pH 7.4. Its decay to the Mn^{IV}-oxo species can be accelerated by nitrite. When the methyl groups of the pyridinium substituents of the porphyrin ligand are in *ortho* positions instead of *para*, then the manganese(V)-oxo porphyrin complex has a longer lifetime (a few minutes) making possible the acquisition of a diamagnetic proton NMR spectrum [69]. This fact supports a low-spin d^2 electronic structure for the ground state of this manganese(V)-oxo complex. This "singlet" state is probably in equilibrium with a "triplet" state; i.e., an Mn^{IV}-oxyl species with a single oxygen-manganese bond, as proposed by Shaik et al. [38]. In fact, the electronic structure of a metal(V)-oxo complex with a metal-oxygen double bond is probably higher in energy than the "diradical" form, a metal(IV)-oxyl species, as singlet dioxygen, with a double bond, is higher in energy by 23 kcal mol^{-1} compared to its triplet ground state, with a single O—O bond and two unpaired electrons (the name "oxyl" underlines the reduction of the bond order between the metal center and the oxygen atom). For a schematic representation of these metal-oxo vs metal-oxyl species, see structure $\mathbf{F_1}$ and $\mathbf{F_2}$, respectively in Scheme 7. A spin conversion of the singlet to the triplet state of the metal(V)-oxo species will occur during the oxygen atom transfer step [38].

The capacity of these electrophilic high-valent iron-oxo and manganese-oxo porphyrin complexes to transfer an oxygen atom to hydrocarbons or olefins will be described with recent mechanistic details in Sect. 3. The existence of Mn^{V}-oxo entities has now been evidenced both by spectroscopic methods and reactivity data.

2.5 Active Species in Heme-Peroxidase Catalytic Cycles

Several heme-peroxidases from the vegetable or animal kingdoms have been extensively studied: e.g., horseradish peroxidase, chloroperoxidase, ligninase, myeloperoxidase, lactoperoxidase, etc. (see [39] for a book and [70, 71] for recent reviews). Here we will only discuss some recent aspects concerning the reactivity of the different possible intermediates (metal-oxo vs metal-peroxo, electron transfer vs oxygen atom transfer) generated during the catalytic cycles of these heme-peroxidases. We will limit the discussion to horseradish peroxidase HRP and chloroperoxidase CPO.

2.5.1
Horseradish Peroxidase

HRP catalyzes the abstraction of one or two electrons (usually two electrons) via a single electron transfer from an organic substrate, hydrogen peroxide being used as electron acceptor, according to Eq. (2):

$$AH_2 + H_2O_2 \xrightarrow{\text{peroxidase}} A + 2H_2O \qquad (2)$$

The different oxidation states and the properties of this enzyme have been well studied. The prosthetic group of HRP is a ferric-protoporphyrin IX with a histidine as proximal ligand (His-170). The high-valent iron-oxo entity of HRP-Compound I is generated from hydrogen peroxide via its heterolytic cleavage involving the guanidinium function of Arg-38 (acting as a proton source for the formation of the water molecule) and His-42 acting as a base in the deprotonation step of H_2O_2 (Scheme 12) [58, 72].

The oxidation of HRP by hydrogen peroxide generates a reactive intermediate (Compound I) containing two redox equivalents above the resting state of the native enzyme. The one-electron reduction of Compound I by the substrate generates Compound II which is still able to abstract one electron from the substrate. Both high-valent iron species have been characterized by several physico-chemical methods. Compounds I and II have a broad Soret band at 400 nm and 420 nm, respectively. Compound I consists of an iron(IV)-oxo species with a radical cation on the porphyrin ring (Scheme 13).

Compounds I and II have been characterized by X-ray absorption and Raman spectroscopies. Both methods confirmed the presence of an iron(IV)-oxo entity in both Compounds I and II with a short Fe=O bond of 1.6 Å

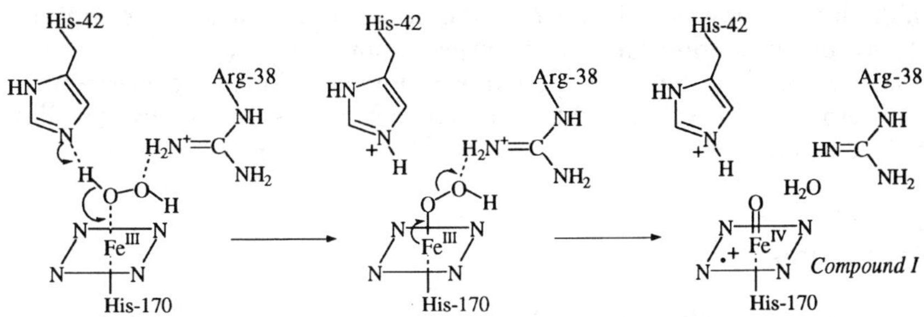

Scheme 12. Essential amino acids involved in the activation of hydrogen peroxide by horseradish peroxidase

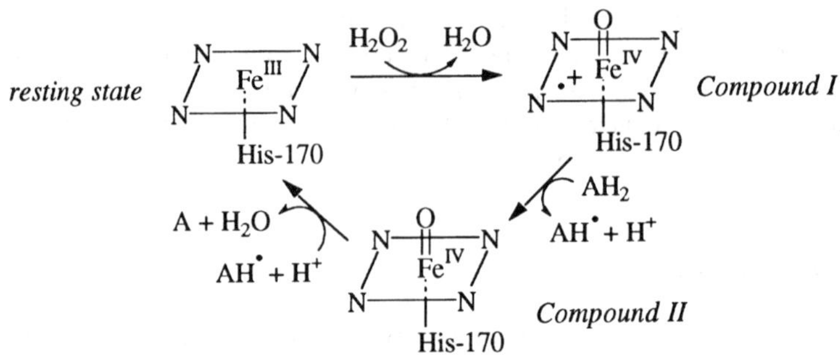

Scheme 13. Catalytic cycle of horseradish peroxidase (AH_2 being a 2-electron donor)

consistent with a ferryl structure [73]. More recently, it has been suggested that the Fe—O bond distance of Compound II was longer: 1.9 Å at pH 7 [74]. Resonance Raman studies confirmed the presence of an Fe=O entity for both Compounds I and II with vibrations at 737 cm^{-1} and 776 cm^{-1}, respectively [75]. The π-radical-cation of Compound I has a predominant $^2A_{2u}$. This porphyrin radical is ferromagnetically coupled with the spin $S = 1$ of the ferryl state [76]. The lifetime of HRP-Compound I can be increased up to one hour at −20 °C with a polyethylene glycolated enzyme [77]. Compound III is an inactive intermediate corresponding to an iron(III)-peroxo species resulting from the addition of dioxygen to the ferrous state of HRP or from the reaction of an excess of hydrogen peroxide with the ferric state of the enzyme [78].

The high-valent iron-oxo intermediates of HRP are not directly accessible to the different substrates. Alkylation of the δ-*meso* position of the heme group by alkylhydrazines indicates that substrates approach the active site of HRP from one edge of the prosthetic group [79]. Recent data on the substrate binding site of HRP have been obtained by proton NMR spectroscopy and molecular dynamics [80, 81]. The weak binding site of HRP is an open pocket able to accommodate a large range of substrates. Despite several claims, HRP is unable to catalyze the transfer of an oxygen atom from hydrogen peroxide to a substrate molecule. HRP is unable to oxidize styrene whereas chloroperoxidase is able to catalyze the oxidation of this olefin to a mixture of the corresponding epoxide and phenylacetaldehyde [82, 83]. The oxidation of sulfides by HRP involves the formation of a radical-cation on the sulfur atom followed by water addition, not an oxygen atom transfer [84].

Despite the use of hydrogen peroxide, a suitable oxidant to generate metal-hydroperoxo species, it should be noted that all the known intermediates involved in the HRP catalytic cycle are high-valent iron-oxo entities.

2.5.2
Chloroperoxidase

The fungal chloroperoxidase (CPO) is among the few peroxidases which are able to catalyze the oxidative chlorination of substrates using H_2O_2 and Cl^- (myeloperoxidase is an other example) according to Eq. (3) [85]:

$$AH + H_2O_2 + Cl^- + H^+ \xrightarrow{CPO} ACl + 2\,H_2O \quad (3)$$

This oxidative chlorination is performed on substrates containing an activated carbon-hydrogen bond such as *β*-diketones (chlorodimedone is a classical substrate used in CPO assays).

A noticeable difference between CPO and HRP is the presence of a cysteine residue as proximal ligand (Cys-29 in CPO) instead of a histidine in HRP [86]. The X-ray structure has been determined recently and confirms the presence of a manganese(II) ion bound to a heme propionate and also surrounded by His-105, Glu-104, and Ser-108 (the manganese ion may be the binding site for a chloride ion) [87, 88].

The addition of hydrogen peroxide to the ferric state of the enzyme generates CPO-Compound I, the only detectable intermediate having an $Fe^{IV}{=}O$ bond with a Raman stretching band at 790 cm^{-1} [89]. The radical-cation of Compound I is probably delocalized on the macrocycle and on the axial ligand. The addition of chloride to Compound I might generate an Fe^{III}-OCl entity, a possible candidate to explain the substrate chlorination [90]. An alternative mechanism is the formation of free HOCl. The same mixture of chloroanisole isomers was observed in the chlorination of anisole by free HOCl or by the CPO chlorinating system [91]. A third possible mechanism is to consider that the manganese site of CPO is able to facilitate the binding of Cl^-. The short distance between this Mn-site and the heme should facilitate the electron removal from the chloride ion and then Cl^+ could be transferred to the substrate present near the active site. This hypothesis would explain the absence of formation of free HOCl without invoking the formation of an iron(III)-hypochlorito intermediate during the catalytic cycle of chloroperoxidase.

Chloroperoxidase is inactivated by formation of *N*-alkyl-heme in the oxidation of terminal olefins [92]. These data suggest that different substrates have relatively easy access to the active site of CPO. Unlike HRP, CPO is able to catalyze the epoxidation of different olefins (styrene, propylene, allyl chloride) [93]. The oxygen atom of the epoxide arises from the primary oxidant as evidenced by using ^{18}O-labeled hydrogen peroxide [83]. Recent studies have demonstrated that the CPO-catalyzed epoxidation of *cis*-disubstituted olefins is enantioselective; enantiomeric excesses (ee values) range from 33% to 85% [93]. With a slow addition of hydrogen peroxide to avoid the degradation of CPO, it has been possible to obtain 4200 catalytic cycles with an ee value of 94% in the epoxidation of methylallyl propionate [94]. Chloroperoxidase is also able to catalyze the enantioselective oxidation of sulfides to provide chiral sulfoxides with ees up to 90–95% [95, 96]. It has been proposed from data obtained with a series of *para*-substituted thioanisoles that *S*-oxygenations catalyzed by CPO involve an oxygen atom transfer from Cpd I to the substrate rather than a one-electron oxidation of the sulfide followed by the addition of H_2O on the *S*-radical-cation intermediate, as proposed for sulfide oxidations catalyzed by HRP [97]. Again, no iron-peroxo species has been identified among the active species generated with hydrogen peroxide during the catalytic cycle of chloroperoxidase.

Recent claims mentioned that microperoxidase might be able to oxygenate substrates via an iron-peroxo species generated by addition of an hydroxide to an iron-oxo (this formation of a weak O—O bond is certainly thermodynamically not favorable; in fact this is the reverse reaction of the heterolytic cleavage of the peroxidic bond when hydrogen peroxide is activated by peroxidases) [98]. However, these claims are based on labeling experiments obtained from a rather complicated work-up on few catalytic cycles. This work should be independently confirmed in effective catalytic reactions before claiming that iron-peroxo species are involved in oxygenation reactions mediated by microperoxidase (see footnote 48 in [41]).

3 Oxo-Hydroxo Tautomerism with Water-Soluble Metalloporphyrins

In the chemistry of P450 models, high valent metal-oxo species have been developed as active intermediates in many different oxidation reactions using manganese or iron porphyrin catalysts and oxygen atom donors (PhIO, NaOCl, $KHSO_5$, H_2O_2, etc), or dioxygen associated to a reductant, as oxygen atom source.

When the oxygenation reaction (hydroxylation or epoxidation) is performed in an organic solvent (usually dichloromethane) with hydrophobic metalloporphyrin catalysts, the oxygen atom incorporated within the substrate originates from the oxidant. This has been evidenced for hypochlorite [66, 99], monopersulfate [100], or iodosylbenzene [101 and references therein], but when these metalloporphyrin-catalyzed oxygenations are performed in aqueous solvents, such metal-oxo species are able to transfer an oxygen atom coming from either the oxygen source or from bulk water. Because the intermolecular exchange of metal-oxo with bulk water is slow, an intramolecular exchange of labeled oxygen atoms via the so-called *oxo-hydroxo tautomerism* [102] has been proposed [103]. This mechanism involves a rapid shift of two electrons and one proton from a hydroxo ligand (electron-rich ligand formed by deprotonation of an aqua ligand) to the *trans* oxo species (electron-poor ligand) leading to the transformation of the hydroxo ligand into an electrophilic oxo entity on the opposite side of the initial oxo (see Scheme 14 for a representation of this tautomeric equilibrium).

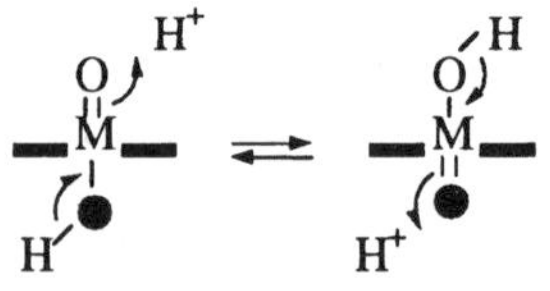

Scheme 14. Key step of the oxo-hydroxo tautomerism

The oxo-hydroxo tautomerism can contribute to a better characterization and an improved understanding of chemical reactivities of these high-valent metal-oxo species, not only important for the knowledge of heme-enzymes catalyzing oxidations, but also in the design of efficient biomimetic or bioinspired oxidation catalysts. At the present stage of the knowledge, the design of metal-oxo complexes able to catalyse oxygenations efficiently is still challenging, since all the parameters involved in the different O-atom transfer steps are not fully understood.

3.1
A New Type of Tautomerism: The Oxo-Hydroxo Tautomerism

We recently discover the *oxo-hydroxo tautomerism* during studies on the origin of the epoxidic oxygen atom in an olefin epoxidation performed in aqueous media with a water-soluble metalloporphyrin catalyst [103].

Using isotopically labeled oxidants or labeled water, it has been shown that the oxygen of the metal-oxo porphyrin complex can be quickly exchanged with water via the axial hydroxo ligand with reaction rates depending on the experimental conditions (pH, temperature, composition of the medium, nature of the axial ligands, etc.). We shall see that, using the label distribution in oxidation products modulated by this oxo-hydroxo tautomerism, it is possible to unambiguously distinguish between oxygenation reactions occurring via an O-transfer from a high-valent metal-oxo complex or via an autoxidation mechanism (see [104] for a review on this controversial debate) in metalloporphyrin-catalyzed oxygenations carried out in the presence of $H_2{}^{18}O$. The main consequence of the oxo-hydroxo tautomerism is the incorporation within the substrate of an oxygen atom coming from either the oxidant or from bulk water, respectively in the ratio 1:1 (Scheme 15). The degree of ^{18}O-incorporation observed into the product (epoxide, alcohol, etc.) is then mechanistically informative.

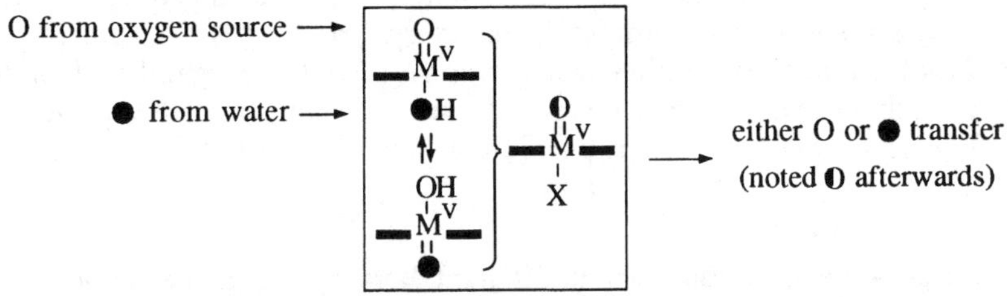

Scheme 15. The oxo-hydroxo tautomerism mediates the incorporation into the oxidation product of 50% of oxygen coming from the primary oxidant and 50% from water. X = hydroxo ligand

3.2
Some Previous Data on O-Exchange of Metal-Oxo Species with Bulk Water

Over the last 15 years, few reports have mentioned the use of ^{18}O-labeling experiments in order to characterize high-valent metal-oxo species or to elucidate the mechanism of O-transfer by these reactive species. For reactions performed in the presence of water, results range from 0% to 100% of O-incorporation from water, depending on the experimental conditions. Most of these results can be understood by reference to the oxo-hydroxo tautomerism (for detailed comments, see the paragraph on requested conditions to observe oxo-hydroxo tautomerism) or on the basis of a direct, intermolecular O-exchange of the oxo ligand with the bulk water. It should be noted that the rate of this intermolecular exchange is slower than the rate of the oxo-hydroxo tautomerism which has been estimated to have a rate constant of $10^3\ s^{-1}$ [68].

A resonance Raman investigation on Cpd II of HRP at pH 7, based on isotopic shift of the $Fe^{IV}{=}O$ stretching mode of Cpd II (the $\nu_{Fe=O}$ was observed at 774 cm^{-1} after activation in $H_2{}^{16}O$ with either $H_2{}^{16}O_2$ or $H_2{}^{18}O_2$, at 740 cm^{-1} for activation in $H_2{}^{18}O$ with either $H_2{}^{16}O_2$ or $H_2{}^{18}O_2$) provided evidence for the oxygen atom exchange between the heme-$Fe^{IV}{=}O$ and bulk water [105]. In a controversial manner, this exchange was not observed in the case of stable species of diacetylheme or manganese substituted HRP: the Raman spectrum of Cpd II generated with $H_2{}^{16}O_2$ or $H_2{}^{18}O_2$ presented lines at 781 cm^{-1} or 745 cm^{-1} (diacetylheme HRP) and at 626 cm^{-1} or 596 cm^{-1} (Mn-HRP), respectively. No isotope-induced change was observed at neutral pH for 1 h at 4 °C in the presence of $H_2{}^{16}O$ or $H_2{}^{18}O$, indicating no appreciable exchange of the iron-oxo entity with bulk water [106].

In studies performed at −80 °C on a ferryl porphyrin π-cation-radical derived from the synthetic Fe^{III}(TMP)Cl, the $\nu_{Fe=O}$ band was observed at 828 cm^{-1} (activation with ^{16}O-*m*-CPBA) or 792 cm^{-1} (activation with ^{18}O-*m*-CPBA), the first band remaining unshifted in the presence of $H_2{}^{18}O$ indicating that the oxygen atom of the iron-oxo was not easily exchanged with water under these conditions [107].

Incorporation of ^{18}O from bulk water was initially reported for epoxidation of norbornene with *m*-CPBA catalyzed by Fe^{III}(TMP)Cl in an organic medium containing 1% $H_2{}^{18}O$ [55] and was further noticed in the course of epoxidation of β-methylstyrene by manganese(V)-oxo porphyrin in CH_2Cl_2 saturated in $H_2{}^{18}O$ [108], indicating a rather fast exchange of the oxo ligand with water. This ^{18}O-exchange was clearly slower in the case of manganese(IV) species and was inhibited by the presence of pyridine as axial ligand [109].

3.3 Oxo-Hydroxo Tautomerism Observed in Different Metalloporphyrin-Catalyzed Oxygenation Reactions

3.3.1 *Epoxidation Reactions*

We initially reported the oxo-hydroxo tautomerism to explain isotopic results observed in aqueous phase during $KHSO_5$ epoxidation of carbamazepine (CBZ), an analgesic and anticonvulsant drug, catalyzed by a cationic water-soluble manganese porphyrin [103]. In such reaction performed at pH 5 in aqueous solution with various contents of $H_2{}^{18}O$, it was shown that half of the oxygen atoms incorporated in the epoxide came from the solvent (Fig. 1a). It

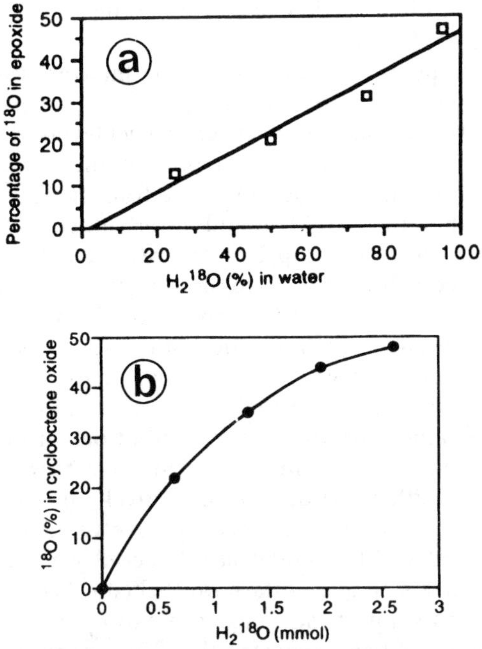

Fig. 1. **a** The amount of labeled oxygen found in CBZ oxide correlates with half the content of ^{18}O-label of water present in the reaction mixture. In abscisse: content (%) of $H_2{}^{18}O$ in water (reproduced with permission from [103]; see also [111] for a similar correlation). **b** Dependence of the oxo-hydroxo tautomerism on a sufficient concentration of water in a non-aqueous solvent observed in cyclooctene catalyzed epoxidation. In abscisse: 0.5, 1, 1.5, 2, and 2.5 mmol of $H_2{}^{18}O$ correspond to 1, 2, 3, 4, and 5 molar concentrations of $H_2{}^{18}O$, respectively. Substrate concentration was 20 mmol l^{-1} (reproduced with permission from [111])

was checked that neither CBZ-oxide nor $KHSO_5$ [100, 110] exchanged oxygen atoms with water in the reaction conditions.

To explain the ratio of 0.5 for the incorporation of oxygen from the solvent, we proposed an oxo-hydroxo tautomerism (previously named "redox tautomerism" in [103, 110–114]) involving a coordinated water molecule on the manganese(III) porphyrin precursor **1** (Schemes 14 and 16). It must be noted that a constant 50% O-exchange corresponds only to the oxo-hydroxo tautomerism involving the hydroxo ligand *trans* to the high-valent metal-oxo

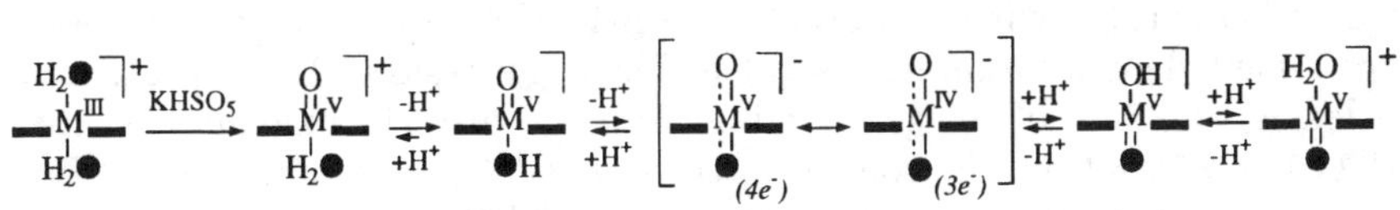

1 **2** **3** **4** **4'** **5** **6**

Scheme 16. General scheme for the oxo-hydroxo tautomerism complex, but not to a direct exchange of the oxo ligand with bulk water that can lead to 100% O-exchange.

Mn^{III}(TMPyP), the pentaacetate of the diaqua-manganese(III) derivative of *meso*-tetrakis(1-methyl-pyridinium-4-yl)porphyrin, can exist in aqueous medium with one or two metal-bound water molecules as axial ligands [**1** in Scheme 16; see [115] for an X-ray structure of the bis-aqua-Mn(TMPyP) complex]. Increasing the metal oxidation state from III to V (going from the Mn^{III} complex **1** to the Mn^{V}-oxo **2**) should lower sufficiently the pK_a value of the ligated water to allow, at the pH of the reaction, its conversion into a hydroxo ligand (**3**; see [116] for a discussion on the pK_a values of aqua and hydroxo ligands in high-valent metalloporphyrins). Removal of a proton from this hydroxo ligand results in the formation of the stabilized anion **4** with $4e^-$ delocalized on both metal-oxygen bonds (**4′** is a mesomeric form with $3e^-$ delocalized and manganese at the formal oxidation state IV). This anion can be protonated with the same probability at the end of one of the two metal-oxo-like bonds, giving rise to either form **3** or **5**, which reacts with CBZ to produce CBZ-oxide containing either ^{16}O or ^{18}O, respectively, in the ratio 1 to 1. The conversion of **3** to **5** does not necessarily involve **4** and **4′** as discrete deprotonated intermediates but might also proceed via a hydrogen-bonded water molecule in a more concerted fashion (Scheme 14).

Other recent reports support the concept of oxo-hydroxo tautomerism [68, 111, 114]. Groves et al. [68] also reported the ^{18}O-incorporation in CBZ-oxide in an Mn(TMPyP)-catalyzed oxidation of CBZ via an oxo-hydroxo interconversion. Lee and Nam described similar ^{18}O-incorporation in cyclooctene epoxidation by either *m*-CPBA, H_2O_2 or *t*-BuOOH catalyzed by *meso*-tetrakis(pentafluoro-phenyl)porphyrinato-iron(III) chloride $Fe^{III}(F_{20}TPP)Cl$, the reaction being performed in a CH_3OH/CH_2Cl_2 mixture containing 10% H_2O [111]. Even at low pH values, CBZ epoxidation data obtained in aqueous solutions by H_2O_2, *t*-BuOOH, or $KHSO_5$ in the presence of Fe^{III}(TDCPPS) indicate that an oxo-hydroxo tautomerism was involved [114]. These latter experiments strongly supported that a common high-valent iron-oxo species was generated from the different oxidants and was the active species responsible for olefin epoxidation.

3.3.2
Hydroxylation Reactions

The oxo-hydroxo tautomerism was further characterized in the oxidation of deoxyribose C—H bonds of DNA by the Mn(TMPyP)/$KHSO_5$ system (see Scheme 17) [112]. Hydroxylation at carbon-1′ of deoxyribose gave in several steps 5-methylene-2-furanone (5-MF) as final sugar residue. In the presence of labeled $H_2{}^{18}O$, 50% of oxygen coming from the primary oxidant (^{16}O from $KHSO_5$) and 50% from the solvent (^{18}O from $H_2{}^{18}O$) were incorporated in 5-MF, strongly supporting a metal-oxo mediated DNA cleavage with an oxo-hydroxo tautomerism to explain the ^{18}O-incorporation in the desoxyribose oxidation product.

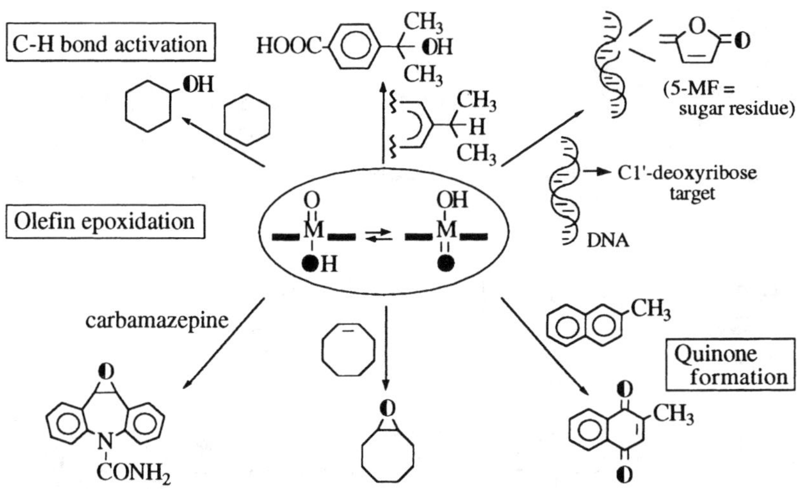

Scheme 17. Examples of oxo-hydroxo tautomerism from literature data. O: unlabeled oxygen; ●: labeled oxygen; ◐: mixed labeled oxygen

Another example came from the monopersulfate oxidation of 4-isopropylbenzoic acid performed in $H_2{}^{18}O$ and catalyzed by the same water-soluble metalloporphyrin Mn(TMPyP) [110]. In the primary hydroxylation product, 4-(1-hydroxy-1-methylethyl)benzoic acid, nearly half of the oxygen atoms incorporated in the alcohol function came from water. In the cyclohexane hydroxylation by *m*-CPBA catalyzed by $Fe(F_{20}TPP)Cl$ [111], or by H_2O_2 catalyzed by $[Fe(TF_4TMAP)](CF_3SO_3)_5$ [117], the percentage of ^{18}O incorporated in cyclohexanol was also found to be 40–50%, with a reaction mixture containing only 7–10% of $H_2{}^{18}O$.

Oxo-hydroxo tautomerism was recently observed in both epoxidation and hydroxylation reactions catalyzed by metalloporphyrins using activation by the sulfite/dioxygen system. Oxidation of sulfite catalyzed by the water-soluble Mn(TMPyP) allows generation of $Mn^V{=}O$ complex [see Scheme 18(A)]. This alternative, biocompatible oxidation system compared to the preformed oxidant $KHSO_5$ allows incorporation of one labeled oxygen atom coming from either water or dioxygen [Scheme 18(B)] as was illustrated in the metalloporphyrin-catalyzed hydroxylation of a benzylic C—H bond [Scheme 18(C)] or in epoxidation of carbamazepine [118].

3.3.3
Quinone Formation

In an aqueous solution, the metalloporphyrin-catalyzed oxidation of 2-methylnaphthalene to *p*-quinones involves two consecutive oxygen transfers from an intermediate metal-oxo entity responsible for 30–55% indirect incorporation of ^{18}O from water into the generated quinones (Scheme 17) [113].

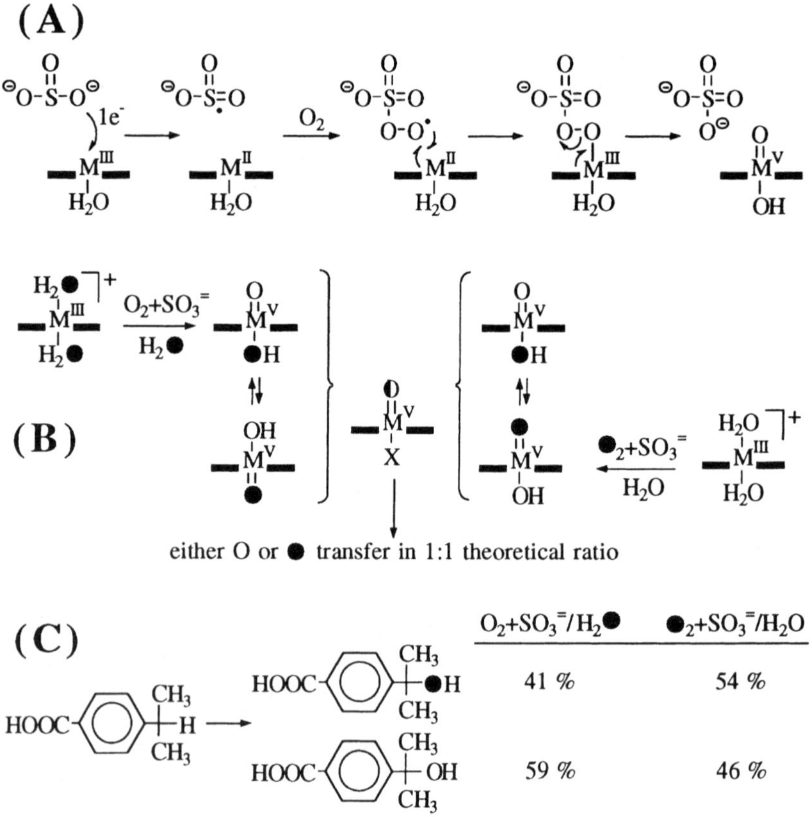

Scheme 18A–C. Activation of metalloporphyrins with $O_2 + SO_3^{2-}$: **A** mechanism of formation of high-valent metal-oxo species; **B** the oxo-hydroxo tautomerism allows one to incorporate one labeled oxygen atom from either water or dioxygen. X = hydroxo ligand; **C** illustration of both labeling routes described in B when performing a metalloporphyrin-catalyzed hydroxylation of a benzylic C—H bond [118]. ○: unlabeled oxygen; ●: labeled oxygen; ◐: mixed labeled oxygen

3.4
Required Conditions to Observe an Oxo-Hydroxo Tautomerism

3.4.1
An Axial Ligand Competitive to the Hydroxo Ligand Inhibits Oxo-Hydroxo Tautomerism

In heme-enzymes, the presence of a cysteinato ligand (cytochrome P-450 or chloroperoxidase) or an imidazole from an histidine (peroxidase) prevents water from coordinating to the iron center. This implies that there is no possible intramolecular exchange between high valent metal-oxo intermediates and water through an oxo-hydroxo tautomerism.

With synthetic metalloporphyrins the situation is different. Iron and manganese porphyrins are known to form imidazole and pyridine complexes when these heterocycles are present in reaction mixtures. In the epoxidation of *cis*-β-methylstyrene with *m*-CPBA catalyzed by Mn^{III}(TMP)Cl, the presence of pyridine completely prevents isotopic enrichment of epoxide products when the reaction is performed in the presence of $H_2{}^{18}O$ [108]. In the epoxidation of cyclooctene by Fe(F_{20}TPP)Cl and H_2O_2, it was found that effectively the ^{18}O-incorporation into the product diminished as the amount of 5-chloro-1-methylimidazole added to the reaction mixture increased [111]. So when the axial position opposite to the oxo group is blocked by a ligand (not derived from water), the oxo-hydroxo tautomerism cannot occur and consequently the oxygen incorporated in the oxidation product is 100% from the oxidant, instead of 50% from the oxidant and 50% from water.

In other respects, when the *trans* ligand is a water molecule instead of a hydroxo ligand, as is probably the case for metal(IV)-oxo species (see hereafter for a discussion on this point), the tautomerism is largely reduced. This may explain why the oxo-hydroxo tautomerism was not observed during the oxidation of polycyclic aromatic hydrocarbons catalyzed by iron tetrasulfophthalocyanine performed in the presence of $H_2{}^{18}O$ (the reactive species was shown to be $Fe^{IV}{=}O$) [119].

3.4.2
A Competitive Autoxidation Route Lowers Incorporation of Oxygen from Water

Groves and Stern have noticed that during the cyclooctene [108] or *cis*-β-methylstyrene [109] epoxidation by manganese(IV)-oxo porphyrin under aerobic conditions, dioxygen was intimately involved in the oxidation process with oxidation products partially resulting from an autoxidation process. The consequence is a decrease of isotopic enrichment from solvent. During the monopersulfate oxidation of ketoprofen catalyzed by Mn(TMPyP), trapping of the intermediate C-centered radical on the substrate by molecular oxygen was competing with the oxygen rebound mechanism, explaining the observed reduction of ^{18}O-incorporation from solvent in the final product when ketoprofen was oxidized under air by the system Mn(TMPyP)/$KHSO_5$/$H_2{}^{18}O$ system [110].

3.4.3
Temperature Effect

In order to characterize the high-valent reactive species formed by oxidative activation of metalloporphyrins, many experiments have been performed at low temperature with some of them concerning ^{18}O-incorporation into the products in the presence of $H_2{}^{18}O$. In the more extensive study, Lee and Nam [111] described the cyclooctene epoxidation by H_2O_2 or *m*-CPBA in the presence of Fe(F_{20}TPP)Cl at different temperatures. The ^{18}O-enrichment in the epoxide gradually increased as the reaction temperature raised from −78 °C to

45 °C. The authors suggested that a putative Fe^{III}—OOR species was involved at low temperature, with a rate of the O—O bond cleavage (leading to the high-valent iron-oxo porphyrin complex) lower than the oxygen atom transfer rate ($k_1 < k_3$ in Scheme 19).

At higher temperature the formation of the high-valent iron oxo porphyrin should be favored (increase of k_1). However, since a tautomeric equilibrium is highly dependent on temperature, the present results might alternatively be re-interpreted as a fast formation of the iron-oxo species ($k_1 > k_3$), even at low temperature, but with a slow prototropy ($k_2 < k_4$) at this temperature and a faster one ($k_2 > k_4$) at higher temperature (k_3 and k_4 = oxygen atom transfer rates; the best conditions to observe the oxo-hydroxo tautomerism correspond to $k_1 > k_3$ and $k_2 > k_4$). We must note that even at low temperature (−78 °C) iron-oxo porphyrin species have been detected and characterized [55, 120, 121].

3.4.4
Differences in Exchange Kinetics for Metal(IV)-Oxo and Metal(V)-Oxo

From experiments conducted on *cis-β*-methylstyrene with manganese(V)-oxo and manganese(IV)-oxo porphyrin complexes, Groves and Stern concluded from the ^{18}O results that, in addition to differences in oxygen transfer occurring with retention or loss of the stereochemistry, the manganese(IV)-oxo slowly exchanged its oxo ligand with $H_2{}^{18}O$, while the exchange was very fast for the manganese(V)-oxo complex [108]. These data are consistent with Schemes 16 and 20 and with literature data on the proton acidity of coordinated water in high-valent species [116, 122, 123]. Going from **3** to **5** (Scheme 16) only requires a prototropy in the case of a manganese(V)-oxo species (Scheme 20; the oxo-hydroxo being the major form), whereas in the

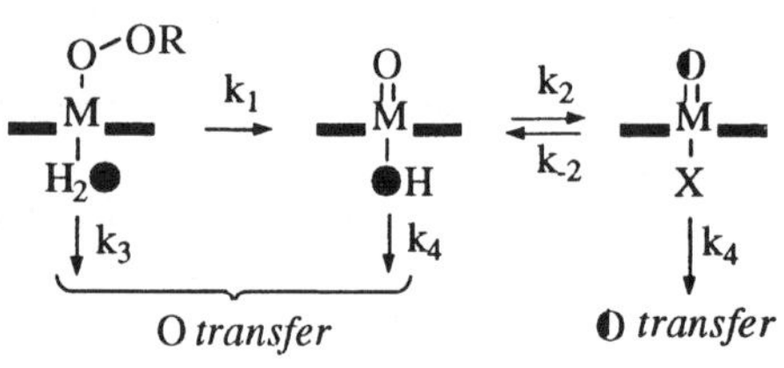

Scheme 19. Kinetic parameters depending on temperature. X = hydroxo ligand

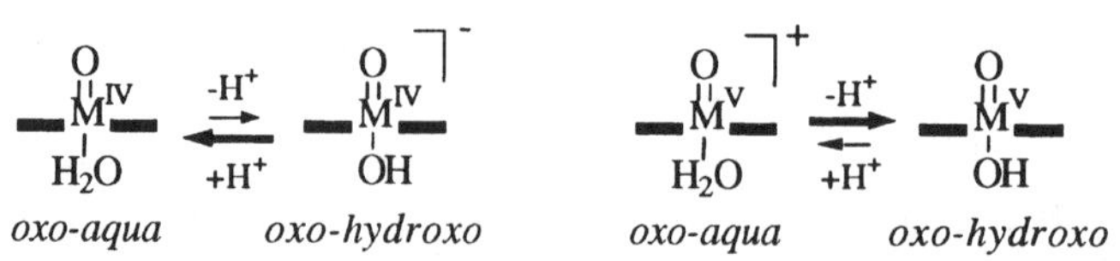

Scheme 20. Equilibria between oxo-aqua and oxo-hydroxo forms depending on the oxidation state of the metal center

case of manganese(IV)-oxo the ligand *trans* to the oxo is mainly a water molecule due to the lower acidity of the ligated water when the metal oxidation state is reduced. So, for manganese(IV)-oxo complexes, the oxo-hydroxo tautomerism can only affect the small fraction of metal-oxo species with a hydroxo ligand, the oxo-aqua form being dominant in this case (Scheme 20).

3.4.5
Role of the Ratio Water/Substrate Concentrations

The percentage of ^{18}O incorporated in oxidation products might be governed by the relative rate to reach the tautomerism equilibrium (k_2 in Scheme 19) which is function of the water concentration, and the rate of oxygen transfer (k_4) which depends on the substrate concentration. For reaction performed in an essentially aqueous medium (a 90% aqueous solution is 50 mol l^{-1} in water) and a substrate rather diluted (1 mmol l^{-1} for example), the competition between the tautomerism and the oxygen transfer was not observed: the water/substrate molar ratio ($\approx 5 \times 10^4$) was largely in favor of the oxo-hydroxo tautomerism [103]. In an organic medium containing only a small amount of water, and at high substrate concentrations, the situation is the opposite and can affect the tautomerism equilibrium and consequently the level of ^{18}O-incorporation [114]. An example of such an extreme condition concerns the epoxidation of 1 mol l^{-1} cyclohexene solution in CH_2Cl_2/CH_3OH containing 5% of $H_2{}^{18}O$ with a metalloporphyrin catalyst and H_2O_2, *t*-BuOOH, or *m*-CPBA as oxygen donor. In these conditions, no ^{18}O-incorporation was observed in the epoxide (water/substrate molar ratio ≈ 3) [101]. In intermediate conditions, such as in epoxidation of 20 mmol l^{-1} cyclooctene with H_2O_2 catalyzed by $Fe(F_{20}TPP)Cl$ in an organic medium containing increasing amounts of water, the 50% O-incorporation from water was observed for a reaction mixture containing at least 10% of water (water/substrate molar ratio above 250; Fig. 1b) [111]. In the rare case of a very small concentration of substrate, then the reaction becomes very slow and the aqua ligand (form **2** or **6** in Scheme 16) can exchange with water from solvent and the ^{18}O incorporation from solvent can rise above 50% [111].

3.4.6
Other Parameters (pH, etc.)

Several other parameters may probably influence the oxo-hydroxo tautomerism but no systematic study has been done up to now. Among them, the pH value of the reaction mixture surely plays a key role in this prototropy mechanism as also do the nature of the metal and the ligand in the metalloporphyrins through their capacity to form differently coordinated complexes. In addition, some reported variations of the ^{18}O rate of incorporation around 50% suggest that the oxo-hydroxo tautomerism equilibrium is probably tuned by small variations of kinetic parameters, including solvent effects or differences in the kinetic parameters of the oxygen transfer from the metal-oxo intermediate to the organic substrate.

4 Conclusion

More than 40 years after the discovery of cytochrome P450 enzymes, the exact nature of the active species (the putative high-valent iron-oxo intermediate) is still a matter of intensive debates. These scientific debates are in fact highly fruitful: enzymologists and structural biologists had to talk with chemists and vice versa, and data obtained with chemical models of these monooxygenases had to be compared with data obtained with the enzymes. Site-directed mutagenesis studies in relation with structural studies allowed one to check different hypotheses. All these studies have enriched the knowledge in molecular enzymology and the coordination chemistry of metal-peroxo and high-valent metal-oxo species.

Among the different methods for studying the mechanism of oxidation mediated by high-valent metal-oxo complexes, the recent discovery of the "oxo-hydroxo tautomerism" provides an additional useful tool to discuss the mechanism of catalytic O-atom transfer reactions, the nature of the axial ligand *trans* to the oxo species, and the oxidation state of these metal-oxo entities.

Acknowledgements. The authors are deeply indebted to the work of collaborators and co-workers whose names are listed in several references of this chapter.

5 References

1. Nelson DR, Koymans L, Kamataki T, Stegeman JJ, Feyereisen R, Waxman DJ, Waterman MR, Gotoh O, Coon MJ, Estabrook RW, Gunsalus IC, Nebert DW (1996) Pharmacogenetics 6: 1
2. Ortiz de Montellano PR (ed) (1996) Cytochrome P450: structure, mechanism and biochemistry. Plenum, New York
3. Ullrich V (1979) Topics in Current Chemistry 83: 67
4. Taube H (1965) J Gen Physiol 49: 29
5. Hayaishi O (ed) (1974) Molecular mechanism of oxygen activation. Academic Press, New York
6. Groves JT, Han YZ (1996) In: Ortiz de Montellano PR (ed) Cytochrome P450: structure, mechanism and biochemistry. Plenum, New York, chap 1, pp 3–48
7. Meunier B (1992) Chem Rev 92: 1411
8. Mansuy D (1987) Pure Appl Chem 59: 759
9. Ostovic D, Bruice TC (1992) Acc Chem Res 25: 314
10. Dolphin D, Traylor TG, Xie LY (1997) Acc Chem Res 30: 259
11. Vaz ADN, McGinnity DF, Coon MJ (1998) Proc Natl Acad Sci USA 95: 3555
12. Meunier B (2000) Biomimetic oxidations mediated by metal complexes. Imperial College Press, London
13. Bernadou J, Meunier B (1998) Chem Commun 2167
14. Poulos TL, Finzel BC, Howard AJ (1986) Biochemistry 25: 5314
15. Poulos TL, Finzel BC, Howard AJ (1987) J Mol Biol 195: 687
16. Griffin BW, Peterson JA (1972) Biochemistry 11: 4740
17. Atkins WA, Sligar SG (1990) Biochemistry 29: 1271

18. Poulos TL, Cupp-Vickery J, Li H (1996) In: Ortiz de Montellano PR (ed) (1996) Cytochrome P450: structure, mechanism and biochemistry. Plenum, New York, chap 4, pp 125–150
19. Ravichandran KG, Boddupalli SS, Hasemann CA, Peterson JA, Deisenhofer J (1993) Science 261: 731
20. Gleria KD, Nickerson DP, Hill HAO, Wong LL, Fülöp V (1998) J Am Chem Soc 120: 46
21. Raag R, Poulos TL (1989) Biochemistry 28: 7586
22. Hanson LK, Eaton WA, Sligar SG, Gunsalus IC, Gouterman M, Connell CR (1976) J Am Chem Soc 98: 2672
23. Caron C, Mitschler A, Rivere G, Ricard L, Schappacher M, Weiss R (1979) J Am Chem Soc 101: 7401
24. Ueno T, Kuosumi Y, Yoshizawa-Kumagaye K, Nakajima K, Ueyama N, Okamura TA, Nakamura A (1998) J Am Chem Soc 120: 12,264
25. Egawa T, Ogura T, Makino R, Ishimura Y, Kitagawa T (1991) J Biol Chem 266: 10,246
26. MacDonald IDG, Sligar SG, Christian JF, Unno M, Champion PM (1999) J Am Chem Soc 121: 376
27. El-Kasmi D, Tetreau C, Lavalette D, Momenteau M (1995) J Am Chem Soc 117: 6041
28. Benson DE, Suslick KS, Sligar SG (1997) Biochemistry 36: 5104
29. Vidakovic M, Sligar SG, Li H, Poulos TL (1998) Biochemistry 37: 9211
30. Vaz ADN, Roberts ES, Coon MJ (1991) J Am Chem Soc 113: 5886
31. Roberts ES, Vaz ADN, Coon MJ (1991) Proc Natl Acad Sci USA 88: 8963
32. Caspi E, Arunachalam T, Nelson PA (1986) J Am Chem Soc 108: 1847
33. (a) Ortiz de Montellano PR (1996) In: Ortiz de Montellano PR (ed) (1996) Cytochrome P450: structure, mechanism and biochemistry. Plenum, New York, chap 8, pp 245–303; (b) Graham-Lorenze S, Amarneh B, White RE, Peterson JA, Simpson ER (1995) Protein Science 4: 1065
34. Moali C, Boucher JL, Sari MA, Stueher DJ, Mansuy D (1998) Biochemistry 37: 10,453
35. Ortiz de Montellano PR (1998) Acc Chem Res 31: 543
36. Gerber NC, Sligar SG (1992) J Am Chem Soc 114: 8742
37. Raag R, Martinis SA, Sligar SG, Poulos TL (1991) Biochemistry 30: 11,420
38. Shaik S, Filatov M, Schröder D, Schwarz H (1998) Chem Eur J 4: 193
39. Everse J, Everse KE, Grisham MB (eds) (1991) Peroxidases in chemistry and biology (2 volumes). CRC Press, Boca Raton
40. Atkinson JF, Ingold KU (1993) Biochemistry 32: 9209
41. Toy PH, Newcomb M, Hollenberg PF (1998) J Am Chem Soc 120: 7719
42. Groves JT, McClusky GA, White RE, Coon MJ (1978) Biochem Biophys Res Commun 81: 154
43. Iyer KR, Jones JP, Darbyshire JF, Trager WF (1997) Biochemistry 36: 7136
44. Manchester JI, Dinnocenzo JP, Higgins LA, Jones JP (1997) J Am Chem Soc 119: 5069
45. Vaz ADN, McGinnity DF, Coon MJ (1998) Proc Natl Acad Sci USA 95: 3555
46. Toy PH, Newcomb M, Coon MJ, Vaz ADN (1998) J Am Chem Soc 120: 9718
47. Siegbahn PEM, Crabtree RC (1999) J Am Chem Soc 121: 117
48. Linde C, Akermark B, Norrby PO, Svensson M (1999) J Am Chem Soc 121: 5083
49. Sisemore MF, Selke M, Burstyn JN, Valentine JS (1997) Inorg Chem 36: 979
50. Selke M, Valentine JS (1998) J Am Chem Soc 120: 2652
51. Wertz DL, Sisemore MF, Selke M, Driscoll J, Valentine JS (1998) J Am Chem Soc 120: 5331
52. Ozette K, Leduc P, Palacio M, Bartoli JF, Barkigia KM, Fajer J, Battioni P, Mansuy D (1997) J Am Chem Soc 119: 6442
53. Gisdakis P, Antonczak S, Köstlmeier S, Herrmann WA, Rösch N (1998) Angew Chem Int Ed 37: 2211
54. Chen K, Que L (1999) Angew Chem Int Ed 38: 2227
55. Groves JT, Haushalter RC, Nakamura M, Nemo TE, Evans BJ (1981) J Am Chem Soc 103: 2884

56. Meunier B, Robert A, Pratviel G, Bernadou J (2000) In: Kadish K, Smith K, Guilard R (eds) The porphyrin handbook. Academic Press, San Diego, vol 4, chap 31, pp 119–187
57. Mandon D, Weiss R, Jayaraj K, Gold A, Terner J, Bill E, Trautwein AX (1992) Inorg Chem 31: 4404
58. Gadjhede M, Schuller DJ, Henriksen A, Smith AT, Poulos TL (1997) Nature Structural Biol 4: 1032
59. Urano Y, Higuchi T, Hirobe M, Nagano T (1997) J Am Chem Soc 119: 12,008
60. Robert A, Loock B, Momenteau M, Meunier B (1991) Inorg Chem 30: 706
61. Collins TJ, Gordon-Wylie SW (1989) J Am Chem Soc 111: 4511
62. Collins TJ (1994) Acc Chem Res 27: 279
63. Bottomley LA, Neely FL (1997) Inorg Chem 37: 5435
64. Du Bois J, Tomooka CS, Hason J, Carreira EM (1997) Acc Chem Res 30: 364
65. Bortolini O, Meunier B (1983) JCS Chem Commun 1364
66. Bortolini O, Ricci M, Meunier B, Friant P, Ascone I, Goulon J (1986) Nouv J Chim 10: 39
67. Ayougou K, Bill E, Charnock JM, Garner CD, Mandon D, Trautwein AX, Weiss R, Winkler H (1995) Angew Chem Int Ed Engl 34: 343
68. Groves JT, Lee J, Marla SS (1997) J Am Chem Soc 119: 6269
69. Jin N, Groves JT (1999) J Am Chem Soc 121: 2923
70. English AM, Tsaprailis G (1995) Adv Inorg Chem 43: 79
71. Meunier B (2000) In: Meunier B (ed) Biomimetic oxidations catalyzed by transition metal complexes. Imperial College Press, London (in press)
72. Dawson JH (1988) Science 240: 433
73. Penner-Hahn JE, Eble KS, McMurry TJ, Renner M, Balch AL, Groves JT, Dawson JH, Hodgson KO (1986) J Am Chem Soc 108: 7819
74. Chang CS, Yamazaki I, Sinclair R, Khalid S, Powers L (1993) Biochemistry 32: 923
75. Paeng KJ, Kincaid JR (1988) J Am Chem Soc 110: 7913
76. Chuang WJ, Van Wart HE (1992) J Biol Chem 267: 13,293
77. Ozaki SI, Inada Y, Watanabe Y (1998) 120: 8020
78. Nakajima R, Yamazaki I (1987) J Biol Chem 262: 2576
79. Ator MA, David SK, Ortiz de Montellano PR (1987) J Biol Chem 262: 14,954
80. Chang YT, Veitch NC, Leow GH (1998) J Am Chem Soc 120: 5168
81. Gajhede M, Schuller DJ, Henriksen A, Smith AT, Poulos TL (1997) Nat Struct Biol 4: 1032
82. Klibanov AM, Berman Z, Alberti BN (1981) J Am Chem Soc 103: 6263
83. Ortiz de Montellano PR, Choe YS, DePillis G, Catalano CE (1987) J Biol Chem 262: 11,641
84. Baciocchi E, Lanzalunga O, Malandrucco S (1996) J Am Chem Soc 118: 8973
85. Griffin BW (1991) In: Everse J, Everse KE, Grisham MB (eds) Peroxidases in chemistry and biology, vol II. CRC Press, Boca Raton, pp 85–137
86. Blanke SR, Hager LP (1988) J Biol Chem 263: 18,739
87. Sundaramoorthy M, Terner J, Poulos TL (1995) Structure 3: 1367
88. Sundaramoorthy M, Terner J, Poulos TL (1998) Chem & Biol 5: 461
89. Egawa T, Miki H, Ogura T, Makino R, Ishimura Y, Kitagawa T (1992) FEBS Let 305: 206
90. Wagenknecht HA, Woggon WD (1997) Chem & Biol 4: 367
91. Geigert J, Lee TD, Dalietos DJ, Hirano DS, Neidleman SL (1986) Biochem Biophys Res Commun 136: 778
92. Debrunner PG, Dexter AF, Schulz CE, Xia YM, Hager LP (1996) Proc Natl Acad Sci USA 93: 12,791
93. Allain EJ, Hager LP, Deng L, Jacobsen EN (1993) 115: 4415
94. Dexter AF, Lakner FJ, Campbell RA, Hager LP (1995) 117: 6412
95. Colonna S, Gaggero N, Manfredi A, Casella L, Gullotti M, Carrea G, Pasta P (1990) Biochemistry 29: 10,465
96. Colonna S, Gaggero N (1994) Phosph Sulf Silicon 95–96: 103
97. Kobayashi S, Nakano M, Kimura T, Schaap AP (1987) Biochemistry 26: 5019

98. Van Handel MJH, Primus JL, Teunis C, Boersma MG, Osman AM, Veeger C, Rietjens IMCM (1998) Inorg Chem Acta 275/276: 98
99. Meunier B, Guilmet E, De Carvalho ME, Poilblanc R (1985) J Am Chem Soc 106: 6668
100. Robert A, Meunier B (1988) New J Chem 12: 885
101. Nam W, Valentine JS (1993) J Am Chem Soc 115: 1772
102. The denomination "oxo-hydroxo tautomerism" which is reminiscent of the keto-enol tautomerism appears to be more precise than "redox tautomerism" that we initially proposed in [103]
103. Bernadou J, Fabiano AS, Robert A, Meunier B (1994) J Am Chem Soc 116: 937
104. Ingold KU, MacFaul PA (2000) In: Meunier B (ed) Biomimetic oxidations catalyzed by transition metals. Imperial College Press, London, chap 2 (in press)
105. Hashimoto S, Tatsuno Y, Kitagawa T (1986) Proc Natl Acad Sci USA 83: 2417
106. Makino R, Uno T, Nishimura Y, Iizuka T, Tsuboi M, Ishimura Y (1986) J Biol Chem 261: 8376
107. Hashimoto S, Tatsuno Y, Kitagawa T (1987) J Am Chem Soc 109: 8096
108. Groves JT, Stern MK (1987) J Am Chem Soc 109: 3812
109. Groves JT, Stern MK (1988) J Am Chem Soc 110: 8628
110. Balahura RJ, Sorokin A, Bernadou J, Meunier B (1997) Inorg Chem 36: 3488
111. Lee KA, Nam W (1997) J Am Chem Soc 119: 1916
112. Pitié M, Bernadou J, Meunier B (1995) J Am Chem Soc 117: 2935
113. Song R, Sorokin A, Bernadou J, Meunier B (1997) J Org Chem 62: 673
114. Yang SJ, Nam W (1998) Inorg Chem 37: 606
115. Prince S, Körber F, Cooke PR, Lindsay Smith JR, Mazid MA (1993) Acta Cryst C49: 1158
116. Jeon S, Bruice TC (1992) Inorg Chem 31: 4843
117. Nam W, Goh YM, Lee YJ, Lim MH, Kim C (1999) Inorg Chem 38: 3238
118. Wietzerbin K, Muller JG, Jameton RA, Pratviel G, Bernadou J, Meunier B, Burrows CJ (1999) Inorg Chem 38: 4123
119. Sorokin A, Meunier B (1998) Eur J Inorg Chem 1269
120. Balch AL, Chan YW, Cheng RJ, La Mar GN, Latos-Grazynski L, Renner MW (1984) J Am Chem Soc 106: 7779
121. Groves JT, Gross Z, Stern MK (1994) Inorg Chem 33: 5065
122. Kaaret TW, Zhang GH, Bruice TC (1991) J Am Chem Soc 113: 4652
123. Rachlewicz K, Latos-Grazynski L (1996) Inorg Chem 35: 1136

Nucleophilicity of Iron-Peroxo Porphyrin Complexes

Diana L. Wertz[1], Joan Selverstone Valentine[2]

Department of Chemistry and Biochemistry, University of California Los Angeles, Los Angeles, CA 90095-1589 USA

[1] *E-mail: wertz@chem.ucla.edu*

[2] *E-mail: jsv@chem.ucla.edu*

For the past 20 years, cytochrome P450 researchers have sought to identify and to characterize the reactive intermediates in reactions of these enzymes. This review focuses on one of those postulated intermediates, the ferric heme peroxo complex, $[(porphyrin)Fe(III)(O_2^{2-})]^-$, a species which has been postulated to be formed transiently in the P450 catalytic cycle. Ferric peroxo porphyrin complexes, inorganic complexes that model the peroxo species, have been synthesized and their chemical reactivities characterized for comparison with the enzymes. Such studies have identified certain peroxo porphyrins as remarkably strong nucleophiles capable of oxidizing a variety of electron-poor molecules. While the ferric heme peroxo intermediate, in the majority of P450 enzymes, rapidly converts to an oxoferryl species, some enzymes, e.g., aromatase, lanosterol 14α-demethylase, progesterone 17α-hydroxylase/17,20-lyase, and NO synthase, appear to use this intermediate as the active oxidant. Additionally, studies of ferric peroxo porphyrin complexes have increased our understanding of the nature of the P450 catalytic cycle and of the mechanisms of generation of other reactive intermediates used in P450 enzymes.

Keywords: Iron-Peroxo, Superoxide, Cytochrome P450, Metal Peroxo, Oxoferryl

List of Abbreviations

CH_3CN Acetonitrile
DMSO Dimethyl sulfoxide
$[Fe(F20TPP)O_2]^-$ Peroxoferritetra(pentafluorophenyl)porphyrin
$[Fe(OEP)O_2]^-$ Peroxoferrioctaethylporphyrin
$[Fe(PPIXDME)O_2]^-$ Peroxoferriprotoporphyrin IX dimethyl ester
$[Fe(TMP)O_2]^-$ Peroxoferritetramesitylporphyrin
$[Fe(TPP)O_2]^-$ Peroxoferritetraphenylporphyrin
K222 Kryptate 222
KO_2^- Potassium superoxide
$[Mn(TPP)O_2]^-$ Manganic peroxo tetraphenylporphyrin
NOS Nitric oxide synthase
P^{2-} Porphyrin ligand
$P^{\bullet -}$ Porphyrin cation radical formed from oxidation of the porphyrin ligand
PPh_3 Triphenylphosphine
THF Tetrahydrofuran

1 Introduction

The cytochrome P450 enzymes comprise a large family of heme-containing enzymes with a cysteine in the active site that acts as an axial ligand to the heme iron. To date, over 470 individual cytochrome P450 sequences have been reported [1]. Cytochrome P450 enzymes are monooxygenases, i.e., they catalyze the incorporation of one oxygen atom of dioxygen into substrate while the other oxygen atom is reduced by two electrons (usually from NADPH) to give water (Eq. 1) [2]:

$$\text{Substrate} + O_2 + 2e^- + 2H^+ \xrightarrow{\text{P450}} \text{Substrate(O)} + H_2O \qquad (1)$$

There is great diversity in reactivity among the different P450 enzymes with over 40 different known types of transformations [3, 4]. The most often encountered processes are hydroxylation, epoxidation, heteratom dealkylation, heteratom oxidation, and alcohol oxidation. This diversity makes the P450 enzyme family one of the body's most important resources for the biotransformation of naturally occurring biomolecules and for the oxidative metabolism of xenobiotics. In fact, Coon et al. recently predicted that each mammalian species will be found to have up to one hundred P450 isoforms, a thousand or more inducers of P450, and a million potential substrates [3]. The ability of one enzyme family to participate in such varied transformations can be understood by examining the P450 catalytic cycle, the process by which the active oxidants are formed. As we will see, slight changes in the enzyme active site may control the formation of one active oxidant vs another and thus provide some measure of flexibility to the P450 family [5].

1.1 The P450 Catalytic Cycle

The catalytic cycle of cytochrome P450 enzymes begins with the binding of substrate to the six-coordinate low-spin ferric resting form of the enzyme (Fig. 1) [2, 5]. Binding of substrate displaces the water ligand and induces a change in spin state from the low-spin resting form to the high-spin ferric form, a form in which reduction is more thermodynamically favorable. This high-spin ferric species is then reduced to the ferrous state by an external reductant, most often a closely associated P450 reductase. Binding of dioxygen to the ferrous porphyrin species generates a low-spin ferrous-dioxygen complex, $[(P^{2-})Fe(II)(O_2)]$, or a ferric superoxide complex, $[(P^{2-})Fe(III)(O_2^-)]$. It is unclear at this time which of the two resonance structures contributes most to the resonance hybrid of this intermediate, and it may well be that the charge distribution of this intermediate will vary depending on the specific active site of each individual cytochrome P450 enzyme.

The catalytic cycle continues as the ferrous-dioxygen complex accepts another electron to form a ferric peroxo species, $[(P^{2-})Fe(III)(O_2^{2-})]^-$, an intermediate which is presumably usually transient but which may, in a subset of the enzymes, serve as a nucleophilic oxidant in direct reactions with the enzyme-bound substrate (see below). The nucleophilic reactivity of the ferric peroxo species might be rationalized by considering it is analogous to

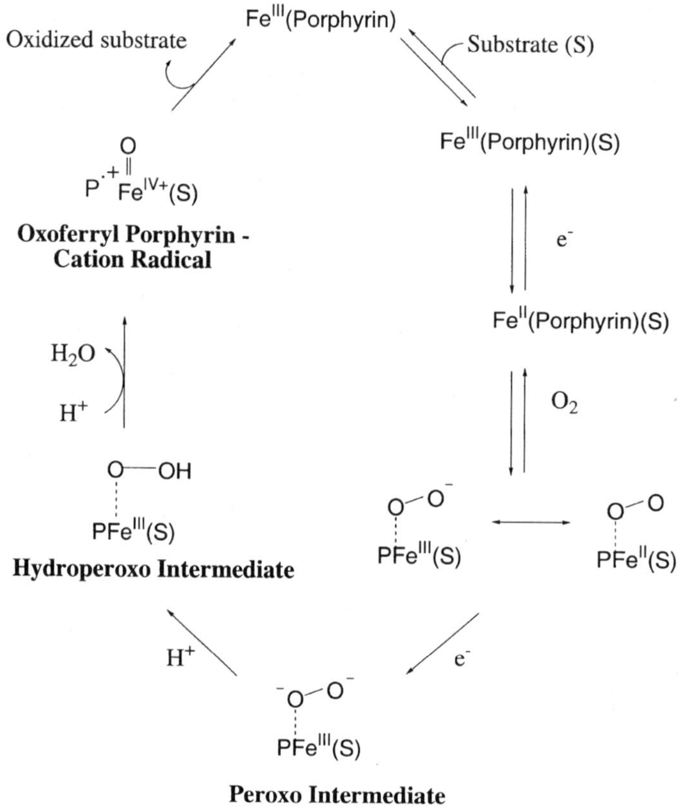

Fig. 1. A simplified version of the P450 catalytic cycle. The timing of proton donation events and axial ligand coordination probably control which species acts as the active oxidant in a particular enzymatic system

HOO^- where the η^1 coordination to the ferric ion is analogous to the H—O bond of deprotonated peroxide. In most cases, however, the ferric peroxo complex is assumed to be rapidly protonated to give the ferric hydroperoxo intermediate, $[(P^{2-})Fe(III)(HOO^-)]$. Again this intermediate is usually considered to be transient in nature, although it has also at times been proposed to react directly with substrate [6, 7]. The reactivity of this species, $[(P^{2-})Fe(III)(HOO^-)]$, would be expected to be analogous to the reactivity of H_2O_2, with the ferric ion having been substituted for a proton; thus, the ferric hydroperoxide species would be expected to be an electrophilic oxidant rather than a nucleophilic oxidant. The delivery of a second proton to the ferric hydroperoxo species results in heterolytic oxygen-oxygen bond cleavage to give an oxoferryl porphyrin cation radical species, $[(P^{\bullet-})Fe(IV)(O^{2-})]^+$. The exact mechanism for the delivery of the two protons to the ferric peroxo intermediate is currently unknown [5]. An alternative to the sequential proton delivery pathway presented above is a concerted pathway in which the protons are delivered contemporaneously to give the oxoferryl porphyrin cation radical species directly.

A great deal of effort has been directed towards understanding the oxoferryl porphyrin cation radical intermediate because most of the transformations of the P450 enzyme family are accomplished through the reaction of substrate with this species. Many excellent reviews that discuss oxoferryl species reactivity and model complexes are available to the reader [2, 8]. The early intermediates, such as the ferric peroxo, have received less attention due in part to the general perception that these intermediates are not the primary active oxidants used in P450 catalysis. Recently this view has changed as several enzymes now appear to use early intermediates to achieve oxidative chemistry [2, 6, 7].

2 Ferric Peroxo Intermediate in P450 Systems

This review will discuss efforts that have been directed towards the study of enzymes using the ferric peroxo species as the active oxidant. We will then describe attempts that have been made to model this active site species and discuss what is known about the reactivity and properties of iron peroxo porphyrin model complexes. These biological and model studies have furthered our understanding of the modes of reactivity available to the ferric peroxo species and have provided us with a framework from which the P450 catalytic cycle and P450 enzymes in general can be better understood.

2.1 Ferric Peroxo Species as an Active Oxidant in Biological Systems

While in most P450 enzymes the peroxo intermediate is rapidly converted to the oxoferryl species upon proton donation (Fig. 2), a subset of P450 enzymes have active sites in which proton donation to the peroxo species may be inhibited. In these enzymes it then becomes possible for the peroxo intermediate, with a now extended lifetime, to act as an active oxidant when

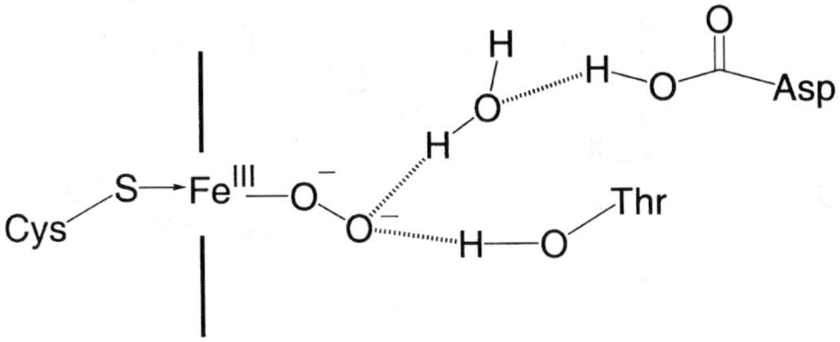

Fig. 2. The active site arrangement of a P450 enzyme where oxoferryl formation would be favored

given the appropriate substrate. For example, one mechanism that may effectively inhibit proton donation is the disruption of hydrogen bonding between the active site water and hydroxyl group and the oxygen atoms of the peroxo by the proper placement of a substrate molecule (Fig. 3) [9].

Evidence for the ferric peroxo species as an active oxidant has been obtained through the study of wild type P450 enzymes as well as of P450 mutants in which the appropriate proton-donating residue is mutated. Several enzymes that appear to use ferric peroxo species to accomplish direct oxidations are now known; experiments with these enzymes that supported the feasibility of the ferric peroxo species as the reactive oxidizing intermediate are described below.

2.1.1
P450 Aromatase

Cytochrome P450 aromatase is perhaps the most thoroughly studied of the enzymes that appear to use the iron peroxo species as an active intermediate. Interest in aromatase stems from the uniqueness of the transformation that it catalyzes, i.e., the conversion of androgens to estrogens [10]. Additionally, aromatase is studied for the biological significance that it holds for humans. Studies indicate that aromatase levels are higher in breast cancer tissue vs normal breast tissue [3]. The production of estrogen may also be important for the development and maintenance of memories [11].

Aromatase catalyzes the conversion of human androgens to estrogens in three steps as shown in Fig. 4 [10]. Each transformation requires one equivalent each of NADPH and dioxygen. The first step involves the hydroxylation of the C19 methyl group to give the C19 alcohol. The second step involves a second hydroxylation of the C19 carbon to give the C19 diol, which presumably loses water rapidly to become the C19 aldehyde. Both the first and second steps are thought to be typical P450 type hydroxylations mediated by the oxoferryl porphyrin cation radical species. The unusual third step, however, requires the cleavage of the C19 aldehyde to give an aromatized A ring [12]. The active oxidant that performs this remarkable transformation

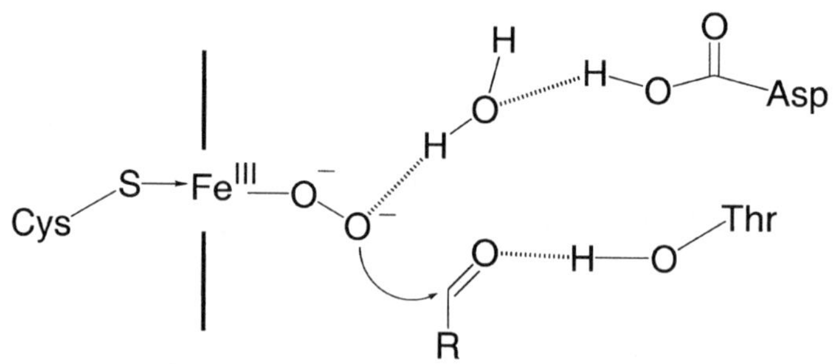

Fig. 3. The active site arrangement of a P450 enzyme where the ferric peroxo intermediate would be the favored active oxidant

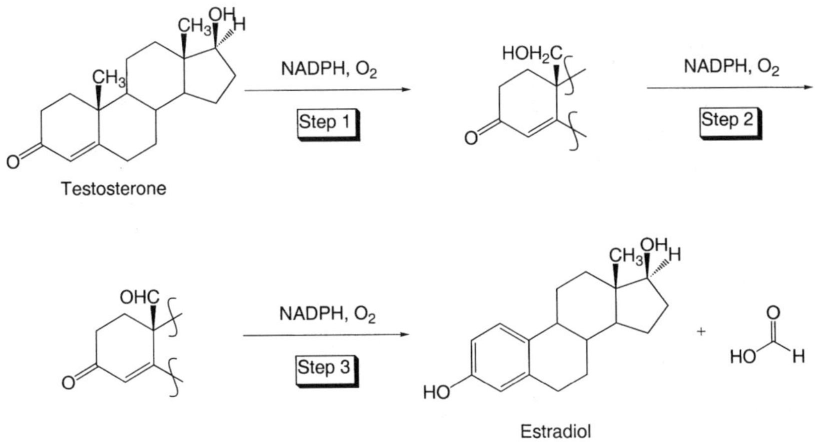

Fig. 4. The conversion of androgens to estrogens by P450 aromatase

has been the subject of debate among P450 researchers for some time [6, 12–15]. Proposed active oxidants for the third step have included the oxoferryl species, the ferric hydroperoxo species, and the ferric peroxo species. However, much evidence (see below) has accumulated that points to the ferric peroxo porphyrin species as the most likely candidate for the active oxidant of aromatase's third catalytic step.

The first researcher to propose the ferric peroxo species as the active oxidant was Akhtar et al. in 1981 [10, 12]. Through elegant isotopic labeling studies, Akhtar et al. established that one atom of the dioxygen taken up by the enzyme for the third step is incorporated into the formic acid product [12]. In the same study, the hydrogen of formic acid was shown to be the original hydrogen of the C19 aldehyde. Additional labeling studies demonstrated that the other oxygen in formic acid originated from the dioxygen required for the first hydroxylation step [16].

The availability of several P450 crystal structures, including the first by Poulos of P450cam [17], allowed Graham-Lorence et al. to construct a three-dimensional hypothetical model of the aromatase active site [9]. One active site residue that they identified as being potentially important in determining the mechanistic pathway of the enzyme is Thr310. That threonine residue is positioned in the active site so that hydrogen bonding to the oxygen of the C19 aldehyde is possible. This hydrogen bonding increases the electrophilicity of the aldehyde. The fact that Thr310 may function to activate the substrate suggests that proton donation to the ferric peroxo species may be prohibited. Graham-Lorence et al. suggest that this active site setup may serve to increase the nucleophilicity of the ferric peroxo species and prevent formation of the ferric hydroperoxo intermediate and its subsequent degradation to the oxoferryl species. One potential mechanism for the third step of aromatase was proposed by Graham-Lorence et al. and is shown in Fig. 5 [9].

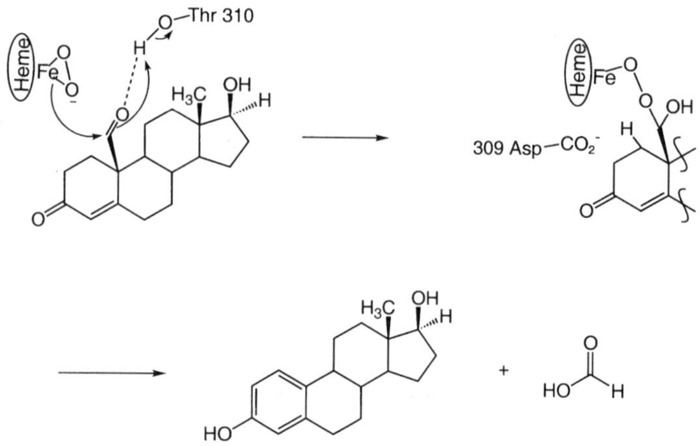

Fig. 5. One potential mechanism of the third step in the conversion of testosterone to estradiol by P450 aromatase

Akhtar, Groves and others propose that the ferric peroxo species attacks the C19 aldehyde to give an iron peroxohemiacetal intermediate [6, 8, 18]. The peroxohemiacetal may then decompose in a concerted or stepwise fashion [19, 20]. Although the degradation pathway of the acetal intermediate is not known, Townsley has established that the 1β (β = above) proton is selectively removed and that enolization of the α,β-unsaturated ketone moiety is necessary in order to produce a fully aromatized steroid A ring [21]. Substrate analogs made and studied by Oh and Robinson suggest that enolization prior to deformylation is favorable [18]. Additionally, the active site modeling studies of Graham-Lorence et al. indicate that Asp309 is in close enough proximity to the C2 protons to act as a base capable of generating the enolized A ring [9].

2.1.2
Cytochrome P450 2B4 and Cytochrome P450 2E1

Convincing evidence for the existence of an iron peroxo intermediate as an active oxidant in biological systems has arisen from mutagenesis studies of several microsomal P450 enzymes [3]. Cytochrome P450 2B4 and cytochrome P450 2E1 are members of the P450 family that are able to hydroxylate substrates as well as deformylate aldehydes [22–26]. Studies of P450 2B4 and P450 2E1 mutants have provided evidence for the use of the peroxo and hydroperoxo intermediates if protons in the active site are not available to promote O—O bond cleavage [7].

In the enzyme P450 2B4, an active site threonine residue is believed to act as a proton donor. Mutation of that residue to alanine (T302A) resulted in greatly decreased hydroxylation of substrates and a five- to tenfold increase in deformylation of cyclohexanecarboxaldehyde through what is believed to be the peroxo intermediate [25]. Vaz et al. had shown in 1994 that microsomal P450 2B4 is able to aromatize an isotopically labeled truncated version of the aromatase natural substrate with selective removal of the 1β hydrogen, as is seen in P450 aromatase [22]. Therefore, it appears that P450 2B4 has an ability to act in a manner similar to aromatase. Removal of the Thr302 active site residue presumably shifts the active oxidant to predominately the ferric peroxo species and, therefore, deformylation is greatly enhanced. Remarkably, the T302A mutation of P450 2E1 resulted in increased epoxidative reactivity towards styrene and decreased allylic hydroxylation activity towards cyclohexene and butene [7]. In cytochrome P450 2E1, trends in the ratios of the rates of epoxidation and hydroxylation support the conclusion that the active oxidant is the ferric hydroperoxo-iron species. Therefore, the removal of the Thr302 amino acid active site proton may shift the protonation equilibrium towards the ferric peroxo intermediate in the case of P450 2B4 and towards a ferric hydroperoxide intermediate in the case of P450 2E1.

The studies discussed above illustrate the idea that an early iron peroxo species will have a sufficient lifetime to act as an active oxidant if proton delivery to the oxygens of the peroxo is inhibited [7, 27–29]. It is important to note that many other researchers have studied mutated P450 enzymes in order to understand how active site proton donating ability affects reactivity; however, we have only presented studies which have direct relevance to the mechanism of aromatase [28–30]. Additionally, Sligar et al. have recently reported the spectroscopic characterization of a reduced ferrous dioxygen species in P450 cam [31]. Density functional calculations by Waskell et al. corroborate the assignment of the observed species as the peroxo intermediate [32].

2.1.3
Lanosterol P450-14α-demethylase

The enzyme lanosterol-14α-demethylase performs a three step transformation of lanosterol that is reminiscent of the transformation performed by aromatase (Fig. 6) [4, 33]. Consequently, it is tempting to assume the mechanisms are similar [6]. Like aromatase, the first two steps in the oxidation of lanosterol are typical P450-type hydroxylations. It is at this point, however, that the two

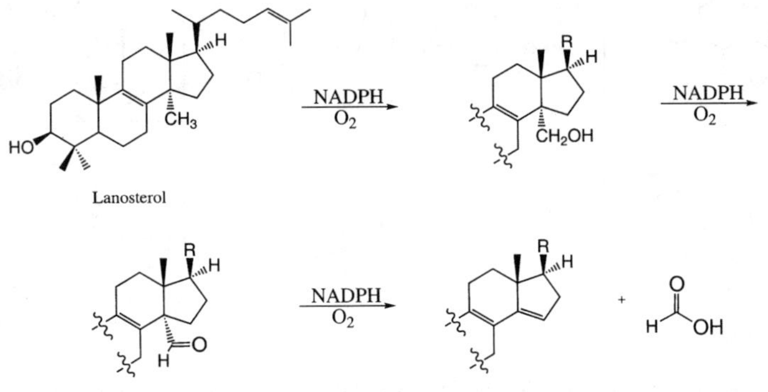

Fig. 6. The deformylation of lanosterol by lanosterol 14α-demethylase

enzymes diverge mechanistically. Instead of direct attack of the ferric peroxo complex onto the aldehyde and direct decomposition as is seen in aromatase, lanosterol-14α-demethylase probably forms an ester intermediate as shown in Fig. 7 [34]. There is much support for the proposal that the third step involves a ferric peroxo mediated Baeyer-Villiger type reaction. The ester shown in Fig. 7 has been isolated and has been shown to proceed to the olefin product in the presence of lanosterol 14α-demethylase [35].

2.1.4
Progesterone 17α-Hydroxylase/17,20-Lyase

The enzyme progesterone 17α-hydroxylase/17,20-lyase catalyzes the cleavage of the C17 side chain of progesterone (or pregnenolone) to form androstenedione (or dehydroepiandrostene) [33]. This unique enzyme is able to follow a hydroxylase/lyase pathway (as shown by the top pathway in Fig. 8) or an aldehyde cleavage pathway (as shown at the bottom of Fig. 8). Recent studies by Akhtar et al. [36] and by Swinney and Mak [37] used isotopic labeling to establish that the reaction does not proceed via an oxoferryl species as the

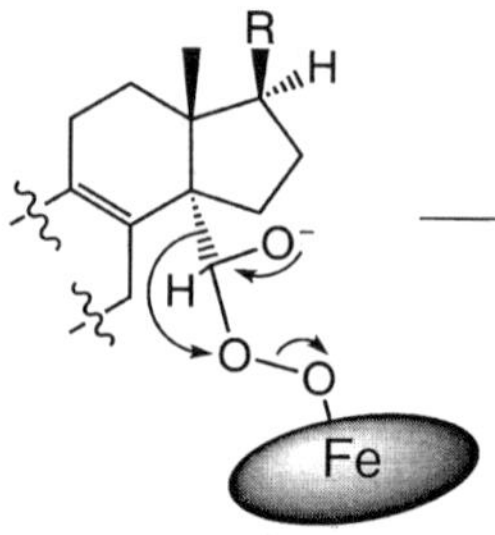

Fig. 7. The biological Baeyer-Villiger type reaction proposed as a part of the mechanism for the deformylation of the D ring of lanosterol by lanosterol 17α-demethylase

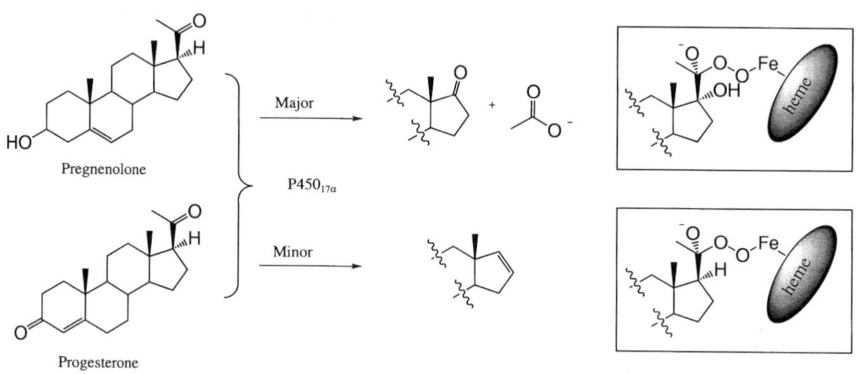

Fig. 8. The two potential oxidative pathways catalyzed by progesterone 17α-hydroxylase/17,20-lyase

active oxidant. Most evidence supports a ferric peroxo porphyrin species being the active oxidant in both pathways. In the hydroxylation/lyase sequence, the substrate is hydroxylated alpha to the carbonyl of the ketone by an oxoferryl species, as is typical of hydroxylation reactions. In the lyase step, however, the ferric peroxo porphyrin is proposed to attack the ketone moiety to form a peroxyhemiketal type intermediate [38]. The mode of decomposition of this intermediate is currently unknown, but it is most likely radical in nature (although a Baeyer-Villiger route is also possible) [39]. A minor pathway as shown on the bottom of Fig. 8 is also possible if the peroxyhemiketal intermediate forms before hydroxylation [4].

2.1.5
NO Synthase

NO synthase (NOS) catalyzes the five-electron oxidation of L-arginine to L-citrulline and nitric oxide, as shown in Fig. 9 [8]. Spectroscopic evidence presented by Degray et al. [40] and Harris et al. [41] suggested that NOS can be classified as a P-450 enzyme. The transformation of L-arginine to L-citrulline is divided into two separate steps. Crane proposes that in the first oxidative step a ferric peroxo porphyrin intermediate removes a proton from a guanidinium moiety of L-arginine [45]. Upon protonation of the ferric peroxo porphyrin complex, the intermediate then decomposes to an oxoferryl cation radical porphyrin species which is then capable of hydroxylating the nitrogen of the guanidinium. Before the next step, an electron transfer must occur, presumably from the substrate to the enzyme [8]. Then, in the second step, it is proposed that a ferric peroxo porphyrin species attacks the carbon of the urea oxime moiety as shown in Fig. 10 [43]. At this time, the mechanism and active oxidant identity during each transformation are subjects of intense debate, but the recent crystal structures of NOS coupled with new research findings may help resolve these issues [42, 44].

3
Inorganic Model Complexes of the Ferric Peroxo Intermediate

Researchers have been actively studying ferric peroxo porphyrin model complexes for over 20 years because of their relevance to the above enzymatic

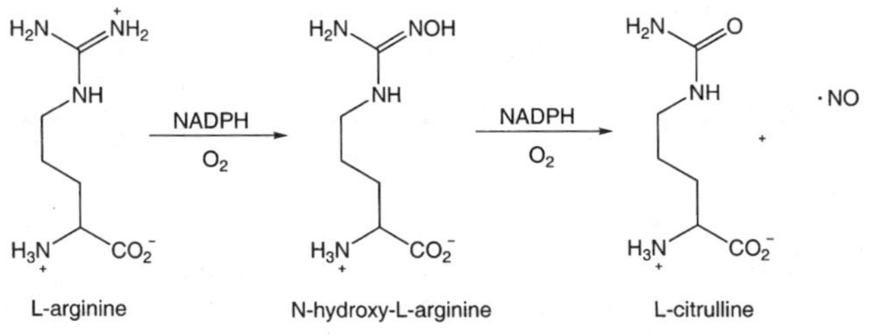

Fig. 9. The conversion of L-arginine to L-citrulline which is catalyzed by NO synthase

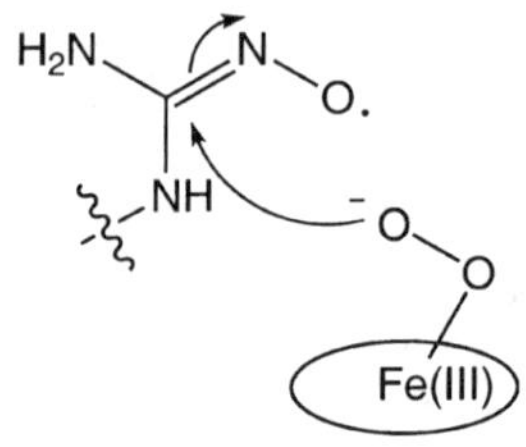

Fig. 10. One mechanism that has been proposed for the carbon-nitrogen bond cleavage step catalyzed by NO synthase

transformations. This section covers the synthesis and characterization of various ferric peroxo porphyrin complexes. In later sections, we review characterization of the reactivity of ferric peroxo porphyrin complexes as well as consider studies which attempt to model the basic transformations that occur in the above-mentioned enzymes.

3.1 Synthesis of Ferric Peroxo Porphyrin Complexes

The synthesis of a ferric peroxo porphyrin complex, $[Fe(TPP)O_2]^-$, was first reported in 1978 [45, 46]. Valentine et al. reported a preparation that took advantage of the solubility of potassium superoxide in the presence of crown ethers [47]. The synthesis, as shown in Fig. 11, is similar to the P450 catalytic cycle in that it provides two electrons and a dioxygen molecule to give an iron(III) peroxo porphyrin. The synthesis is possible only in aprotic solvents such as acetonitrile and toluene due to the disproportionation of superoxide in protic solvents. Khenkin et al. also reported the synthesis of a compound with the spectroscopic properties of the ferric peroxo porphyrin complex, but it was assumed to be a ferrous superoxo porphyrin complex at the time [46].

Another synthetic route to ferric peroxo porphyrin complexes was reported in 1981 when Welborn et al. [48] reported the one-electron reduction of a porphyrin ferrous dioxygen complex to give a ferric peroxo porphyrin complex. A third synthetic route reported by Reed in 1982 illustrated that

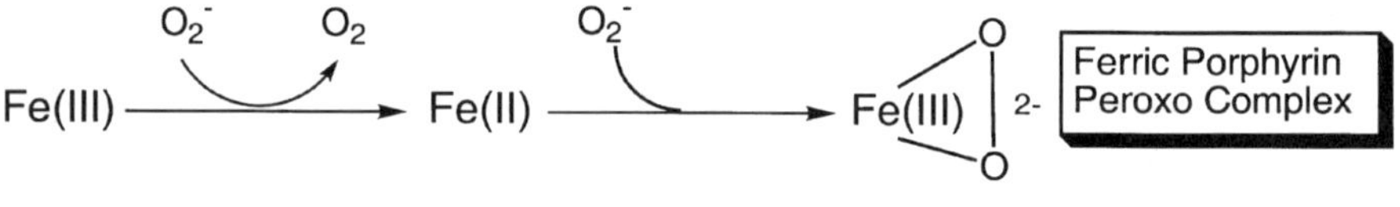

Fig. 11. The synthesis of ferric porphyrin peroxo complexes is possible by the addition of two equivalents of superoxide to a ferric porphyrin complex. The first equivalent of superoxide reduces the ferric porphyrin to the ferrous porphyrin. Oxidative addition of a superoxide molecule to the ferrous porphyrin produces the ferric η^2-peroxo porphyrin complex

dioxygen addition to an iron(I) porphyrin compound also yielded the ferric peroxo porphyrin complex [49].

3.2 Characterization of Ferric Peroxo Porphyrin Complexes

The $[Fe(OEP)O_2]^-$ complex was reported in 1980, along with spectroscopic characterization of both $[Fe(OEP)O_2]^-$ and $[Fe(TPP)O_2]^-$ [47]. In order to establish whether the peroxide ligand was bound in an η^1 or an η^2 fashion, the frequency of the O—O bond in $[Fe(OEP)O_2]^-$ was measured by IR spectroscopy. η^1-Superoxo type metal complexes typically exhibit oxygen-oxygen bond frequencies in the 1030–1180 cm^{-1} region while those of η^2-peroxo type metal complexes are in the 750–950 cm^{-1} region [50]. The 806 cm^{-1} vibration observed for $[Fe(OEP)O_2]^-$, along with the observed shift in frequency to 759 cm^{-1} upon substitution with $^{18}O_2$, suggest that the actual structure is a ferric η^2-peroxo complex [47]. Further evidence for a η^2-peroxo type metal complex is provided by the EPR spectrum of $[Fe(OEP)O_2]^-$. Signals at $g = 4$, $g = 8$ (weak), and $g = 2$ (weak) indicate the presence of a high spin rhombic ferric complex. The observed rhombicity supports the hypothesis that the peroxo ligand in $[Fe(OEP)O_2]^-$ is bound to iron in an η^2 fashion. The deuterium NMR characterization of the $[Fe(TPP)O_2]^-$ complex by Shirazi and Goff is also consistent with a high spin ferric species [51].

Although a crystal structure of a ferric peroxo porphyrin is still not available, EXAFS studies on $[Fe(TPP)O_2]^-$ are consistent with a peroxide bound in an η^2 fashion [52]. Additionally, the $[Mn(TPP)O_2]^-$ complex has been crystallized and is known to contain a peroxo ligand bound to manganese in an η^2 fashion, as shown in Fig. 12 [53]. Mossbauer, EPR, and magnetic susceptibility data for $[Fe(TPP)O_2]^-$ are all consistent with a geometry for the ferric peroxo complex that is similar to that observed for $[Mn(TPP)O_2]^-$ [54]. Therefore, the crystal structure of the manganese peroxo porphyrin probably reflects the nature of the ferric peroxo porphyrin as well.

4 Reactivity of Ferric Peroxo Porphyrin Complexes

After the initial synthesis of $[Fe(TPP)O_2]^-$, many additional ferric peroxo porphyrin complexes were synthesized and studied in order to determine whether they exhibited nucleophilic or electrophilic properties. Attempts to react ferric peroxo porphyrins with electron-rich substrates such as styrene and triphenylphosphine have resulted in little or no reactivity. For example, Welborn et al. [48] observed no epoxidation of styrene while a recent study showed virtually no oxidation of triphenylphosphine in THF [55, 56]. These studies established that ferric peroxo porphyrin complexes are not, by nature, *electrophilic*. The attention of researchers then shifted in order to explore the potential for nucleophilic reactivity.

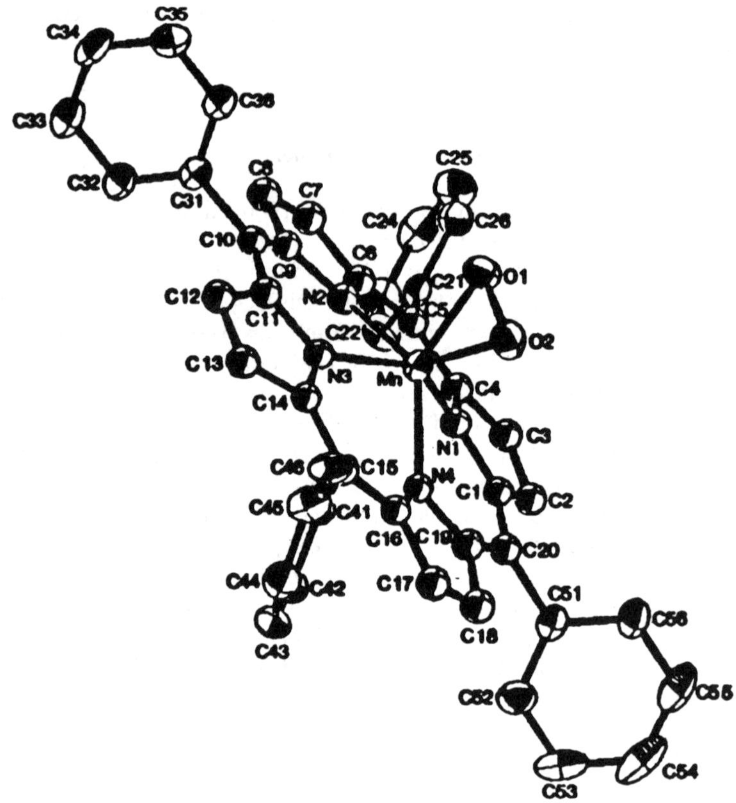

Fig. 12. The crystal structure of $[Mn(TPP)O_2]^-$ complex as determined by Van Atta et al. [53] The associated cation, potassium K222, has been removed for this figure

4.1
Demonstration of the Nucleophilic Nature of Ferric Peroxo Porphyrin Complexes

The first demonstration of the *nucleophilic* nature of a ferric peroxo complex was in 1982 when Khenkin and Shteinmann reacted $[Fe(TPP)O_2]^-$ with acetic anhydride or acetyl chloride and observed the oxidation of cyclohexane to cyclohexanol [57]. They postulated that the ferric peroxo porphyrin complex reacts with the acyl anhydride at –70 °C to form, presumably, a peracetic acid iron(III) porphyrin complex. This species then degrades above –70 °C to give the high valent iron oxo species which is capable of alkane hydroxylation. Khenkin and Shteinmann later published a full report on the subject in 1984 and provided further evidence to support the formation of the peracetic acid iron(III) porphyrin complex [58].

Shappacher and Weiss [59] and Groves and Watanabe [60] observed additional examples of ferric peroxo porphyrins reacting with electrophiles to produce iron(IV) oxo porphyrin species. In 1985 Schappacher and Weiss demonstrated that ferric peroxo picket fence porphyrins react with carbon dioxide to give an iron(IV) oxo porphyrin adduct [59]. In this case a monoperoxy carbonate iron(III) porphyrin intermediate was not detected but presumably formed. Groves and Watanabe also reported the reaction of ferric peroxo complexes with acid anhydrides and acyl halides to produce the iron(IV) oxo porphyrin adduct [60]. Therefore, by 1986 it was established that ferric peroxo complexes could be activated by addition of an acyl group, and presumably a proton, to give high valent iron oxo species. This type of reaction is similar to protonation of the ferric peroxo porphyrin and subsequent degradation of the hydroperoxo species as seen in Fig. 1.

It is also possible to generate lower valent porphyrin products by reacting the ferric peroxo complex with electrophiles. For example, Mitzkal et al. determined that $[Fe(TPP)O_2]^-$ reacted with SO_2 to form the [Fe(II)(TPP)] $(DMSO)_2$ complex as the porphyrin product [61]. SO_4^{2-} was presumed to be the primary product of SO_2 oxidation, although, the authors actually observed $S_2O_7^{2-}$, a product of sulfate condensation.

These early studies of ferric peroxo porphyrin reactivity were important for several reasons. First, the ferric peroxo porphyrin could now be classified as a nucleophilic oxidant. Second, the peroxyacetate ferric porphyrin intermediate generated in the reaction of the ferric peroxo porphyrin with acyl anhydrides and acetic anhydride was also of interest to researchers because oxygen-oxygen bond cleavage was shown to occur to give higher valent iron oxo porphyrin compounds. While these studies were the first models of the events that occur in P450 actives sites to generate the oxoferryl intermediate, they did not necessarily help to establish the ferric peroxo species as a viable active oxidant in biological systems. Later work would provide support for the idea that the ferric peroxo species may be an important active oxidant of electrophilic substrates in certain select P450 enzymes. Several of the studies that helped to establish the viability of the ferric peroxo porphyrin species as a primary oxidant are next considered.

4.2
Epoxidation of Electron-Poor Enones by Ferric Peroxo Porphyrin Complexes

An important step towards establishing the oxidizing ability of ferric peroxo porphyrin complexes came when Sisemore et al. demonstrated that the very nucleophilic $[Fe(TMP)O_2]^-$ epoxidized electron-deficient olefins [62]. The epoxidation, shown in Fig. 13, was the first demonstration of direct oxygen insertion into an olefinic bond by a ferric peroxo porphyrin complex. In the proposed mechanism, the authors suggest that the ferric peroxo species attacks the electron-deficient β-carbon of the enone to yield an epoxide product. This process is similar to the epoxidation of certain alkenes by basic hydrogen peroxide.

In a later study, Selke et al. [63] synthesized and characterized $[Fe(PPIXDME)O_2]^-$, the ferric peroxo complex of the biological porphyrin. The $[Fe(PPIXDME)O_2]^-$ was found to be as effective at oxygen transfer to

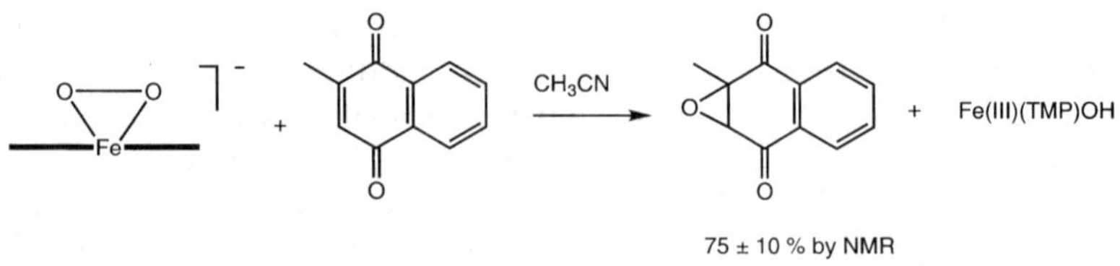

Fig. 13. The epoxidation of menadione by $[Fe(TMP)O_2]^-$ in CH_3CN

menadione as $[Fe(TMP)O_2]^-$ (75% conversion to epoxide). This study established that the heme group used in most P450 enzymes is capable of affecting direct nucleophilic attack on substrates in the ferric peroxo form. Remarkably, in the same study triphenylphosphine was found to bind weakly to the electron-deficient ferric peroxo complex $[Fe(F_{20}TPP)O_2]^-$. This weak binding is reminiscent of the axial ligation that occurs in P450 active sites by the active site cysteinyl thiolate.

Selke and Valentine [64] later demonstrated the importance of axial ligation in activating ferric peroxo porphyrin species when they reported the "switching-on" of the formerly unreactive $[Fe(F_{20}TPP)O_2]^-$. In acetonitrile, $[Fe(F_{20}TPP)O_2]^-$ was unreactive towards menadione, but epoxidation occurred when the reaction was carried out in DMSO, a solvent capable of axial ligation. As shown in Fig. 14, the yield of epoxidation increased from 0% to 75%. This result suggests the possibility that the nucleophilicity of a ferric peroxo porphyrin intermediate may be enhanced in an enzyme's active site by coordination of the thiolate side chain of cysteine.

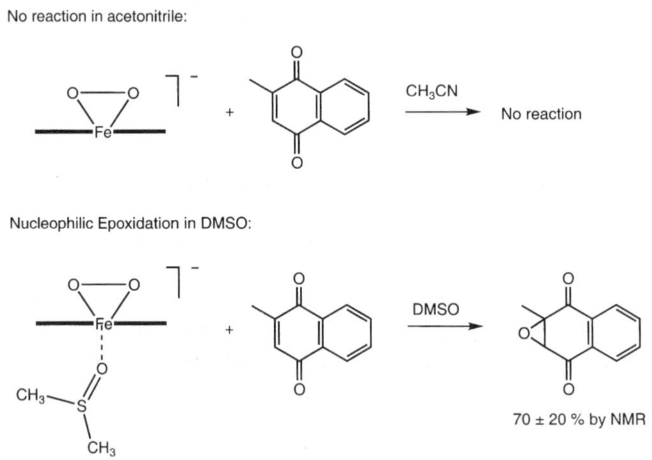

Fig. 14. Comparison of the extent of epoxidation of menadione by $[Fe(F_{20}TPP)O_2]^-$ in acetonitrile vs DMSO. Epoxidation in DMSO may be explained by the ability of DMSO to bind as an axial ligand to the iron and increase the nucleophilic character of the peroxo oxygens

5
Nucleophilic Reactivity of the Ferric Peroxo Porphyrin Species in Biological Systems

The above studies served to establish the nucleophilic nature of ferric peroxo porphyrin complexes. Since they also demonstrated that ferric peroxo porphyrins could react with biologically known functional groups, the studies established the ferric peroxo as a biologically viable P450 active oxidant. The next section considers how ferric peroxo porphyrin model complexes have been used to investigate the mechanisms of enzymes thought to react via the ferric peroxo species.

5.1
Ferric Peroxo Porphyrin Model Complexes as Mimics of the Third Catalytic Step of Aromatase

Model studies aimed at determining the mechanism of aromatase have helped researchers arrive at an accepted catalytic mechanism. In 1990, Watanabe reported that the natural substrate of aromatase reacted with an oxoferryl porphyrin to give oxidation of the aldehyde to the carboxylic acid rather than aromatization of the A ring [65]. This important result allows us to conclude that an oxoferryl species is not likely as the active oxidant of aromatase. A later study by Goto et al. [66] demonstrated that ferric porphyrin peroxo complexes, such as $[Fe(TMP)O_2]^-$, were mediators of carbon-carbon bond cleavage in aliphatic aldehydes (Fig. 15). This reaction mimicked the first part of the aromatization process catalyzed by aromatase and thus provided support for the concept of a ferric peroxo species as an active oxidant.

Cole and Robinson obtained additional evidence for the peroxo intermediate as an active oxidant when they reported that the enolized version of the natural substrate reacted slowly with hydrogen peroxide to give the aromatized product [14, 19]. This chemical model of aromatase also supported the importance of enolization of the substrate prior to deformylation as the unenolized derivative of the natural substrate failed to react with hydrogen peroxide.

In a recent model study, Wertz et al. [67] were able to show that $[Fe(TMP)O_2]^-$ reacts rapidly with an enolized mimic of the A ring of androstenedione to give an aromatized product (Fig. 16A). In this reaction

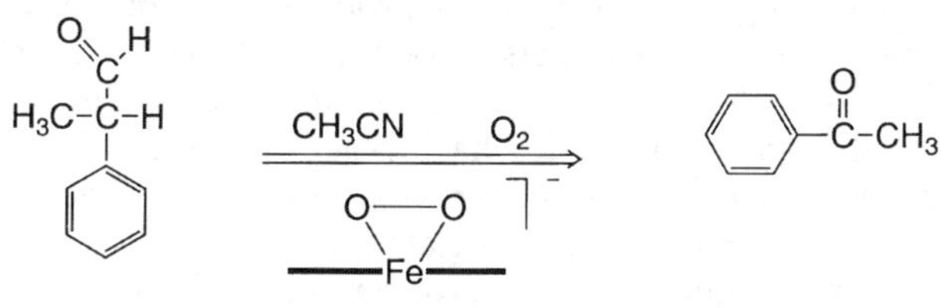

Fig. 15. The carbon-carbon bond cleavage of phenylpropionaldehyde by $[Fe(TMP)O_2]^-$ in acetonitrile

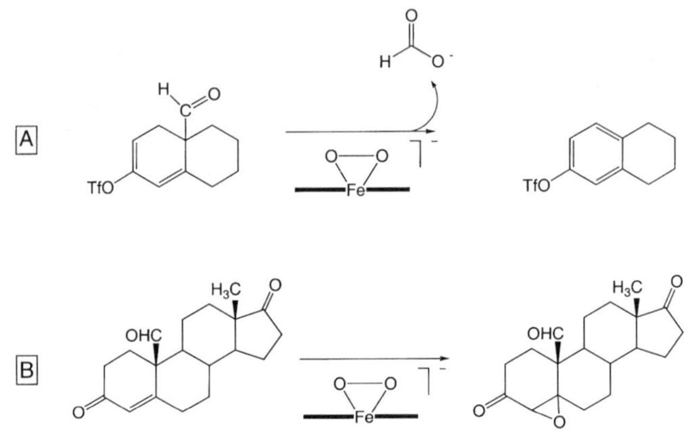

Fig. 16A,B. Comparison of the reactivity of a ferric peroxo porphyrin with the natural substrate of aromatase shown in B vs an enolized mimic of the A and B ring. The epoxidation observed in B is explained by the greater electrophilicity of the β-carbon compared to the carbonyl carbon of the aldehyde

both carbon-carbon bond cleavage and aromatization of the A ring mimic occur to yield a product analogous to what is observed in the third step of aromatase. Reaction with the non-enolized natural substrate resulted in epoxidation of the α,β-unsaturated ketone of the A ring rather than deformylation (Fig. 16B). This finding is consistent with Cole and Robinson's observation that only enolized mimics react with hydrogen peroxide to give deformylation [19]. Calculations by Graham-Lorence et al. also suggest that enolization is achieved in the enzymatic active site before deformylation occurs [9]. Taken together these results lend strong support for the ferric peroxo porphyrin species as the active oxidant in the third catalytic step of aromatase.

5.2 Ferric Peroxo Porphyrin Complexes as Mimics of Lanosterol-14α-demethylase, Progesterone 17α-hydroxylase/17,20-Lyase, and NO Synthase

Theoretically, it should be possible for ferric peroxo porphyrin complexes to mimic the applicable steps in lanosterol-14α-demethylase and progesterone 17α-hydroxylase/17,20-lyase. In the case of lanosterol-14α-demethylase, a ferric peroxo porphyrin complex would have to react with an aldehyde-containing molecule to give a formyl ester product, presumably via a Baeyer-Villiger type mechanism. Researchers have turned to the use of ketones to explore the potential of the peroxo complex for Baeyer-Villiger chemistry because, in recent studies, aldehydes have reacted to yield deformylated or carbon-carbon bond cleaved products[66, 67]. Studying ketone reactivity has the additional benefit of being an appropriate functional group for the study of the mechanism of progesterone 17α-hydroxylase/17,20-lyase.

Direct nucleophilic attack of a ferric peroxo porphyrin complex onto the carbonyl group of a ketone and subsequent rearrangement should be within the range of chemistry possible for the ferric peroxo. However, reactivity studies involving ketone oxidation by ferric peroxo porphyrin complexes have not resulted in Baeyer-Villiger type chemistry. Goto et al. observed carbon-carbon bond cleavage, not ester formation, when 2-phenyl-butyrophenone was reacted with a ferric peroxo porphyrin complex [66]. Additionally, reactivity studies with $[Fe(TMP)O_2]^-$ and various simple ketones did not result in ester formation even with strained ketones such as cyclobutanone, a ketone that would be expected to ring open rapidly to form a stable five-membered ring lactone [68]. The reason for the lack of Baeyer-Villiger type chemistry remains puzzling at this time. One might postulate that the C and D ring of the steroid and active-site energetic factors may convey an unknown thermodynamic driving force that is not present in the model system. Future investigations of ferric peroxo porphyrin reactivity with substrate-like ketones and α-hydroxy-ketones might yield functional mimics of lanosterol-14α-demethylase and progesterone 17α-hydroxylase/17,20-lyase.

The use of ferric peroxo model complexes to model the first and second steps of NO synthase may yield interesting results. The carbon of the urea oxime in *N*-hydroxy-L-arginine should be electrophilic enough to encourage nucleophilic attack of a ferric peroxo model complex. We are unaware of any studies that use ferric peroxo porphyrin complexes to model either of the pertinent steps of NO synthase.

6 Failure to Convert the Synthetic Ferric Peroxo Porphyrin Complex to an Electrophilic High Valent Oxoferryl Species in a Fashion Analogous to Cytochrome P450

The ferric heme peroxo intermediate, $[(P^{2-})Fe(III)(O_2^{2-})]^-$, in the cytochrome P450 mechanism is protonated twice and then undergoes heterolytic O—O bond cleavage to form the oxoferryl porphyrin cation radical species, $[(P^{\bullet -})Fe(IV)(O^{2-})]^+$. We know from the X-ray crystallography of P450 enzymes that two hydrogen bond donors, a water molecule and a hydroxyl group from an amino acid side chain, are appropriately positioned to interact with the peroxo ligand of the intermediate when it is formed and that O—O bond heterolysis occurs quickly and without the formation of observable intermediates such as the hydroperoxo species [17]. As ferric peroxo complexes seem to be able to mimic some aspects of P450 biological reactivity, one might also expect that it would be possible to observe oxygen-oxygen bond heterolysis of synthetic ferric peroxo complexes upon the addition of two protons. However, attempts to mimic this seemingly straightforward enzymatic step by protonating the synthetic ferric porphyrin peroxo complexes with a wide variety of protic and Lewis acids of varying strengths and trapping the oxoferryl intermediate with olefins have failed utterly [69]. Dissociation of hydrogen peroxide apparently is favored over O—O bond cleavage. In this section, we consider why we have been unable to model this critical step of the enzymatic chemistry using the synthetic porphyrins.

The synthetic ferric peroxo complexes are high spin, $S = 5/2$, and the peroxo ligand is thought to be coordinated in an η^2 fashion. In order for these complexes to achieve O—O bond heterolysis, it is necessary for the peroxo complex to be protonated, to isomerize to a monodentate configuration, and to undergo a spin conversion as the low spin oxoferryl porphyrin cation radical species is formed (Fig. 17, pathway A). Our failure to induce O—O bond heterolysis by protonating the synthetic ferric porphyrin peroxo complex suggests to us that this pathway is not facile and is not used by the enzyme.

Deoxy ferrous heme proteins and models are high spin ($S = 2$), but they react with dioxygen to give low spin oxy complexes (either low spin ferrous-dioxygen or low spin ferric superoxide complexes). An alternative possibility to pathway A is conversion from the low spin oxy heme intermediate, ferrous-dioxygen or ferric superoxide, to the low spin oxoferryl porphyrin cation radical species via a pathway that does not require high spin intermediates. Such a pathway is feasible if the low spin oxy heme intermediate is reduced and protonated simultaneously, i.e., via proton-coupled electron transfer (pathway B).

The mechanism described above in pathway B is analogous to that proposed by Babcock and coworkers for heterolytic O—O bond cleavage in cytochrome c oxidase [70]. In that case, hydrogen atom transfer from a tyrosine side chain to the oxy heme intermediate is concerted with O—O bond cleavage to produce the oxoferryl species $[(P^{2-})Fe(IV)(O^{2-})]$ and the tyrosyl radical, which is analogous to the intermediate proposed for cytochrome P450, $[(P^{\bullet -})Fe(IV)(O^{2-})]^{+}$, except that the electron deficiency is in the tyrosine radical rather than in the porphyrin radical. A high spin ferric peroxo complex need never be invoked.

7 Conclusions

Unlike the early transition metal peroxo complexes, ferric peroxo porphyrin complexes are nucleophilic. Their reactivity with anhydrides, acyl halides, carbon dioxide, and sulfur dioxide is well documented. The electronic characteristics of the porphyrin macrocycle influence the degree to which the peroxo moiety acts as a nucleophile. For example, electron-rich $[Fe(TMP)O_2]^-$ epoxidizes menadione while the electron-poor peroxo complex $[Fe(F20TPP)O_2]^-$ does not. Certain small molecules such as DMSO and PPh_3 may serve as axial ligands to the model complexes and increase nucleophilic reactivity. This axial ligation may mimic the effect that the active site thiolate has on the nucleophilicity of the peroxo intermediate in the enzyme.

Fig. 17. Two potential pathways leading to the heterolytic oxygen-oxygen bond cleavage of a ferric superoxide species. Pathway A is the route a synthetic ferric peroxo porphyrin complex would be expected to traverse as it is converted to the oxoferryl porphyrin cation radical intermediate. Pathway B depicts the expected spin state changes in the case of proton-coupled electron transfer

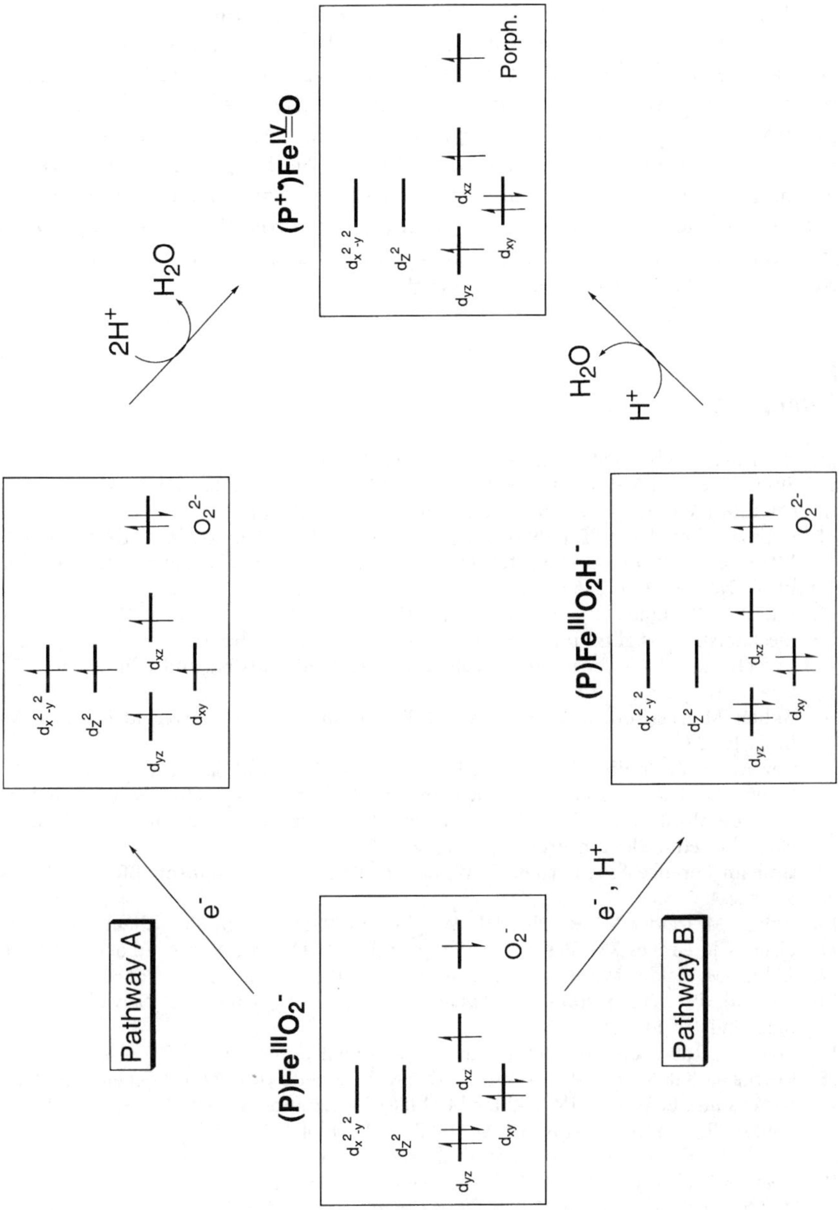
(P)Fe^{III}O$_2^-$
(P)Fe^{III}O$_2^{2-}$
(P)Fe^{III}O$_2$H$^-$
(P$^{+•}$)Fe^{IV}=O
Pathway A
Pathway B
e$^-$
e$^-$, H$^+$
2H$^+$
H$_2$O
H$^+$
H$_2$O
d$_{x^2-y^2}$
d$_{z^2}$
d$_{yz}$
d$_{xz}$
d$_{xy}$
O$_2^-$
O$_2^{2-}$
Porph.

Although the ferric peroxo species was once thought to be only an intermediate necessary for the generation of the oxoferryl porphyrin cation radical species, biological and model studies have established the importance of the ferric peroxo as an active oxidant. Reactivity studies of ferric peroxo porphyrin complexes with biologically-relevant electrophiles have provided mechanistic information concerning the use of the peroxo species by P450 aromatase. The continued study of peroxo porphyrin model complexes should eventually lead to a greater understanding of the role that peroxo species play in other P450 enzymes such as lanosterol-14α-demethylase, progesterone 17α-hydroxylase/17,20-lyase, and NO synthase.

8 References

1. Information obtained from P-450 sequence databank
2. Sono M, Roach MP, Coulter ED, Dawson JH (1996) Chem Rev 96: 2841
3. Coon MJ, Vaz ADN, Bestervelt LL (1996) Faseb J 10: 429
4. Ortiz de Montellano PR (1995) Oxygen activation and reactivity. In: Ortiz de Montellano PR (ed) Cytochrome P-450, structure, mechanism and biochemistry, 2nd edn. Plenum Press, New York, p 245
5. Mueller EJ, Loida PJ, Sligar SG (1995) Twenty-five years of P450cam research: mechanistic insights into oxygenase catalysis. In: Ortiz de Montellano PR (ed) Cytochrome P-450, structure, mechanism and biochemistry. 2nd edn. Plenum Press, New York, p 83
6. Akhtar M, Lee-Robichaud P, Akhtar ME, Wright JN (1997) J Steroid Biochem Molec Biol 61: 127
7. Vaz ADN, McGinnity DF, Coon MJ (1998) Proc Natl Acad Sci USA 95: 3555
8. Groves JT, Han Y-Z (1995) Models and mechanisms of cytochrome P450 action. In: Ortiz de Montellano PR (ed) Cytochrome P-450, structure, mechanism and biochemistry, 2nd edn. Plenum Press, New York, p 3
9. Graham-Lorence S, Amarneh B, White RE, Peterson JA, Simpson ER (1995) Protein Science 4: 1065
10. Akhtar M, Calder MR, Corina DL, Wright JN (1981) J Chem Soc, Chem Comm 3: 129
11. Fisher CR, Graves KH, Parlow AF, Simpson ER (1998) Proc Natl Acad Sci USA 95: 6965
12. Akhtar M, Calder MR, Corina DL, Wright JN (1982) Biochem J 201: 569
13. Morand P, Williamson DG, Layne DS, Lompa-Krzymien L, Salvador J (1975) Biochemistry 14: 635
14. Cole PA, Robinson CH (1986) J Chem Soc, Chem Comm 1651
15. Korzekwa KR, Trager WF, Smith SJ, Osawa Y, Gillette JR (1991) Biochemistry 30: 6155
16. Stevenson DE, Wright JN, Akhtar M (1988) J Chem Soc, Perkin Trans 1: 2043
17. Poulos TL, Finzel BC, Howard AJ (1987) J Mol Biol 195: 687
18. Oh SS, Robinson CH (1993) J Steroid Biochem Mol Biol 44: 389
19. Cole PA, Robinson CH (1991) J Am Chem Soc 113: 8130
20. Watanabe Y, Ishimura Y (1989) J Am Chem Soc 111: 8047
21. Townsley JD, Broodie HJ (1968) Biochemistry 7: 33
22. Vaz ADN, Kessell KJ, Coon MJ (1994) Biochemistry 33: 13,651
23. Raner GM, Chiang EW, Vaz ADN, Coon MJ (1997) Biochemistry 36: 4895
24. Roberts ES, Vaz ADN, Coon MJ (1991) Proc Natl Acad Sci USA 88: 8963
25. Vaz ADN, Pernecky SJ, Raner GM, Coon MJ (1996) Proc Natl Acad Sci USA 93: 4644
26. Peng H-M, Raner GM, Vaz ADN, Coon MJ (1995) Arch Biochem Biophys 318: 333
27. Imai M, Shimada H, Watanabe Y, Matsushima-Hibiya Y, Makino R, Koga H, Horiuchi T, Ishimura Y (1989) Proc Natl Acad Sci USA 86: 7823

28. Raag R, Martinis SA, Sligar SG, Poulos TL (1991) Biochemistry 30: 11,420
29. Counts Gerber NC, Sligar SG (1994) J Biol Chem 269: 4260
30. Poulos TL, Raag R (1992) FASEB J 6: 674
31. Benson DE, Suslick KS, Sligar SG (1997) Biochemistry 36: 5104
32. Harris D, Loew G, Waskell L (1998) J Am Chem Soc 120: 4308
33. Goodman LS, Gilman A (1996) The pharmacological basis of therapeutics, 9th edn. McGraw-Hill, New York
34. Shyadehi AZ, Lamb DC, Kelly SL, Kelly DE, Schunck W-H, Wright JN, Corina D, Akhtar M (1996) J Biol Chem 271: 12,445
35. Fischer RT, Trzaskos JM, Magolda RL, Lo SS, Brosz CS, Larsen B (1991) J Biol Chem 266: 6124
36. Akhtar M, Corina D, Miller S, Shyadehi AZ, Wright JN (1994) Biochemistry 33: 4410
37. Swinney DC, Mak AY (1994) Biochemistry 33: 2185
38. Lee-Robichaud P, Shyadehi AZ, Wright JN, Akhtar ME, Akhtar M (1995) Biochemistry 34: 14,104
39. Lee-Robichaud P, Akhtar ME, Akhtar M (1998) Biochemical J 330: 967
40. Degray JA, Lassmann G, Curtis JF, Kennedy TA, Marnett LJ, Eling TE, Mason RP (1992) J Biol Chem 267: 23,583
41. Harris RZ, Newmyer SL, Ortiz de Montellano PR (1993) J Biol Chem 268: 1637
42. Crane BR, Arvai AS, Ghosh DK, Wu C, Getzoff ED, Stuehr DJ, Tainer JA (1998) Science 279: 2121
43. Pufahl RA, Wishnok JS, Marletta MA (1995) Biochemistry 34: 1930
44. Crane BR, Arvai AS, Gachhui R, Wu C, Ghosh DK, Getzoff ED, Stuehr DJ, Tainer JA (1997) Science 278: 425
45. Valentine JS, McCandlish E (1978) Reactions of superoxide with metalloporphyrins. In: Dutton P, Leigh JS, Scarpa A (eds) Frontiers of biological energetics. Academic Press, New York, p 933
46. Kol'tover VK, Koifman OI, Khenkin AM, Shteinman AA (1982) Izvest Akad Nauk SSSR, Ser Khim 7: 1690
47. McCandlish E, Miksztal AR, Nappa M, Sprenger AQ, Valentine JS, Stong JD, Spiro TG (1980) J Am Chem Soc 102: 4268
48. Welborn CH, Dolphin D, James BR (1981) J Am Chem Soc 103: 2869
49. Reed CA (1982) Iron(I) and iron(IV) porphyrins. In: Kadish KM (ed) Electrochemical and spectrochemical studies of biological redox components. American Chemical Society, Washington DC, p 333
50. Ahmad S, McCallum JD, Shiemke AK, Appelman EH, Loehr TM, Sanders-Loehr J (1988) Inorganic Chemistry 27: 2230
51. Shirazi A, Goff HM (1982) J Am Chem Soc 104: 6318
52. Friant R, Goulon J, Fischer J, Ricard L, Schappacher M, Weiss R, Momenteau M (1985) Nouv J Chim 9: 33
53. Van Atta RB, Strouse CE, Hanson LK, Valentine JS (1987) J Am Chem Soc 109: 1425
54. Burstyn JN, Roe JA, Miksztal AR, Shaevitz BA, Lang G, Valentine JS (1988) J Am Chem Soc 110: 1382
55. Burstyn JN (1986) Ph.D. Thesis, University of California, Los Angeles
56. Sisemore MF, Selke M, Burstyn JN, Valentine JS (1997) Inorganic Chemistry 36: 979
57. Khenkin AM, Shteinman AA (1982) Kinet Katal 23: 219
58. Khenkin AM, Shteinman AA (1984) J Chem Soc, Chem Comm 18: 1219
59. Schappacher M, Weiss R (1985) J Am Chem Soc 107: 3736
60. Groves JT, Watanabe Y (1986) J Am Chem Soc 108: 7834
61. Miksztal AR, Valentine JS (1984) Inorganic Chemistry 23: 3548
62. Sisemore MF, Burstyn JN, Valentine JS (1996) Angew Chem Int Ed Engl 35: 206
63. Selke M, Sisemore MF, Valentine JS (1996) J Am Chem Soc 118: 2008
64. Selke M, Valentine JS (1998) J Am Chem Soc 120: 2652
65. Watanabe Y, Takehira K, Shimizu M, Hayakawa T, Orita H (1990) J Chem Soc, Chem Comm 13: 927

66. Goto Y, Wada S, Morishima I, Watanabe Y (1998) J Inorg Biochem 69: 241
67. Wertz DL, Sisemore MF, Selke M, Driscoll J, Valentine JS (1998) J Am Chem Soc 120: 5331
68. Wertz DL (1999) Ph.D. Thesis, University of California, Los Angeles
69. Valentine Lab, unpublished results
70. Proshlyakov DA, Pressler MA, Babcock GT (1998) Proc Natl Acad Sci USA 95: 8020

Characterization of High-Valent Oxo-Metalloporphyrins

Yoshihito Watanabe[1], Hiroshi Fujii[2]

Institute for Molecular Science, Okazaki 444-8585, Japan
[1] *E-mail: yoshi@ims.ac.jp*
[2] *E-mail: hiro@ims.ac.jp*

High-valent oxo metalloporphyrins have been known as reactive intermediates in the catalytic cycles of many heme enzymes and in the oxidation reactions mediated by synthetic metalloporphyrins. In this review, we survey studies of high-valent oxo metalloporphyrins in the past decade. In Sect. 2, electronic structures and magnetic properties of compound I and compound II species in various heme enzymes are discussed on the basis of recent studies of synthetic model complexes of the compound I and compound II. In Sect. 3, spectroscopic properties of oxo manganese porphyrin complexes and their application to biological systems are summarized. In the Sects. 4 and 5, synthesis and characterization of oxo chromium and oxo ruthenium porphyrin complexes are reviewed. In the last section, we have summarized the synthesis and characterization of Mo, Nb, Ti, and V oxo metalloporphyrins complexes.

Keywords: Cytochrome P450, Peroxidase, Heme, Oxo, Metalloporphyrin

1 Introduction

The biological use of Fe=O structures for oxidative metabolisms has been intensively studied for over three decades. Readily available examples are heme enzymes including cytochrome P-450, peroxidases, oxidases, and so on [1–5]. Among those heme enzymes, cytochrome P-450 has received much attention due to its unusually high catalytic hydroxylation activity [2, 3, 6–8]. Cytochrome P-450cam is able to hydroxylate *d*-camphor to 5-*exo*-hydroxy-*d*-camphor at a turnover number of >1000/min [9]. While, the active intermediate(s) of cytochrome P-450 responsible for the oxidations is still controversial, an oxo-ferryl porphyrin π-cation radical (O=Fe(IV) $Por^{+\bullet}$), equivalent to compound I of peroxidases, has been considered to be the active species [1–8]. Early studies mimicking cytochrome P-450-type reactions by metalloporphyrins were carried out by Tabushi and his coworkers in the late 1970s [10–12]. For example, they employed a Mn(III) hematoporphyrin/NaOCl system for the oxidation of hydrocarbons, alcohols and ethers [10, 11]. Though Mn(IV) intermediates were proposed for the active species, no high-valent Mn complexes were characterized at the time. Efforts for the preparation of high-valent iron porphyrin complexes related to compounds I and II of peroxidases and catalase resulted in the preparation of $O{=}Fe^{IV}(TPP)$ by electrochemical oxidation of $[Fe^{III}(TPP)]_2O$ in CH_2Cl_2 in 1971 [13]. In 1979, Groves et al. reported the hydroxylation and epoxidation by PhIO in the presence of Fe(TPP) as a catalyst [14]. A few months later, Chang and Kuo were the first to observe a transient intermediate similar to compound I of HRP by the reaction of Fe(III)(OEP) and excess Ar—IO in CH_2Cl_2 at −45 °C [15]. Since then, many efforts have been devoted to the identification and application of oxo-metalloporphyrins and related intermediates.

In this review article, the authors have focused mainly on the formation and characterization of oxo-metalloporphyrin complexes both in biological and synthetic model systems. Thus, oxidations catalyzed by metalloporphyrin complexes are not described very much unless the reactions are highly related to mechanistic aspects of the high valent intermediate formation. For the convenience of readers who are interested in the metalloporphyrin catalyzed oxidations, we have listed the review articles in References [16–21].

2
O=Fe Porphyrins and Related Complexes

2.1
Physical Properties and Characteristics of High Valent Intermediates in Heme Enzymes

High valent iron porphyrins have been detected in the catalytic cycles of peroxidases [22]. When peroxidase reacts with 1 equivalent of hydrogen peroxide, a metastable green intermediate named compound I is formed. The compound I oxidizes organic substrates by one electron to form the second red intermediate named compound II (Eq. 1).

$$\begin{aligned}&\text{Resting State (ferric high spin)} + H_2O_2 \longrightarrow \text{Compound I}\\&\text{Compound I} + \text{Substrate} \longrightarrow \text{Compound II} + \text{Oxidized Product}\\&\text{Compound II} + \text{Substrate} \longrightarrow \text{Resting State} + \text{Oxidized Product}\end{aligned} \quad (1)$$

The compound I of horseradish peroxidase(HRP) exhibits a unique absorption spectrum; a broad Soret band (400 nm) with loss of intensity and new broad peak at 651 nm [23]. The spectral feature of the compound I is similar to that of metalloporphyrin π-cation radical [24]. The absorption spectrum of the HRP compound II shows peaks at 418, 527 and 554 nm [23]. The Mössbauer spectra of the HRP compound I and II indicate that heme irons in compound I and II are in a ferryl oxidation state [25]. Based on these results, the electronic structures of compound I and II are proposed as oxo-iron(IV) porphyrin π-cation radical and oxo-iron(IV) porphyrin, respectively.

Two types (a_{1u} and a_{2u}) of porphyrin π-cation radical states have been characterized from EPR and theoretical studies of metalloporphyrin π-cation radical complexes (Fig. 1) [26, 27]. The HRP compound I had been thought the a_{2u} radical state from its absorption spectral feature [28]. However, recent studies have brought question the a_{1u}/a_{2u} assignment based on the absorption spectral feature and pointed out, rather, the a_{1u} radical state for HRP compound I on the basis of several physicochemical measurements (Fig. 1)

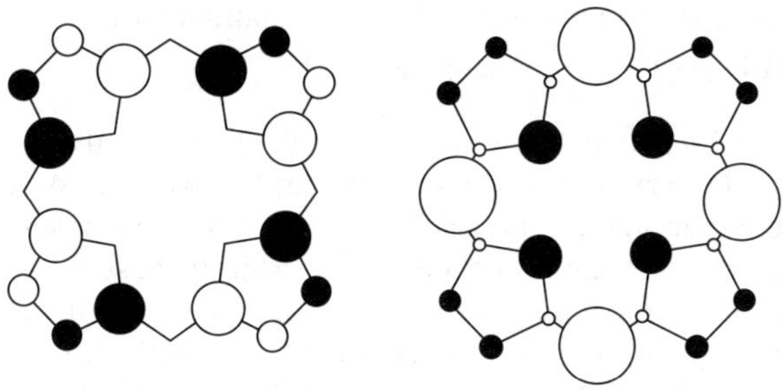

Fig. 1. Electron spin distribution of porphyrin atomic orbitals with a_{1u} (*left*) and a_{2u} (*right*) symmetries. *Black and white circles* represent signs of the upper lobe of the p_π AOs

[29–31]. An EPR spectrum of HRP compound I exhibits a broad signal around $g = 2$ resulting from weak magnetic interaction between ferryl iron spins and porphyrin π-cation radical spin [32, 33]. Even in the porphyrin π-cation radical state, the HRP compound I exhibits well-resolved paramagnetic ^{1}H-NMR signals of prophyrin peripheral protons [34–36]. The porphyrin π-cation radical state of HRP compound I is also supported by ^{1}H- and ^{14}N-ENDOR spectra [37]. The presence of oxo ligands in HRP compound I is confirmed by its ^{17}O-ENDOR spectra [38]. The EXAFS measurements of the HRP compound I and II show the iron-oxygen distance to be 1.6 Å [39]. While the resonance Raman (RR) spectrum of HRP compound II shows the ν(Fe=O) signal at 774 cm^{-1} [30, 40], the RR band of ν(Fe=O) for HRP compound I has not been determined and still in controversy [30, 41–43].

The compounds I and II of cytochrome *c* peroxidase (C*c*P) are characterized by various spectroscopic methods. The absorption spectrum of C*c*P compound I is different from that of HRP compound I, but rather similar to that of HRP compound II [44, 45]. The EPR mesurement of C*c*P compound I exhibits the presence of a protein radical [46]. The Mössbauer spectrum of the C*c*P compound I is in good agreement with a ferryl oxidation state [47]. On the basis of these results, C*c*P compounds I and II are assigned as oxo-iron(IV) porphyrin with a protein radical and oxo-iron(IV) porphyrin, respectively. The ENDOR study of C*c*P compound I has identified the primary site of the protein radical at tryptophan-191 [48]. The ν(Fe=O) band is observed at 767 cm^{-1} for the compound I [49, 50].

The compound I of chloroperoxidase (CPO), which has a cystein thiolate axial ligand instead of the histdyl imidazole ligand in HRP, has absorption peaks at 367 and 689 nm [51]. Compared with HRP compound I, these peaks are blue and red-shifts, respectively. The EPR spectrum of the CPO compound I shows $g = 1.64$, 1.73, and 2.00, which indicate antiferromagnetic coupling between the ferryl iron and porphyrin π-cation radical spins [52]. This is in contrast to the weak magnetic interaction in HRP compound I. The RR spectrum of CPO compound I shows a ν(Fe=O) vibration at 790 cm^{-1} [53].

In the catalytic cycle of catalase (CAT), the compound I is also an active intermediate which can disproportionate hydrogen peroxide to molecular oxygen and water (Eq. 2) [54].

$$\begin{aligned} &\text{Resting State (ferric high spin)} + H_2O_2 \longrightarrow \text{Compound I} \\ &\text{Compound I} + H_2O_2 \longrightarrow \text{Resting State} + H_2O + O_2 \end{aligned} \tag{2}$$

The compound I intermediate is not detected in the catalytic cycle because hydrogen peroxide serves as both oxidant and substrate. When catalase is oxidized by peracetic acid, a metastable compound I is observed [55]. The absorption spectrum of CAT compound I exhibits absorption at 400 and 662 nm, which is a typical spectrum of porphyrin π-cation radical complex [56]. Thus the CAT compound I is an oxo-iron(IV) porphyrin π-cation radical, as HRP compound I. Although the peak positions of CAT compound I are close to those of HRP compound I, the spectral feature of CAT compound I is obviously different from that of HRP compound I. This is interpreted as an effect of the axial tyrosyl ligand [57]. The EPR spectrum of compound I from *Micrococcus luteus* (MLC) showed EPR signals at $g = 3.45$ and 2.0, indicating weak ferromagnetic interaction of ferryl iron and porphyrin spins [58]. The EPR spectrum of CAT compound I from *Proteus mirabilis* (PMC) is also similar to that of MLC compound I at low pH, but changes to the signal at $g = 2$ like HRP compound I at neutral pH [59]. The RR band of ν(Fe=O) has not been observed for CAT compound I, but observed at 775 cm^{-1} at neutral pH for CAT compound II from bovine liver [60]. From these spectral results, the CAT compound I is also assigned to the a_{1u} radical state.

Compound I and II species are also proposed in the catalytic cycle of cytochrome c oxidase (C*c*O), which is a terminal oxidase and catalyzes reduction of molecular oxygen with four electrons arriving through the respiratory chain [61, 62]. The high valent intermediates of C*c*O have been detected in flow-flash experiments [63, 64]. Two intermediates having absorption peaks at 607 and 580 nm are reported and assigned as a peroxy species (compound I level) and ferryl species (compound II level), respectively [63]. The RR spectra of these intermediates assigned for ν(O=Fe) show bands at 804 and 785 cm^{-1} [64].

2.2 Studies on the Preparation and Characterization of Oxo-Ferryl Porphyrin Model Complexes Related to Cytochrome P-450 and Peroxidases

In 1981, Groves and coworkers found the reaction of Fe^{III}(TMP)Cl (TMP: tetramesityl porphyrin) with *m*-chloroperbenzoic acid (*m*CPBA) in CH_2Cl_2-MeOH at −78 °C gave a green intermediate [65]. According to this color change, the β-pyrrole hydrogen resonance at 120 ppm in ^1H NMR for Fe^{III}(TMP)Cl was shifted to −27 ppm. The chemical shift of the intermediate obeyed the Curie law between −26 and −89 °C. The magnetic susceptibility determined by Evans method was 4.2 μ_B, indicating $S = 3/2$. The Mössbauer spectrum of the intermediate showed a quadrupole doublet centered at 0.05 mm/s (ΔEq = 1.49 mm/s). In addition, the UV-Vis spectrum of the intermediate showed a less intense Soret band with a broad band around 600 to 700 nm. All these physical properties support the green intermediate to be $O{=}Fe^{IV}TMP^{+\bullet}$. The green complex shows EPR signals at $g = 4.3$, 3.9, and 1.99 at 4 K, indicating strong ferromagnetic coupling between ferryl iron ($S = 1$) and the porphyrin π-cation radical spin ($S = 1/2$) at $J > +40$ cm^{-1} [66]. This is in contrast to weak magnetic couplings observed for compounds I of peroxidases. Evidence for the O=Fe ligation was provided by EXAFS spectroscopic studies on $O{=}Fe^{IV}TMP^{+\bullet}$[67]. For example, O=Fe distances for HRP compound I and $O{=}Fe^{IV}TMP^{+\bullet}$ were reported to be 1.61–1.64 and 1.62–1.66 Å, respectively [67]. Furthermore, Hashimoto et al. prepared $^{16}O{=}Fe^{IV}TMP^{+\bullet}$ and $^{18}O{=}Fe^{IV}TMP^{+\bullet}$ in CH_2Cl_2/MeOH by using ^{16}O- and ^{18}O-labeled *m*CPBA [68]. The resonance Raman (RR) bands for $\nu_{O=Fe}$ were observed at 828 and 792 cm^{-1} for ^{16}O and ^{18}O, respectively. In addition [54], Fe substituted samples gave the RR band at 832 cm^{-1}. These isotope shifts are consistent with theoretical isotope frequency shifts expected for an O=Fe oscillator. The O=Fe RR band is sensitive to the 6th ligand of $O{=}Fe^{IV}$ Porphyrin$^{+\bullet}$ [69], thus, Kincaid et al. observed the O=Fe RR band of $O{=}Fe^{IV}TMP^{+\bullet}$ at 801 cm^{-1} in CH_2Cl_2 [70].

Detailed mechanistic studies on the reaction of $Fe^{III}TMP$ with *m*CPBA by Groves and Watanabe showed the transient formation of acylperoxo-$Fe^{III}TMP$ followed by the heterolytic O—O cleavage giving $O{=}Fe^{IV}TMP^{+\bullet}$ in polar solvent such as CH_2Cl_2 [71, 72]. Replacement of the solvent to less polar toluene or benzene changed the nature of O—O bond cleavage from the heterolysis to the homolysis (Scheme 1) [72, 73]. Substituent effects of porphyrin ring, peracid and 6th ligand on the homolysis and heterolysis were systematically examined and their relevance to cytochrome P-450 and peroxidases were discussed [74–75].

2.2.1
A_{1u} and A_{2u}

Change in the electron-negativity of *meso*-substituent alters the reactivity and electronic state of the oxo-iron(IV) porphyrin π-cation radical complex. Iron porphyrins having electron-withdrawing substituents such as tetra-2,6-dchlorophenylporphyrin (TDCPP) and tetra-pentafluorophenylporphyrin (TPFPP) (Fig. 2) are shown to be efficient catalysts of oxygen atom transfer

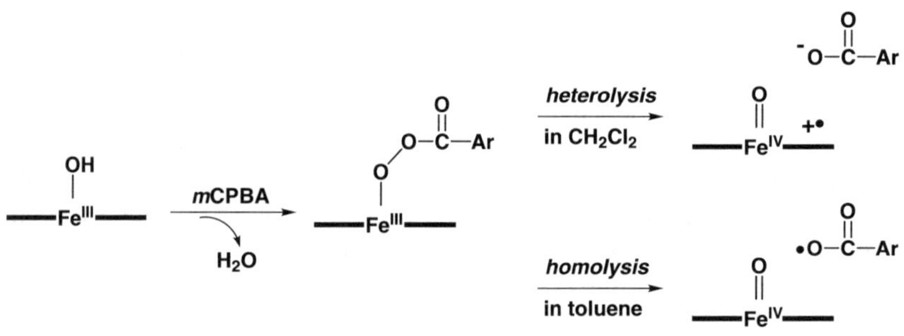

Scheme 1. The heterolysis and homolysis of acylperoxo-Fe(III) porphyrins

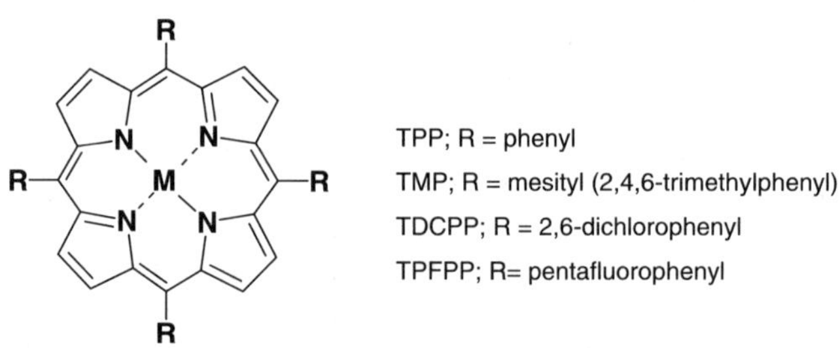

Fig. 2. The structures of *meso*-tetraarylporphyrins employed for synthesis of high valent oxo porphyrin complexes

reactions from iodosobenzene, hydrogen peroxide and so on [76]. $O{=}Fe^{IV}(TDCPP^{+\bullet})$ is prepared by the oxidation of $Fe^{III}(TDCPP)(ClO_4)$ with ozone in acetonitile at −35 °C [77]. The EPR spectrum of $O{=}Fe^{IV}(TDCPP^{+\bullet})$ exhibits the signals at $g = 2$ in acetonitrile [78] and at $g = 4.26, 3.62, 1.98$ in dichloromethane-methanol (6:1) [66]. The EPR spectrum in dichloromethane-methanol is similar to that of $O{=}Fe^{IV}(TMP^{+\bullet})$ and interpreted by ferromagnetic interaction of the ferryl iron and the porphyrin π-cation radical with $J = +38\ cm^{-1}$ [66]. However, ^{1}H-NMR spectrum of $O{=}Fe^{IV}(TDCPP^{+\bullet})$ shows much more upfield shift of pyrrole proton (−66 ppm) than that (−27 ppm) of $O{=}Fe^{IV}(TMP^{+\bullet})$ [31], suggesting the change in the electronic state of the porphyrin radical. The comprehensive studies of the electron-negativity of *meso*-substituents revealed that the pyrrole proton signal shifts upfield with increase in electron-negativity of *meso*-substituents; 2,4,6-trichlorophenyl > 2,6-dichlorophenyl > 2-methyl-6-chlorophenyl > mesityl [31]. The spectral change is interpreted as the mixing of the a_{1u} and a_{2u} radical states by a vibronic coupling. Electron-withdrawing substituent at the *meso*-positions should lower the energy of the a_{2u} orbital relative to the a_{1u} orbital via the interaction of the aryl and porphyrin π-orbitals, since electron density is concentrated at the *meso*-position for the a_{2u} orbital while the a_{1u} orbital has a node (Fig. 1). This leads to a decrease in the a_{2u}-a_{1u} energy separation (Fig. 4) and stronger mixing of the a_{1u} radical state to the a_{2u} state would be expected with increase in the electron-negativity of *meso*-substituent. EPR spectra of these electron-negative π-cation radical complexes are similar to that of $O{=}Fe^{IV}(TMP^{+\bullet})$, indicating ferromagnetic coupling between the ferryl iron and the porphyrin radical spins [79]. When the *meso*-substituents are much more electron-negative such as pentafluorophenyl, HOMO of the porpyhrin orbital is switched to the a_{1u} orbital from the original a_{2u} orbital. Thus, $O{=}Fe^{IV}(TPFPP^{+\bullet})$ shows large up-shift (−96 ppm) of pyrrole proton signal, consistent with the a_{1u} radical state [80]. The EPR spectrum of $O{=}Fe^{IV}(TPFPP^{+\bullet})$ exhibits a broad signal around $g = 2$ as observed for HRP compound I suggests a weak interaction of the ferryl iron and the porphyrin π-cation radical spins [80]. This is also reasonable to the a_{1u} radical state. The spin interaction of the a_{1u} radical with the iron spins would be weak because of no spin density at the pyrrole nitrogen atoms in the a_{1u} orbital, but that of the a_{2u} would be strong because of large spin density at the pyrrole nitrogen atoms in the a_{2u} orbital. Drastic change in the magnetic interaction with the change from the a_{2u} radical to the a_{1u} radical is also observed in the copper(II) porphyrin π-cation radical complexes [81].

The position of porphyrin substituents is also important to determine the electronic state of oxo-iron(IV) porphyrin π-cation radical (Fig. 3). While heme enzymes utilize protoporphyrin or its modified forms as the prosthetic group, i.e., pyrrole β-substituted porphyrin, most of model complexes which provide oxo-iron(IV) porphyrin π-cation radical complexes are *meso*-tetraphenylporphyrin derivatives because of their easy preparations. An oxo-iron(IV) porphyrin π-cation radical complex bearing pyrrole β-substituents has been prepared and characterized by Fujii et al. using 2,7,12,17-tetramethyl-3,8,13,18-tetramesityporphyrin (TMTMP) [82]. By changing the substituents

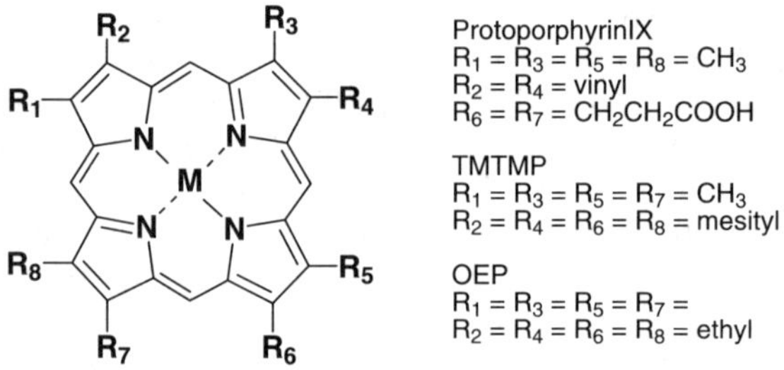

Fig. 3. The structures of pyrrole β-substituted porphyrin complexes

from the *meso*-positions to the pyrrole β-positions, spectroscopic properties are altered to those of compounds I of heme enzymes. The absorption spectrum of O═Fe^{IV}(TMTMP$^{+\bullet}$) is similar to that of HRP compound I, but very different from that of O═Fe^{IV}(TMP$^{+\bullet}$) [31, 82]. The ^{1}H-NMR spectrum of O═Fe^{IV}TMTMP$^{+\bullet}$ exhibits large downward shift of the pyrrole methyl proton at 140 ppm and small shift of the *meso*-proton at 53 ppm at −70 °C, consistent with the spin distribution of the a_{1u} orbital [31, 82]. The EPR spectrum of O═Fe^{IV}TMTMP$^{+\bullet}$ shows signals at g = 3.1 and 2.0, which is close to that of compounds I of *Micrococcus luteus* catalase and ascorbate peroxidase [79]. The EPR spectrum is interpreted by a weak ferromagnetic interaction between ferryl iron and porphyrin radical spins, as expected for the a_{1u} radical state. The Raman spectrum of O═Fe^{IV}TMTMP$^{+\bullet}$(ClO_4) shows the ν(Fe═O) band at 801 cm^{-1} [83], which is the same position as that of O═Fe^{IV}TMP$^{+\bullet}$(ClO_4) [84]. This means that the a_{1u}/a_{2u} radical state does not change the Fe═O bond strength. In contrast to the *meso*-substituted porphyrin, the electronic structure is not altered by the electron-withdrawing substituents at pyrrole β-positions (Fig. 4) [31]. This can be also explained by the spin distributions

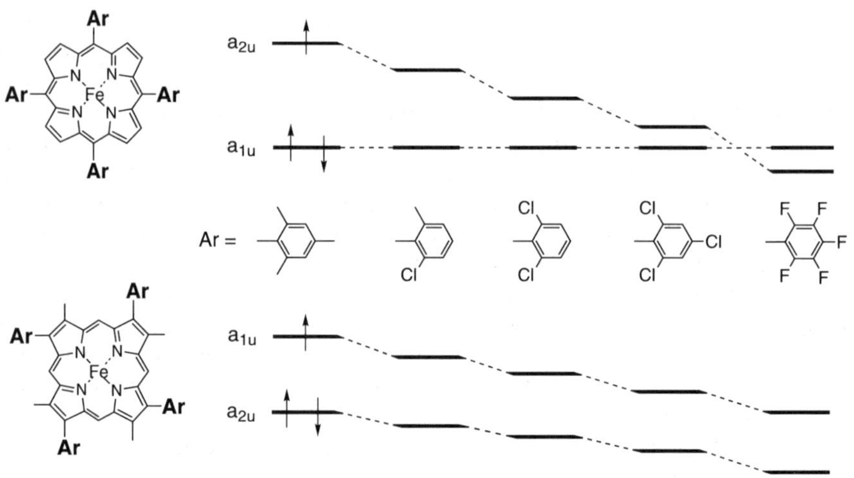

Fig. 4. Change in the a_{1u} and a_{2u} orbitals with increasing in the electron-negativity of the substituent

of the a_{1u} and a_{2u} orbitals (Fig. 1). Since the a_{1u} and a_{2u} orbitals have small spin densities at the pyrrole β-positions, both orbitals are slightly stabilized by electron-withdrawing substituents, as shown in Fig. 4. Thus, pyrrole β-substituents do not drastically alter the orbital occupancy of the pyrrole β-substituted oxo iron(IV) porphyrin π-cation radical, consistent with the a_{1u} radical state. On the basis of these results, the a_{1u} radical state of HRP compound I is proposed [31].

2.2.2
Ligand Effects

The effect of an axial ligand on the reactivities of oxo-iron(IV) porphyrin π-cation radicals has been studied in connection with various axial ligands in heme enzymes. The epoxidation reactivity of oxo-iron(IV) porphyrin π-cation radicals was found to be drastically altered by its axial ligand. $O{=}Fe^{IV}(TMP^{+\bullet})(F)$ rapidly reacted with styrene to form styrene oxide, though $O{=}Fe^{IV}(TMP^{+\bullet})(X)$, in which X = methanol, chloride and acetate, were less reactive. In addition, $O{=}Fe^{IV}(TMP^{+\bullet})(ClO_4)$ did not react with olefins [85]. The RR ν(Fe=O) band is affected by the nature of the *trans*-axial ligand, for example, the RR ν(Fe=O) band was observed near 835 cm^{-1} for perchlorate and trifrate complexes while near 800 cm^{-1} for the *m*-chlorobenzoate, fluoride and chloride complexes [84]. In these studies, alcohols, halides and some other weakly coordinating anions have been examined. The RR spectrum of $O{=}Fe^{IV}(TMP^{+\bullet})(ImH)$ showed ν(Fe=O) vibration at 810 cm^{-1} and that of $O{=}Fe^{IV}(TMP)(ImH)$ appared at 814 cm^{-1}, the Raman band is just 4 cm^{-1} down-shifted by porphyrin ring oxidation [86].

Very recently, oxo-iron(IV) TMTMP π-cation radical complexes bearing imidazole and *p*-nitrophenol have been prepared as models for the compounds I of peroxidase and catalase (Scheme 2) and characterized by absorption and ^{1}H-NMR spectroscopies [87]. The absorption spectra of $O{=}Fe^{IV}TMTMP^{+\bullet}(ImH)$ and $O{=}Fe^{IV}TMTMP^{+\bullet}(p\text{-}NO_2\text{-}PhO)$ were quite close to those of compounds I of peroxidase and catalase, respectively [87]. ^{1}H-NMR

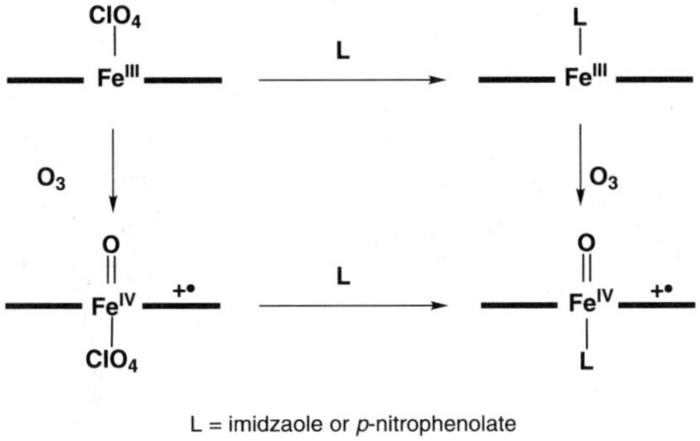

Scheme 2. Introduction of biologically relevant trans-axial ligand into a oxo-iron(IV) porphyrin π-cation radical complex

of $O{=}Fe^{IV}TMTMP^{+\bullet}(ImH)$ exhibits paramagnetic shifts of the pyrrole β-methyl protons similar to those of HRP compound I. $O{=}Fe^{IV}TMTMP^{+\bullet}(ImH)$ is readily assigned as the a_{1u} radical state even with coordinating imidazole as an axial ligand. The preparation of high valent iron porphyrin bearing thiolate ligand as a model for cytochrome P-450 has been hampered by high reactivity of the thiolate ligand with air and oxidants [88, 89]. Recently, air-stable iron(III) porphyrin complexes having thiolate ligands have been synthesized by connecting the thiolate ligands with porphyrin through a covalent bond and/or by surrounding the thiolate ligand with pivaroyl groups (Fig. 5) [90–92]. Because of the strong push-effect from the *trans*-axial thiolate ligand, the oxygenation activities of these thiolate complexes are much higher than those of non-thiolate complexes. Very recently, an intermediate bearing absorption peak at 388 nm has been observed and tentatively assigned as oxo iron(IV) porphyrin π-cation radical state [93].

2.2.3
Chlorin Derivatives

The effect of porphyrin ring reduction (i.e., chlorin) is studied as a model for the compounds I of chlorin-containing heme enzymes such as *Neurospora crassa* catalase [94]. Oxo-iron(IV) tetramesitylchlorin (TMC) π-cation radical, $O{=}Fe^{IV}(TMC^{+\bullet})$ (Fig. 6) is prepared by *m*CPBA oxidation in dichloromethane at −80 °C and characterized by absorption and ^{1}H- and ^{2}D-NMR spectroscopies [95]. The ^{1}H-NMR shifts of pyrrole and mesityl protons of $O{=}Fe^{IV}(TMC^{+\bullet})$ are consistent with an a_2 radical character (a_2 symmetry corresponds to the a_{1u} in the D_{4h} symmetry). This is in contrast to the a_{2u} state assigned to $O{=}Fe^{IV}(TMP^{+\bullet})$. The porphyrin ring reduction destabilizes the energy of the a_2 orbital relative to that of the b_2 orbital (the a_{2u} in D_{4h} symmetry). The RR band of $O{=}Fe^{IV}$(dihydroxy-dimethyl-$TMC^{+\bullet}$) at 832 cm^{-1} is shifted to 800 cm^{-1} on oxidation with ^{18}O-*m*CPBA. Thus, the $\nu(Fe{=}O)$ bond is very close to that of $O{=}Fe^{IV}(TMP^{+\bullet})$ [96]. $O{=}Fe^{IV}$(dihydroxy-dimethyl-$TMC^{+\bullet}$) exhibits EPR signals at g = 4.23, 3.80, and 2.02 arisen from the S = 3/2 state [96]. The Mössbauer parameters of $O{=}Fe^{IV}$(dihydroxy-dimethyl-$TMC^{+\bullet}$)

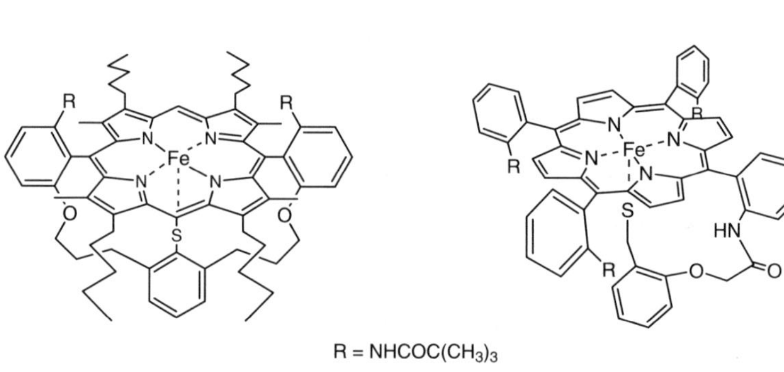

R = $NHCOC(CH_3)_3$

Fig. 5. Air-stable iron(III) porphyrin complexes having thiolate ligands

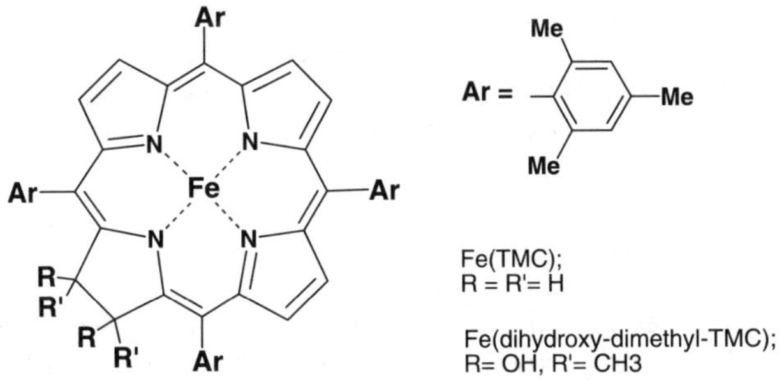

Fig. 6. Fe tetramesitylchlorin, Fe(TMC)

are $\delta = 0.01$ mm/s and $\Delta E_q = +1.25$ mm/s, which is almost identical with those of $O{=}Fe^{IV}(TMP^{+\bullet})$ [96]. Interestingly, the oxidation of norborene by $O{=}Fe^{IV}(TMC^{+\bullet})$ proceeds much slower than that by the corresponding TMP complex [95].

2.2.4
Isoelectronic Forms of Oxo-Iron(IV) Porphyrin π-Cation Radical

Iron(V) porphyrin, which is an isoelectronic form of the oxo-iron(IV) porphyrin π-cation radical, has been prepared by introducing methanol into an oxo-iron(IV) TDCPP π-cation radical as an axial ligand at −90 °C [97, 98]. The iron(V) oxidation state can be prepared by the introduction of strong electron-withdrawing substituents on the porphyrin ring and by the coordination of methoxide. The absorption spectrum of $O{=}Fe^{V}(TDCPP)$ is similar to that of $O{=}Fe^{IV}(TDCPP)$. The iron(V) oxidation state of the complex is confirmed by iodometric titration. Oxo-iron(V) porphyrins are reactive toward olefin and $O{=}Fe^{V}(TDCPP)$ oxidized norborene to norborene oxide in 50% yield at −90 °C.

The other isoelectronic forms of oxo-iron(IV) porphyrin π-cation radical are iron(III) porphyrin *N*-oxide [99–101] and iron(III) porphyrin dication complexes (Fig. 7) [102]. These complexes have been prepared and characterized by absorption, ^{1}H-NMR, EPR and Raman spectroscopies.

2.2.5
Oxo-Iron(IV) Porphyrins

Through the works on the autooxidation of Fe(II) porphyrin complexes, Balch et al. showed the transient formation of Fe(III)—O—O—Fe(III) species [103].

Fig. 7. Fe porphyrin *N*-oxide and Fe porphyrin dication

More importantly, they found that the introduction of *N*-methylimidazole (MeIm) to the Fe(III)—O—O—Fe(III) complex affords O=Fe^{IV}(Por)(MeIm) (Eq. 3) [104, 105]. O=Fe^{IV}(Por)(MeIm) is able to oxidize triphenylphosphine even at −80 °C in toluene over a period of several hours to give triphenylposphine oxide quantitatively [105].

$$(\mathrm{Por})\mathrm{Fe^{III}{-}O{-}O{-}Fe^{III}}(\mathrm{Por}) + 2\ \mathrm{Melm} \longrightarrow 2\ \mathrm{O{=}Fe^{IV}}(\mathrm{Por})(\mathrm{Melm}) \quad (3)$$

Very recently, a series of independent papers have appeared concerning the conversion of Fe(III) porphyrins to the corresponding O=Fe(IV) complexes. Groves et al. prepared O=Fe^{IV}(TMP) from $Fe^{III}(TMP^{+\bullet})(ClO_4)_2$ by passing through basic alumina [106]. Though O=Fe(IV) porphyrins are believed to be poor oxidants for olefin oxidation, they found the epoxidation of substituted styrenes by isolated O=Fe^{IV}(TMP). Kinetic data and product analyses in benzene indicated that the first and rate determining step is attack of the ferryl oxygen of O=Fe^{IV}(TMP) on the olefin π-bond to give a carbon radical intermediate, following by a fast reaction with O_2 to afford benzaldehyde or epoxides with loss of the configuration of the starting olefin. These mechanistic features are quite different from the results by O=$Fe^{IV}(TMP^{+\bullet})$ (Scheme 3).

The first electrochemical oxidation of haloiron(III) porphyrins generally affords porphyrin centered π-cation radicals [107], while the same oxidation of hydroxoiron(III) porphyrins gives oxoiron(IV) porphyrins [108–111]. The $E_{1/2}$ values of eight Fe^{III}(Por)(OH) was plotted against the $E_{1/2}$ values of the corresponding Fe^{III}(Por)(Cl) [106]. The correlation was found to be unity, indicating that first oxidation of Fe^{III}(Por)(OH) proceed through $Fe^{III}(Por^{+\bullet})$(OH) to afford O=Fe^{IV}(Por). A similar conclusion was also obtained by Gray et al. upon photoinduced oxidation of microperoxidase-8, the heme octapeptide derived from enzymatic cleavage of cytochrome *c* [112]. Very recently, Hamachi et al. have succeeded in detecting a transiently formed $Fe^{III}(Por^{+\bullet})$ as a precursor of O=Fe(IV) by using semisynthetic myoglobins having covalently appended $Ru(byp)_3$ [113].

2.3
Development of Heme Protein Engineering

Peroxidases and catalase react with hydrogen peroxide to afford oxo-ferryl porphyrin π-cation radicals so called compound I through the heterolysis of an

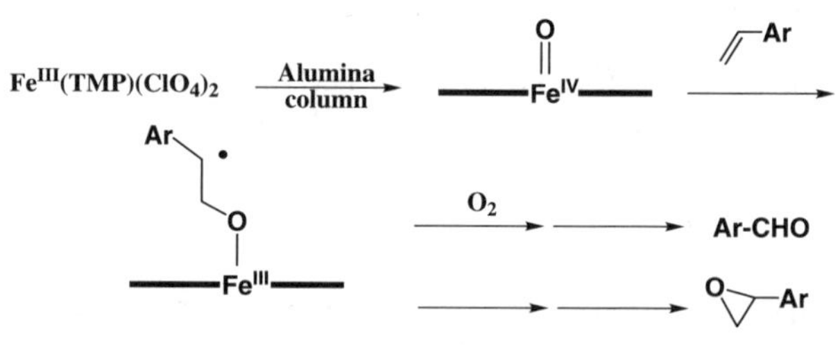

Scheme 3. The oxidation of olefins by O=Fe(IV) porphyrins

iron bound hydroperoxide [4, 5]. Homolytic O—O bond cleavage is usually observed for the reactions of hydroperoxide with many transition metal ions to form hydroxyl radical [114] , thus, general acid-base function of the distal histidine in those enzymes is believed to be crucial for avoiding radical process (Scheme 4) [115].

Quantitative roles of the distal residues of cytochrome *c* peroxidase were examined by replacing these residues with other amino acids by site-directed mutagenesis. For example, mutation of the distal histidine to leucine was found to depress the compound I formation rate from 3×10^{-7} to 2×10^{-2} M^{-1} s^{-1} [116, 117].

In the case of cytochrome P-450, the actual intermediate responsible for the oxygenation is still controversial, however, Ishimura et al. and Sligar et al. observed the important participation of the distal threonine 252 residue in cytochrome P-450cam as a proton donor for the heterolytic O—O bond cleavage as discussed in Scheme 4 [118, 119]. Recent mechanistic studies have shown the presence of a proton network in the distal site of P-450cam to allow the activation of a peroxo intermediate [120]. Very recently, a crystal structure of nitric oxide synthase-oxygenase dimer with pterin and the substrate (L-arginine) has been reported by Crane et al. [121]. Even in this case, L-arginine stays very close to the heme iron and is capable of serving as a proton donor (Scheme 5). These protons could work together with cysteinate ligation for the compound I formation in cytochrome P-450.

By comparison of the distal site structures of Mb [122] and cytochrome *c* peroxidase [123], two major differences are apparent; 1) the location of two histidines and 2) the histidine of C*c*P makes a hydrogen bond with asparagine (Asn 82) while the histidine of Mb stands alone (Fig. 8) [122]. According to the hydrogen bond, the distal histidine in C*c*P is much more basic than that of Mb [122]. As shown in Scheme 4, the proton abstract from hydrogen peroxide is the initial step of compound I formation, suggesting that the basicity of the distal histidine could play an important role. In fact, replacement of asparagine 70 in *HRP*, which interacts with the distal histidine (His 42) of HRP through the hydrogen bond, with valine decreased the compound I formation less than 10% of the native HRP [123].

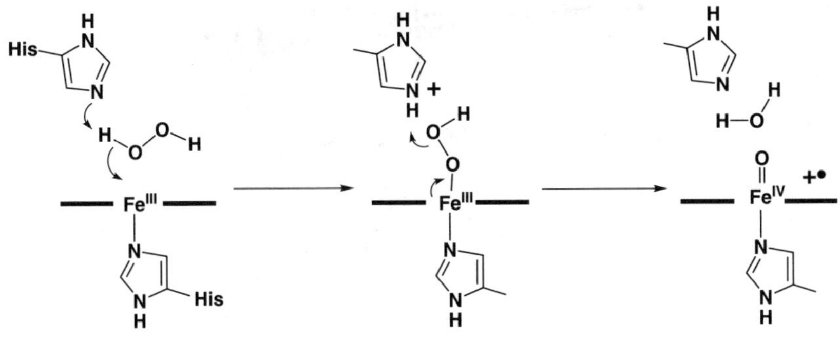

Scheme 4. Roles of the distal histidine for the formation of compound I

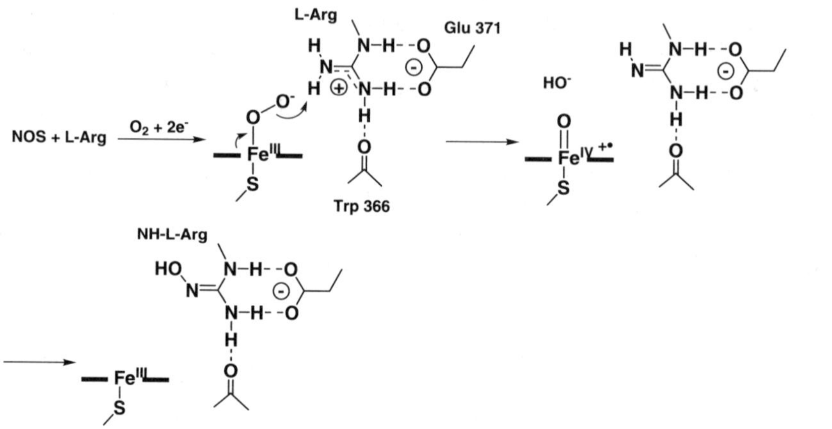

Scheme 5. Proposed proton assisted formation of compound I in the first step of NO synthesis by NOS

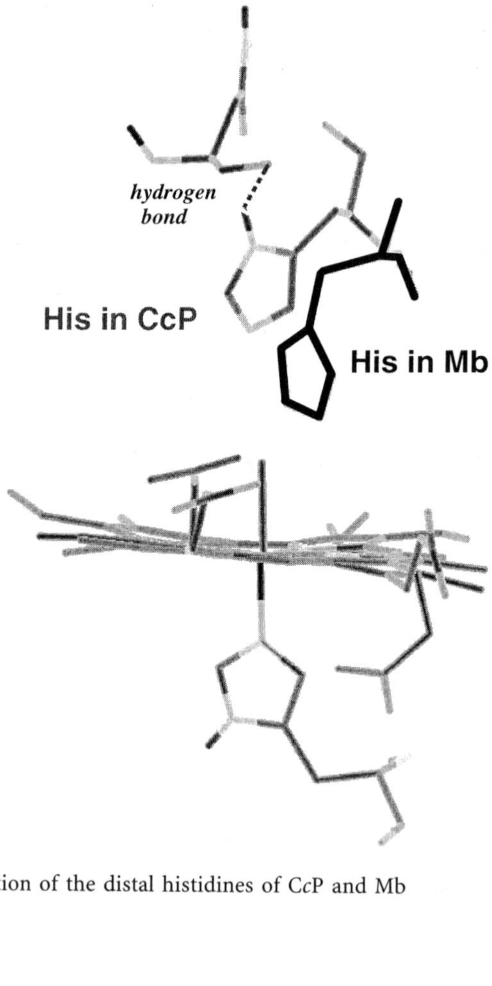

Fig. 8. Superimposition of the distal histidines of CcP and Mb

The location of the distal histidine in the peroxidase should also be very important for the compound I formation. According to the crystal structures of C*c*P and beef liver catalase [124], the distances between the heme iron and Nε of the distal histidine are 5.6 and 4.8 Å, respectively [123, 124]. About 5.0–5.5 Å could be the distance suitable for the hydrogen bond between the protonated distal histidine and the terminal oxygen of the iron bound hydroperoxide (Fig. 9) [125]. On the other hand, the distance between the heme iron and Nε of the distal histidine (His43) in Mb is 4.3 Å [122] and it seems too close to make a peroxidase type hydrogen bond; instead, the protonated distal histidine readily makes the hydrogen bonds with both oxygens of the hydroperoxide (Fig. 9) [126].

To investigate this hypothesis, Watanabe et al. have prepared F43H/H64L Mb, in which the distal His64 was moved to the position of Phe43 [127]. On the basis of the mutant crystal structure, the distance of Nε of the His43 and the heme iron was determined to be about 5.5 Å [128]. F43H/H64L Mb oxidized thioanisole and styrene 200–300 times faster than the wild-type Mb with very high enantio-selectivity [127a, 129]. For example, thioanisole and *trans*-*β*-methylstyrene were oxidized to *R*-sulfoxide and 1*R*, 2*R*-epoxide in e.e. of 85% and 96%, respectively [129]. Furthermore, the reaction of F43H/H64L Mb and *m*CPBA gave a transient intermediate characterized as compound I, which gradually photo-reduced to compound II (Fig. 10) [127b, 128]. Finally, compound I of F43H/H64L Mb has been shown to react with substrates including H_2O_2 (i.e. catalase activity) by stopped flow measurements [127b,c, 129, 130].

3 O=Mn Porphyrins

3.1 Early Work on the O=Mn Porphyrins

Hill and Groves independently reported the oxidation of hydrocarbons by PhIO in the presence of Mn^{III}(TPP)Cl [131, 132]. As shown in Scheme 6, the

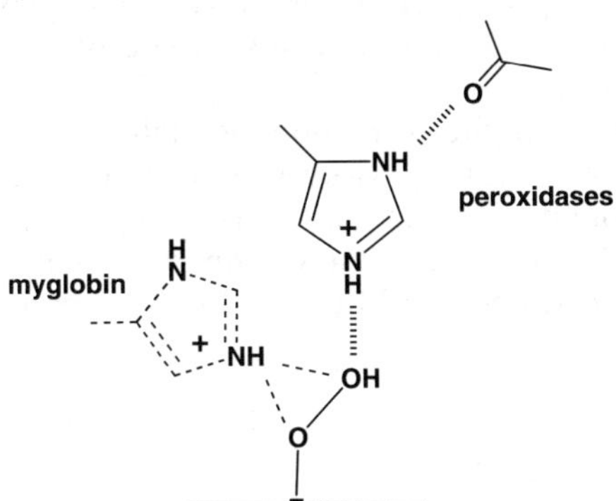

Fig. 9. Schematic drawing of two types of hydrogen bond by the distal histidines of C*c*P and Mb

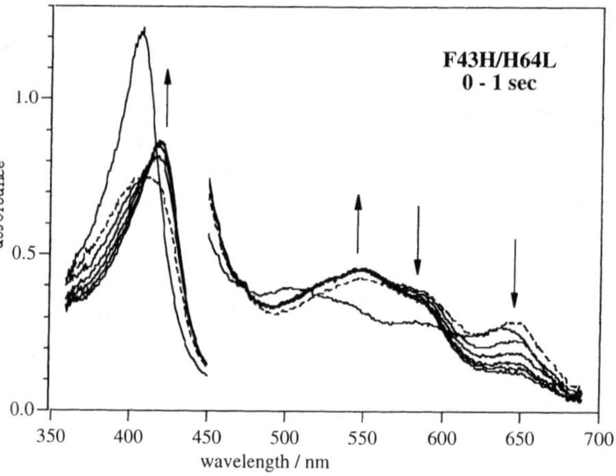

Fig. 10. Time dependent spectral changes of F43H/H64L Mb upon the addition of *m*CPBA

reaction is expected to proceed through the radical mechanism on the basis of the products [131, 132]. Upon the addition of PhIO to a Mn^{III}(TPP)Cl solution, a reactive intermediate tentatively assigned as O=Mn^{V}TPP was observed. Through crystallographic studies, Hill et al. demonstrated a series of Mn(IV) structures (Fig. 11) [133–136], whose UV-Vis spectra are very similar to those observed in earlier work. For the $Mn^{IV}(TPP)(OMe)_2$ complex, bond distances are as follows: two Mn—O bonds, 1.839(2) Å; two Mn—N1 bonds, 1.993(3) Å; two Mn—N2 bonds, 2.031(3) Å; two O—C bonds, 1.387(3) Å [134]. The magnetic susceptibility of $Mn^{IV}(TPP)(OMe)_2$ determined in solution and the solid state at 25 °C gave $\mu_{eff} = 3.9\ \mu_B$, consistent with a high-spin d^3 configuration of Mn(IV). Meunier et al. also isolated Mn(IV) complexes by using NaOCl(*aq.*) as an oxidant [137]. According to X-ray absorption spectroscopies (EXAFS, XANES), one axial ligand was well defined ($R = 1.84 \pm 0.02$ Å) and assigned to a Mn—O single bond and the other bond was much larger (R = ca. 2.3 Å) and assigned to a loosely bound water molecule [138]. EPR spectra of those Mn(IV) complexes showed strong $g_\perp \sim 4.0$ and weak $g_\perp \sim 2.0$ signals characteristic of a high-spin d^3 configuration [139].

Mechanistic studies on the formation of high-valent oxo-Mn porphyrin complexes were carried out by Groves and Watanabe [140]. A peroxo-Mn^{III}(TMP) was found to react with acylating reagents to afford an acylperoxo-Mn^{III}(TMP) in CH_3CN at low temperatures. By employing *m*CPBA, they also observed the acylperoxo-Mn(III) adduct formation in CH_2Cl_2, which

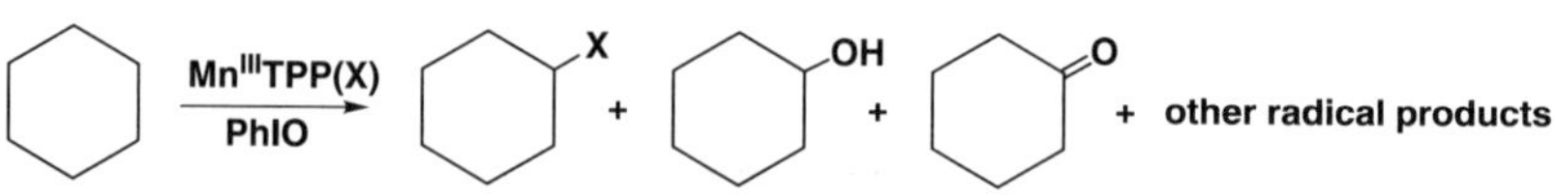

Scheme 6. Mn porphyrin catalyzed oxidation of hydrocarbons

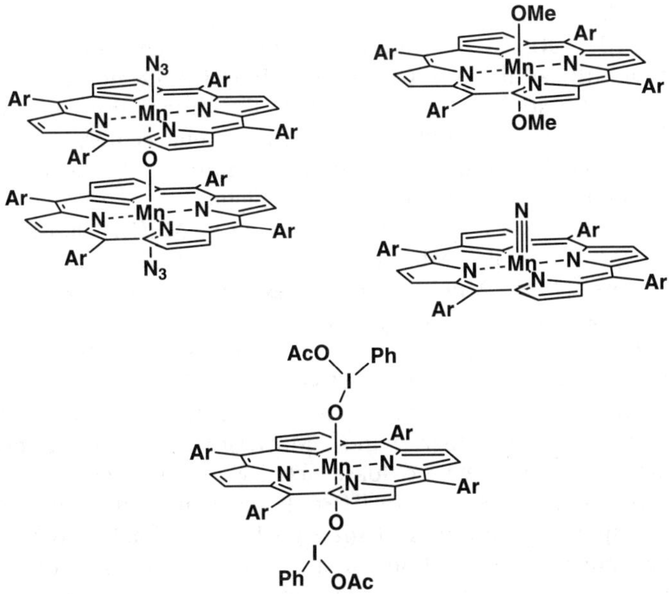

Fig. 11. Mn(IV) porphyrin complexes well-characterized by X-ray crystallography

readily decomposed to afford O═Mn(V) porphyrins [140]. A similar oxygen activation of peroxo-Mn(III) complexes by CO_2 were also reported [139b]. The O═Mn^VTMP is EPR silent immediate after the formation, however, EPR signals assignable to O═Mn^{IV}(TMP) was observed in a few seconds even −78 °C in CH_2Cl_2, implying O═Mn^V(TMP) is very unstable even at low temperatures [141]. Substituent effect of the aromatic ring of peroxybenzoic acid on the formation of O═Mn^V(TMP) indicated heterolytic cleavage of the O—O bond in the acylperoxo-Mn adducts proceed under acidic conditions, while homolysis is a favorable process in the presence of hydroxide anion [141]. The very similar homolysis/heterolysis nature of acylperoxo-Fe(III) porphyrin has been also demonstrated by Groves and Watanabe [71–73]. Schappacher and Weiss also reported both homolytic and heterolytic O—O bond cleavage for Mn^{III}(Por)(OO—CO_2) [139b].

3.2 Application of O═Mn Porphryin Chemistry to Biological Systems

The heme prosthetic group of heme proteins has been replaced by many manganese(III) porphyrins [142–148]. UV-Vis spectra of Mn(III) substituted heme proteins are usually very similar to those of Mn(III) porphyrins in aqueous solutions. Anionic ligands such as N_3^-, F^- and CN^- does not bind to Mn reconstituted myoglobin (Mb) and horseradish peroxidase (HRP) as evidenced by no UV-Vis spectral change upon the addition of these external ligands, while Mn substituted cytochrome *c* peroxidase (C*c*P) gives altered spectra [142]. Modi, Mitra and co-workers examined the binding of SCN^- and CN^- to Mn(III)•HRP by relaxation rate measurements of ^{15}N resonance [147a]. Though the addition of CN^- to Mn(III)•HRP did not alter its UV-Vis spectrum, ^{15}N-T_1 studies indicated that the interaction of CN^- with Mn(III)•HRP is very weak and away from the metal center, ca. 6.8 Å from the Mn(III) center. Non-binding of anionic ligands may be caused by strong binding of a hydroxide ion as the distal site ligand.

Yonetani and Asakura systematically prepared Mn(III) substituted C*c*P, HRP and sperm whale Mb with proto-, hemato-, *meso*- and deuteroporphyrins [144]. Mn(III)•peroxidases reacted with H_2O_2 to form intermediates whose UV-Vis spectra were similar to those of well characterized Mn(IV) porphyrins described above, while these Mn•peroxidases showed very low peroxidase activities. The intermediates are able to oxidize peroxidase substrates and retain 2 oxidizing equivalents per mole of the enzymes. On the other hand, Mn(III)•Mb neither afforded the intermediate nor exhibited peroxidase activity by the reaction with H_2O_2. Upon the addition of H_2O_2, Mn(III)•C*c*P gave an EPR signal at $g = 2$, however, the spin concentration of the signal was estimated to be less than 5% of the molar concentration of the enzyme. Thus, Yonetai et al. concluded that the free radical was not the major component of the intermediate [144].

Later, the axial ligand in the Mn(IV)•HRP prepared by H_2O_2 was postulated to be a hydroxo ligand by Makino et al. [148]. Groves, Spiro and coworkers further studied the reaction of Mn(III)•HRP with *m*CPBA instead of H_2O_2 [146]. The intermediate, they called Mn•HRP-I, showed absorption maximum at 412 nm, typical for Mn(IV) porphyrins, and stable enough to isolate by Sephadex chromatography at 4 °C. The EPR spectrum of the isolated Mn•HRP-I exhibited signals at g 5.25 and 2.08, characteristic of Mn(IV) and a free radical. Titration of Mn•HRP-I to Mn^{III}HRP required 2 equivalent of thiosalicylate. The first equivalent exclusively eliminated the Mn(IV) signal. The second equivalent titrated the g 2 signal. At the same time, 413.1 nm excited RR spectrum of Mn•HRP-I showed a band at 622 cm^{-1}, which shifted to 592 cm^{-1} by the *m*CPBA-^{18}O preparation. Likewise synthetic O=Mn(IV) porphyrins, 622 cm^{-1} band was also observed even in $H_2{}^{18}O$, consistent with slow oxo exchange with solvent H_2O observed for model systems (*vide infra*). Further, the 622 cm^{-1} band was down-shifted by 4 cm^{-1} in D_2O [146].

Gray et al. have prepared manganese(III) substituted microperoxidase-8 (Mn•MP8) and examined its reactions with H_2O_2 and $Ru(byp)_3^{3+}$ [149]. A transient intermediate formed by H_2O_2 reaction exhibited the Soret at 401 nm and was less stable and readily reduced to Mn^{III}•MP8 in seconds. The same product was also observed by the photoinduced oxidation of Mn^{III}•MP8 in the presence of $Ru(byp)_3^{2+}$. On the basis of the spectroscopic characteristics, Gray et al. proposed the intermediate to be O=Mn^{IV}•MP8. More importantly, the low reactivity of Mn(III) substituted heme proteins with peroxides was attributed to reduced "push effect" from the other axial ligand [149].

3.3
Recent Advances in O=Mn Porphryin Chemistry

As mentioned above, isolated high valent Mn porphyrin complexes never had the O=Mn bond in early work. Finally, O=Mn^{IV}(TMP) and O=Mn^{IV}(TMP)(OH) were isolated and characterized [150]. The Mn^{IV}=O stretching frequency for the former complex was identified at 754 cm^{-1} (^{18}O: 722 cm^{-1}) by both RR and IR spectroscopies. On the other hand, 6-coordinate O=Mn^{IV}(TMP)(OH) showed the O=Mn stretching frequency at 712 cm^{-1} while Fe—OH was not apparent. More than 100 cm^{-1} lowering of O=Mn band in HRP was attributed to the imidazolate character of the proximal histidine ligand in HRP. Likewise the O=Fe porphyrins, O=Mn(V) porphyrins and O=Mn(IV) porphyrins exhibit very different reaction profiles as summarized in Scheme 7.

Ayougou et al. prepared a Mn(IV)(T_{piv}PP) complex by KO_2/CO_2 at −70 °C according to Scheme 8 [139]. The sample was applied to X-ray absorption spectroscopy (XAS). The results were consistent with the formation of a five-coordinate O=Mn(IV) complex, in which the Mn—O bond distance was 1.69 ± 0.03.

Due to the instability of O=Mn(V) porphyrins, their formation is usually demonstrated by oxidation of organic substrates such as olefins, alcohols, and

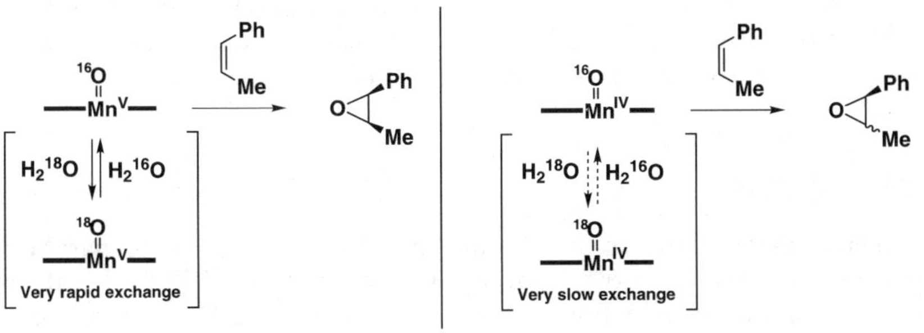

Scheme 7. Comparison of O=Mn(V) and O=Mn(IV) porphyrin catalyzed epoxidation

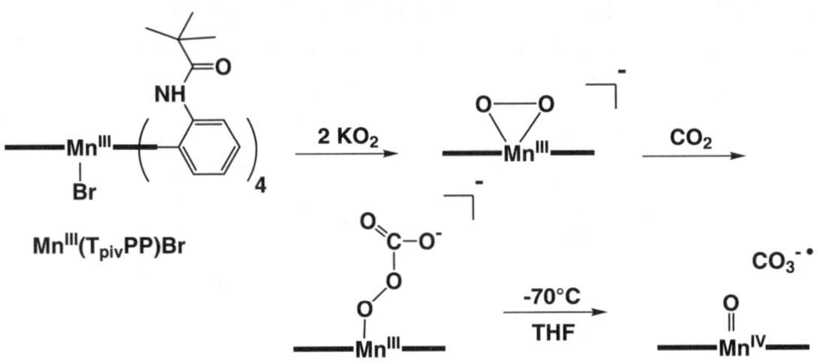

Scheme 8. Preparation of O=Mn^{IV}(T_{piv}PP)

hydrocarbons [151–153]. Stereospecific epoxidation shown in Scheme 7 is also an example of characteristic nature of O=Mn(V) porphyrins [150a]. Thus, O=Mn^V(TMP) was only observed in CH_2Cl_2 as a transient intermediate and reduced in a few seconds to O=Mn(IV) species even at −78 °C [141]. Very recently, Groves et al. successfully prepared an O=Mn(V) porphyrin complex by employing water soluble tetra-(*N*-methylpyridyl)porphyrin (TMPyP) at room temperature [154]. The reaction of Mn^{III}(TMPyP) with *m*CPBA, HSO_5^-, or ClO- has afforded a short lived intermediate with the Soret maximum at 443 nm with second order kinetics, first order in [Mn^{III}(TMPyP)] and [oxidant]. The intermediate decays to O=Mn^{IV}(TMPyP) as a relatively stable species (λ_{max}: 428 nm). The O=Mn(V) intermediate is highly reactive toward olefins while O=Mn(IV) complex is not capable of the reactions under the conditions. In addition, the oxo oxygen was demonstrated to be exchangeable with $H_2{}^{18}O$ as described in Scheme 7.

Unlike the O=Mn(IV) porphyrins, neither crystal structures nor vibration of O=Mn(V) porphyrins are available. However, Collins et al. have reported crystal structures of oxo-Mn(V) complexes having non-porphyrin ligands (Fig. 12) [155]. The oxo-Mn(V) complexes are diamagnetic and the O—Mn bond distances are 1.548–1.555(4) Å. According to the short O—Mn bond distance, they have attributed the oxo-Mn(V) bonds to be a triple bond formulation [156]. Thus, O—Mn stretching bands appears around 970–979 cm^{-1} (933–942 cm^{-1} for ^{18}O), depending on the structure of the ligand. In the case of the oxo-Mn(V) complex having a macrocyclic tetraamide ligand, the Mn atom sits 0.60 Å above the mean plane. Likewise the O=Mn(V) porphyrins, the oxo ligand exchanges in ^{18}O-labeled water [157].

4 O=Cr Porphyrins

The methodology used for the O=Mn and O=Fe porphyrin preparation is applicable for the O=Cr(V) porphyrin synthesis. Cr^{III}(TPP)Cl catalyzes hydroxylation and epoxidation with PhIO in good yields [158]. Treatment of Cr^{III}(TPP)Cl either with PhIO or *m*CPBA gives a red intermediate assigned as O=Cr^V(TPP). A dilute solution of O=Cr^V(TPP) (ca. 10^{-5}M) is stable for several hours at room temperature, addition of olefins or alcohols causes rapid regeneration of Cr^{III}(TPP)Cl. Clear evidence for the formation of O=Cr was provided by IR measurements; $\nu_{^{16}O=Cr}$ = 1026 cm^{-1}, $\nu_{^{18}O=Cr}$ = 982 cm^{-1}.

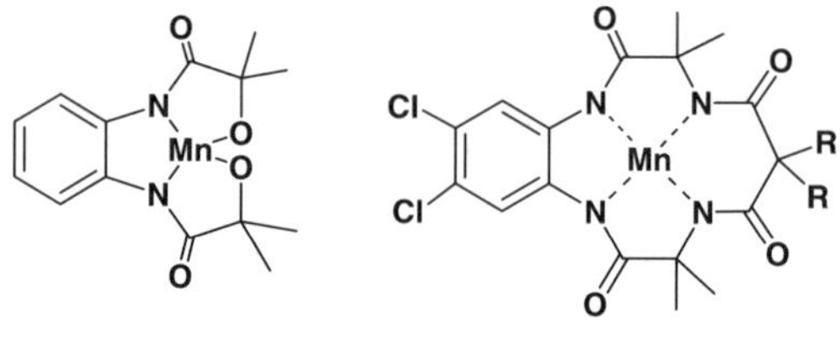

Fig. 12. Non-porphyrin Mn complexes, which are able to form O=Mn(V) complexes

The O=Cr(V) was also demonstrated by the EPR spectra of $O{=}Cr^{V}(TPP)$, in which ^{53}Cr and ^{17}O hyperfine splitting was evident [158, 159].

In 1981 and 1982, three groups independently reported the preparation of $O{=}Cr^{IV}$ porphyrin complexes as diamagnetic species [160, 162]. According to the crystal structures of $O{=}Cr^{IV}(TPP)$, O=Cr was determined to be 1.62(2) by Budge et al. [160] and 1.572(6) by Groves et al. [161] respectively. The $\nu_{O=Cr}$ was observed at 1025 cm^{-1} (981 cm^{-1} for ^{18}O), very similar to that of $O{=}Cr^{V}(TPP)$. A less common preparation of O=Cr(IV) complexes was also reported by Yamaji et al., i.e., the photochemical formation of $O{=}Cr^{IV}(TPP)$ from $ONO{-}Cr^{III}(TPP)$ [163]. $O{=}Cr^{IV}$ porphyrin complexes are less reactive than the corresponding $O{=}Cr^{V}$ complexes, though quantitative conversion of benzyl alcohol to benzaldehyde and triphenyl phosphine to its oxide were reported. Greager and Murray prepared $O{=}Cr^{V}(TPP)$ by electrochemical oxidation of $O{=}Cr^{IV}(TPP)$ [164].

Likewise $Cr^{III}(Por)$, $Cr^{III}(salen)$ complexes react with PhIO to yield $O{=}Cr^{V}(salen)^{+}X^{-}$ complexes, which oxidize olefins in reasonable yields [165]. The $\nu_{O=Cr}$ band appeared in the range of 935–997 cm^{-1} depending on the nature of a counter anion (X^{-}).

5
O=Ru Porphyrins

Because of the stable formation of high oxidation states, a wide variety of O=Ru complexes have been prepared. Most of those complexes are prepared by using non-porphyrin ligands [166–169]. For the ruthenium porphyrin complexes, μ-oxo-bis-ruthenium(IV) complexes, $[(Ru^{IV}(Por)(OR)]_2O$, were prepared and characterized by X-ray crystallography [170, 171].

Leung et al. oxidized $Ru^{III}(OEP)(PPh_3)_2$ with PhIO in CH_2Cl_2 to afford a green intermediate, which was tentatively assigned as $O{=}Ru^{IV}(OEP^{+\bullet})Br$ [172]. A stoichiometric titration with PPh_3 gave $O{=}PPh_3$ and $[Ru^{IV}(OEP)_2O]$ quantitatively. By using $Ru^{III}(OEP)(PPh_3)_2$ as a catalyst, PhIO was able to oxidize hydrocarbons such as norbornene, styrene, cyclohexene and cyclohexane. Groves and Quinn prepared highly oxidized *trans*-bis-oxo-ruthenium porphryin, $(O{=})_2Ru^{VI}(TMP)$ by the reaction of $Ru^{II}(TMP)(CO)$ and *m*CPBA in CH_2Cl_2 at room temperature [173]. Further development of $(O{=})_2Ru^{VI}(TMP)$ preparation by aerobic oxidtion of $Ru^{II}(TMP)$ according to the mechanism shown in Scheme 9 allowed to proceed aerobic epoxidation [174]. Homogeneous catalytic oxidations using ruthenium porphyrin complexes are summarized by James [175].

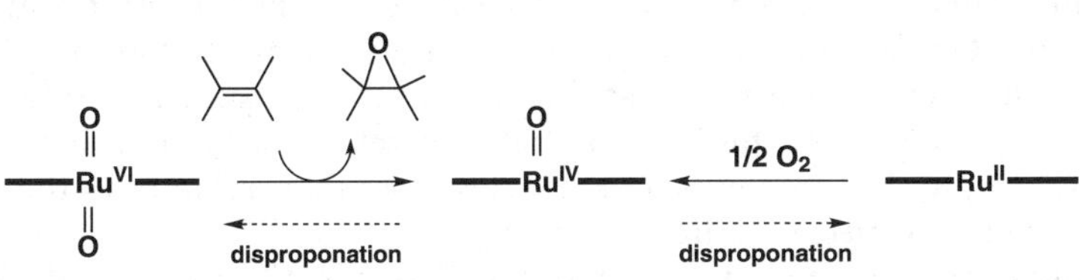

Scheme 9. Aerobic epoxidation catalyzed by Ru(TMP)

As expected, $(O{=})_2Ru^{VI}(TMP)$ complex is diamagnetic and exhibits well-resolved ^1H-NMR signals in diamagnetic region [173]. The complex can be isolated as a solid form at room temperature [176], but the X-ray crystal structure has not been reported. The crystal structure of analogous dioxoosmium(VI) porphyrin has been determined; 1.743 Å for Os=O bond [177]. The infrared spectrum of $(O{=})_2Ru^{VI}(TMP)$ shows a strong band at 821 cm^{-1} which is assigned as ν(Ru=O) band. The ν(Os=O) band for $(O{=})_2Os^{IV}(OEP)$ is assigned as 825 cm^{-1}. $(O{=})_2Ru^{VI}(OEP)$ and $(O{=})_2Ru^{VI}(TPP)$ which are not sterically bulky porphyrin complexes are prepared by the oxidations of $CORu^{II}(OEP)$ and $CORu^{II}(TPP)$ with *m*CPBA in the presence of alcohol (methanol or ethanol), which occupies the vacant coordination site of a ruthenium(IV) intermediate [176]. Later, the ruthenium(IV) intermediate is identified as dihydroxy ruthenium(IV) porphyrin complex, but not an oxo ruthenium(IV) porphyrin [178, 179]. The cyclic voltammogram of $(O{=})_2Ru^{VI}(TMP)$ in dichloromethane shows one-electron reversible oxidation couples around 0.6 V vs SCE [173]. The oxidized product was characterized dioxo ruthenium(VI) porphyrin π-cation radical complex [173, 180] $O{=}Ru^{IV}(TMP)$ complex has been characterized both in the synthesis of $(O{=})_2Ru^{VI}(TMP)$ from $CORu^{II}(TMP)$ and in the reduction of $(O{=})_2Ru^{VI}(TMP)$ with triphenylphosphine [181]. $O{=}Ru^{IV}(TMP)$ is unstable toward disproportionation which affords $(O{=})_2Ru^{VI}(TMP)$ and $Ru^{II}(TMP)$ [181]. While $(O{=})_2Ru^{VI}(TMP)$ oxidizes an olefin to an epoxide, $O{=}Ru^{IV}(TMP)$ does not oxidize the olefin but reacts with triphenylphosphine [180]. Recently, highly efficient oxidizing systems using ruthenium porphyrin have been reported [182]. The systems employ 2,6-dichloropyridine *N*-oxide as the oxidant and oxidize adamantane to adamantan-1-ol (up to 68%), adamantan-1,3-diol (25%) and adamantan-2-one (1%). Tentative formulation of an active intermediate includes oxo-Ru(V) porphyrin, oxo-Ru(IV) porphyrin π-cation radical or oxo-Ru species haiving the pyridine *N*-oxide ligand.

6 Other O=M Porphyrins

In connection with the intermediates of native heme enzymes introduced in the previous section, various oxo metalloporphyrin complexes have been synthesized. Oxo metalloporphyrin complex shows various reactivities depending on the central metal ion and its oxidation state. For example, with oxophilic metals such as titanium and vanadium, oxo metalloporphyrins are inert and unable to transfer an oxygen atom to organic substrates at room temperature. On the other hand, oxo metalloporphyrins of iron, manganese, and chromium are known to be capable of oxidizing organic substrates. This section is focused on synthesis and characterization of various high valent metal oxo porphyrin complexes and also discusses their relation to the nature of active iron-oxo species in heme enzymes.

Metalloporphyrin complexes with groups IVB to VIIB readily binds oxygen atom(s) to form both mononuclear and binuclear complexes. Very early studies by Buchler et al. showed a series of high valent oxo and μ-oxo complexes of Hf(V), Mo(V), Nb(V), W(V) and Re(V) [183]. A crystal structure of $O{=}Mo^{V}{-}O{-}Mo^{V}{=}O$ showed the O=Mo and Mo–O distances to be 1.708(3) and 1.936(3) Å, respectively [184]. Replacement of the O=Mo(V) unit in the $O{=}Mo^{V}{-}O{-}Mo^{V}{=}O$ to Cl–Mo(IV) caused the O–Mo distance in $Mo^{IV}{-}O{-}Mo^{IV}$ to be 1.851(6) [185]. On the other hand, $O{=}Mo^{IV}$(TPP) showed the O=Mo bond distance being 1.656(6) Å [186]. $O{=}Mo^{IV}$(TPP) reacts with O_2 to form $O{=}Mo^{V}{-}O{-}Mo^{V}{=}O$. When sterically hindered porphyrin such as TMP is used instead of TPP, an oxygen adduct of molybdenum(VI) complex, $O{=}Mo^{VI}(TMP)(O_2)$, is characterized from the reaction of $O{=}Mo^{IV}$(TMP) with O_2 [187]. The O=Mo distances for *cis*-dioxo-Mo^{VI}(TPP) were also reported to be 1.709(9) and 1.744(9) Å with a saddle shape porphyrin core (Fig. 13a) [188]. Interestingly, the crystal structure of $O_3[Nb(TPP)]_2$ was quite different from that for Mo derivatives as shown in Fig. 13 [184].

Oxo-metalloporphyrins other than those described above are the O=Ti complexes. For instance, Guilard et al. reported the transformtion of Ti(Por)Cl to peroxo-Ti(Por) via O=Ti(Por) [189]. The structures of O=Ti and (O_2)Ti(Por) were confirmed by their crystal structures.

Oxo-vanadium(IV) porphyrin complex has a stable V=O bond. The oxo ligand of oxo-vanadium(IV) porphyrin is displaced with halides to form bis halide vanadium(IV) porphyrin complex only when it reacts with thionyl halides or oxalyl halide [190]. The *bis* halide complex reacts with water to reproduce oxo-vanadium(IV) porphyrin complex. The reaction is utilized to synthesize ^{18}O-labeled oxo-vanadium(IV) porphyrin complex [191]. One electron oxidation of oxo-vanadium(IV) porphyrin gives stable oxo-vanadium(IV) porphyrin π-cation radical complex [192], which allows to facilitate systematic studies of the effect of solvent and porphyrin oxidation on ν(V=O) vibration as models for compounds I and II. Raman studies shows that the frequency for ν(V=O) decreased with increase in solvent acceptor number, consistent with solvent-induced polarization of the ν(V=O) bond [193]. The downshift of ν(V=O) band upon ligation increased with increasing donor strength of the ligand; OH > imidazolate > imidazole > pyridine > alcohol

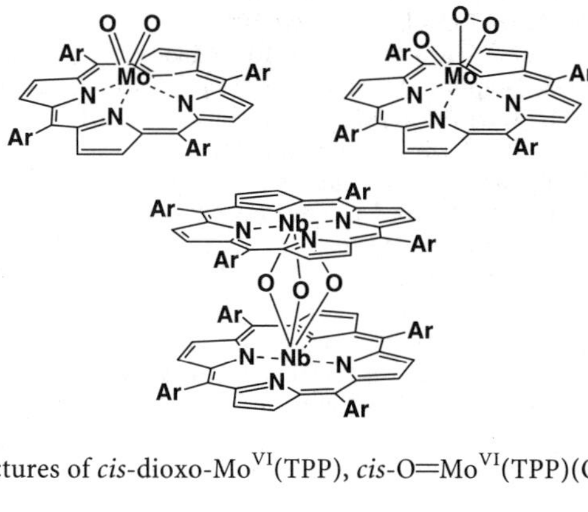

Fig. 13. Crystal structures of *cis*-dioxo-Mo^{VI}(TPP), *cis*-$O{=}Mo^{VI}(TPP)(O_2)$ and $O_3[Nb(TPP)]_2$

and water. An earlier work reported that the V=O band shifted to higher frequency for $^2A_{1u}$ cations, such as $O{=}V(OEP^{+\bullet})$, but to lower frequency for $^2A_{2u}$ cations, such as $O{=}V(TMP^{+\bullet})$ [194]. However, a recent study demonstrated that the shift of the $\nu(V{=}O)$ band is insensitive to radical type, as pointed out for the oxo-ferryl complexes [191]. Oxo-vanadium(IV) porphyrin has unpaired electron in vanadium d_{xy} orbital, which gives sharp EPR signals [195]. ESEEM spectra of oxo-vanadium porphyrins are also reported [196]. Oxo-vanadium(IV) porphyrin π-cation radical is biradical which exhibits a triplet EPR signal at low temperature [192]. Recent study of $O{=}V(OEP^{+\bullet})(H_2O)$ showed that the radical complex forms a dimer both in solid and in solution at low temperature and exhibits intramolecular ferromagnetic coupling and intermolecular antiferromagnetic coupling [197].

7
References

1. Dawson JH, Sono M (1987) Chem Rev 87: 1255
2. Watanabe Y, Groves JT (1992) Molecular mechanism of oxygen activation by cytochrome P-450. In: Boyer PD, Sigman DS (eds) The Enzymes, Academic Press, New York, vol 20, p 405
3. Ortiz de Montellano PR (1986) Cytochrome P450, 2nd edn. Plenum Press, New York
4. Dunford HB (1999) Heme Peroxidases, Wiley-VCH, New York, Chap 4
5. Everse J, Everse KE, Grisham MB (1991) Peroxidases in Chemistry and Biology, CRC Press, Boca Raton, vol I, II
6. Shapiro S, Caspi E (1998) Tetrahedron 54: 5005
7. Guengerich FP, Macdonald TL (1993) Sequential Electron Transfer Reactions Catalyzed by Cytochrome P-450 Enzymes. In: Advances in Electron Transfer Chemistry, vol 3, JAI Press, p 191
8. Mansuy D, Battioni P (1993) Dioxygen Activation at Heme Centers in Enzymes and Synthetic Analogs. In: Reedijk J (ed) Bioinorganic Catalysis, Marcel Dekker, New York, Chap 12, p 395
9. Gunsalus IC, Wagner GC (1978) Bacterial P-450_{cam} Methylene Monooxygenase Components: Cytochrome *m*, Putidaredoxin, and Putidaredoxin Reductase. In: Fleischer S, Packer L (eds) Methods in Enzymology, Academic Press, New York, vol 52, p166
10. Tabushi I, Koga N (1979) Tetrahedron Lett 3681
11. Tabushi I, Koga N (1978) Tetrahedron Lett 5017
12. Tabushi I, Koga N (1979) J Am Chem Soc 101: 6456
13. Felton RH, Owen GS, Dolphin D, Faier J (1971) J Am Chem Soc 93: 6332
14. Groves JT, Nemo TE, Myers RS (1979) J Am Chem Soc 101: 1032
15. Chang CK, Kuo M-S (1979) J Am Chem Soc 101: 3413
16. Matsuura T (1977) Tetrahedron 33: 2869
17. Gunter MJ, Turner P (1991) Coord Chem Rev 108: 115
18. Meunier B (1992) Chem Rev 92: 1411
19. Sheldon RA (1994) Metalloporphyrins in Catalytic Oxidations, Marcel Dekker, New York
20. Collman JP, Zhang X, Lee VJ, Uffelman ES, Braaauman JI (1993) Science 261: 1404
21. Watanabe Y (1997) Model Studies on Heme Monooxygenases. In: Funabiki T (ed) Oxygenases and Model Systems, Kluwer Academic Publishers, Netherlands, vol 19, p 223
22. Dunford HB (1982) Adv Inorg Biochem 4: 41

23. Blumberg WE, Peisach J, Wittenberg BA, Wittenberg AB (1968) J Biol Chem 243: 1854
24. Wolberg A, Manassen J (1970) J Am Chem Soc 92: 2982
25. Moss TH, Ehrenberg A, Bearden AJ (1969) Biochemistry 8: 4159
26. Fajer J, Borg DC, Forman A, Felton RH, Vegh L, Dolphin D (1973) Ann NY Acad Sci 206: 349
27. Loew G, Herman ZS (1980) J Am Chem Soc 102: 6173
28. Dolphin D, Forman A, Borg DC, Fajer J, Felton RH (1971) Proc Natl Acad Sci USA 68: 614
29. Sandusky PO, Salehi A, Chang CK, Babcock GT (1989) J Am Chem Soc 111: 6437
30. Palaniappan V, Terner J (1989) J Biol Chem 264: 16046
31. Fujii H (1993) J Am Chem Soc 115: 4641
32. Schulz CE, Devaney PW, Winkler H, Debrunner PG, Doan N, Chiang R, Rutter R, Hager LP (1979) FEBS Lett 103: 102
33. Schulz CE, Rutter R, Sage JT, Debrunner PG, Hager LP (1984) Biochemistry 23: 4743
34. Morishima I, Ogawa S (1978) J Am Chem Soc 100: 7125
35. La Mar G, de Ropp JS (1980) J Am Chem Soc 102: 395
36. Thanabal V, La Mar G, de Ropp JS (1988) Biochemistry 27: 5400
37. Roberts JE, Hoffman BM, Rutter R, Hager LP (1981) J Biol Chem 256: 2118
38. Roberts JE, Hoffman BM (1981) J Am Chem Soc 103: 7654
39. Penner-Hahn JE, Ebel KS, McMurry TJ, Renner M, Balch AL, Groves JT, Dawson JH, Hodgson KO (1986) J Am Chem Soc 108: 7819
40. Hashimoto S, Tatsuno Y, Kitagawa T (1986) Proc Natl Acad Sci USA 82: 2417
41. Ogura T, Kitagawa T (1987) J Am Chem Soc 109: 2177
42. Paeng K-J, Kincaid JR (1988) J Am Chem Soc 110: 7913
43. Chuang W-J, Van Wart HE (1992) J Biol Chem 267: 13293
44. Yonetani T, Chance B, Kajiwara S (1966) J Biol Chem 241: 2981
45. Yonetani T (1976) The Enzymes, Academic Press, New York, vol 13
46. Yonetani T, Schleyer H, Ehrenberg A (1966) J Biol Chem 241: 3240
47. Lang G, Spartalian K, Yonetani T (1976) Biochim Biophys Acta 451: 250
48. Sivaraja M, Goodin, DB, Smith M, Hoffman BM (1989) Science 245: 738
49. Hashimoto S, Teraoka J, Inubushi T, Yonetani T, Kitagawa T (1986) J Biol Chem 261: 11110
50. Reczek CM, Sitter AJ, Terner J (1989) J Mol Struct 214: 27
51. Palcic MM, Rutter R, Araiso T, Hager LP, Dunford HB (1980) Biochem Biophys Res Comm 94: 1123
52. Rutter R, Hager LP, Dhonau H, Hendrich M, Valentine M, Debrunner P (1984) Biochemistry 23: 6809
53. Egawa T, Miki H, Ogura T, Makino R, Ishimura I, Kitagawa T (1992) FEBS Lett 305: 206
54. Schonbaum GR, Chance B (1976) The Enzymes. Academic Press, New York, vol 13
55. Jones P, Middlemiss DN (1972) Biochem J 130: 411
56. Chance B (1949) J Biol Chem 180: 947
57. Gouet P, Jouve H-M, Williams PA, Anderson I, Andreoletti P, Nussaume L, Hajdu J (1996) Nature Struct Biol 3: 951
58. Benecky MJ, Frew JE, Scowen N, Jones P, Hoffman BM (1993) Biochemistry 32: 11929
59. Ivancich A, Jouve HM, Sartor B, Gaillard J (1997) Biochemistry 36: 9356
60. Chuang W-J, Heldt J, Van Wart HE (1989) J Biol Chem 264: 14209
61. Malmstrom BG (1990) Chem Rev 90: 1247
62. Kitagawa T, Ogura T (1997) Progress in Inorganic Chemistry. Karlin KD (Ed), Wiley, vol 45: 431
63. Morgan JE, Verkhovsky MI, Wikström (1996) Biochemistry 35: 12235
64. Ogura T, Takahashi S, Hirota S, Sinzawa-Itoh K, Yoshikawa S, Kitagawa T (1993) J Am Chem Soc 115: 8527
65. Groves JT, Haushalter RC, Nakamura M, Nemo TE, Evans BJ (1981) J Am Chem Soc 103: 2884

66. Mandon D, Weiss R, Jayaraj K, Gold A, Terner J, Bill E, Trautwein AX (1992) Inorg Chem 31: 4404
67. Penner-Hahn JE, Eble KS, McMurry TJ, Renner M, Balch AL, Groves JT, Dawson JH, Hodgson KO (1986) J Am Chem Soc 108: 7819
68. Hashimoto S, Tatsuno Y, Kitagawa T (1987) J Am Chem Soc 109: 8096
69. Hashimoto S, Mizutani Y, Tatsuno Y, Kitagawa T (1991) J Am Chem Soc 114: 6542
70. Kincaid JR, Schneider AJ, Paeng K-J (1989) J Am Chem Soc 111: 735
71. Groves JT, Watanabe Y (1986) J Am Chem Soc 108: 7834
72. Groves JT, Watanabe Y (1988) J Am Chem Soc 110: 8443
73. Groves JT, Watanabe Y (1986) J Am Chem Soc 108: 7836
74. Yamaguchi K, Watanabe Y, Morishima I (1992) J Chem Soc, Chem Commun 1709
75. Yamaguchi K, Watanabe Y, Morishima I (1993) J Am Chem Soc 115: 4058
76. Dolphin D, Trayler TG, Xie LY (1997) Acc Chem Rev 30: 251
77. Sugimoto H, Tung H-C, Sawyer DT (1988) J Am Chem Soc 110: 2465
78. Bill E, Ding X-Q, Bominaar EL, Trautwein AX, Winkler H, Mandon D, Weiss R, Gold A, Jayaraj K, Hatfield WE, Kirk M (1990) Eur J Biochem 188: 665
79. Fujii H, Yoshimura T, Kamada H (1996) Inorg Chem 35: 2373
80. Fujii H (1994) Chem Lett 1491
81. Fujii H (1993) Inorg Chem 32: 1993
82. Fujii H, Ichikawa K (1992) Inorg Chem 31: 1110
83. Czarnecki K, Proniewicz LM, Fujii H, Kincaid J (1996) J Am Chem Soc 118: 4680
84. Czarnecki K, Nimri S, Gross Z, Proniewicz LM, Kincaid J (1996) J Am Chem Soc 118: 2929
85. Gross Z, Nimri S (1994) Inorg Chem 33: 1731
86. Czarnecki K, Kincaid J, Fujii H (1999) J Am Chem Soc 121: 7953
87. Fujii H, Yoshimura T, Kamada H (1997) Inorg Chem 36: 6142
88. Collman JP, Sorrell TN, Hoffman BM (1975) J Am Chem Soc 97: 914
89. Ogoshi H, Sugimoto H, Yoshida Z (1975) Tetrahedron Lett: 2289
90. Higuchi T, Uzu S, Hirobe M (1990) J Am Chem Soc 112: 7051
91. Aissaoui H, Ghirlanda S, Gmür C, Woggon W-D (1996) J Mol Cat 113: 393
92. Volz H, Holzbecher M (1997) Angew Chem Int Ed Engl 36: 1142
93. Wagenknecht H-A, Woggon W-D (1997) Angew Chem Int Ed Engl 36: 390
94. Jacob GS, Orme-Johnson WH (1979) Biochemistry 18: 2967
95. Ozawa S, Watanabe Y, Morishima I (1994) J Am Chem Soc 116: 5832
96. Jayaraj K, Gold A, Austin RN, Mandon D, Weiss R, Tenner J, Bill E, Müther M, Trautwein AX (1995) J Am Chem Soc 117: 9079
97. Yamaguchi K, Watanabe Y, Morishima I (1992) J Cjem Soc Chem Comm 1721
98. Murakami T, Yamaguchi K, Watanabe Y, Morishima I (1998) Bull Chem Soc Jpn 71: 1343
99. Groves JT, Watanabe Y (1986) J Am Chem Soc 108: 7836
100. Groves JT, Watanabe Y (1988) J Am Chem Soc 110: 8443
101. Mizutani Y, Watanabe Y, Kitagawa T (1994) J Am Chem Soc 116: 3439
102. Tsurumaki H, Watanabe Y, Morishima I (1993) J Am Chem Soc 115: 11784
103. Balch AL, Chan Y-W, Cheng R-J, La Mar GN, Latos-Grazynski L, Renner MW (1984) J Am Chem Soc 106: 7779
104. Chin DH, Balch AL, La Mar GN (1980) J Am Chem Soc 102: 1446
105. Chin DH, La Mar GN, Balch AL (1980) J Am Chem Soc 102: 5945
106. Groves JT, Gross Z, Stern MK (1994) Inorg Chem 33: 5065
107. Phillippi MA, Goff HM (1982) J Am Chem Soc 104: 6026
108. Caldewood TS, Lee WA, Bruice TC (1985) J Am Chem Soc 107: 8273
109. Lee WA, Calderwood TS, Bruice TC (1985) Proc Natl Acad Sci USA 82: 4301
110. Groves JT, Gilbert JA (1986) Inorg Chem 25: 123
111. Swistak C, Mu XH, Kadish KM (1987) Inorg Chem 26: 4360
112. Low DW, Winkler JR, Gray HB (1996) J Am Chem Soc 118: 117
113. Hamachi I, Tsukiji S, Shinkai S, Oishi S (1999) J Am Chem Soc 121: 5500

114. Walling C (1975) Acc Chem Rev 8: 125
115. Poulos TL, Kraut J (1980) J Biol Chem 255: 8199
116. Erman JE, Vitello LB, Miller MA, Kraut J (1992) J Am Chem Soc 114: 6592
117. Erman JE, Vitello LB, Miller MA, Shaw A, Brown KA, Kraut J (1993) Biochemistry 32: 9798
118. Imai M, Shimada H, Watanabe Y, Matsushima-Hibiya Y, Makino R, Koga H, Horiuchi T, Ishimura Y (1989) Proc Natl Acad Sci USA 86: 7823
119. Martinis SA, Atkins WM, Stayton PS, Sligar SG (1989) J Am Chem Soc 111: 9252
120. Shimada H, Sligar SG, Yeom H, Ishimura Y (1997) Heme Monooxygenase. In: Funabiki T (ed) Oxygenases and Model Systems, Kluwer Academic Publishers, Netherlands, pp 195–221
121. (a) Crane BR, Arvai AS, Ghosh DK, Wu C, Getzoff ED, Stuehr DJ, Tainer JA (1998) Science 279: 2121 (b) Crane BR, Arvai AS, Gachhui R, Wu C, Ghosh DK, Getzoff ED, Stuehr DJ, Tainer JA (1997) Science 278: 425
122. Quillin ML, Arduini RM, Olson JS, Phillips GN Jr (1993) J Mol Biol 234: 140
123. (a) Poulos TL, Freer ST, Alden RA, Edwards SL, Skogland U, Takio K, Eriksson B, Xuong N-H, Yonetani T, Kraut J (1980) J Biol Chem 255: 575 (b) Finzel BC, Poulos TL, Kraut J (1984) J Biol Chem 259: 13027
124. Murthy MRN, Reid TJ, Sicignano A, Tanaka N, Rossmann MG (1981) J Mol Biol 152: 465
125. Miller MA, Shaw A, Kraut J (1994) Nature Struct Biol 1: 524
126. Phillips SE, Schoenborn BP (1981) Nature 292: 81
127. (a) Ozaki S, Matsui T, Watanabe Y (1996) J Am Chem Soc 118: 9784 (b) Ozaki S-I, Matsui T, Watanabe Y (1997) J Am Chem Soc 119: 6666 (c) Matsui T, Ozaki S, Watanabe Y (1997) J Biol Chem 272: 32731
128. Matsui T, Ozaki S, Liong E, Phillips GN, Watanabe Y (1999) J Biol Chem 273: 2838
129. Ozaki S-I, Yang H-J, Matsui T, Goto Y, Watanabe Y (1999) Tetrahedron Asymmetry 10: 183
130. Goto Y, Matsui T, Ozaki S-I, Watanabe Y, Fukuzumi S (1999) J Am Chem Soc p 9497
131. Hill CL, Schardt BC (1980) J Am Chem Soc 102: 6374
132. Groves JT, Kruper WJ, Haushalter RC (1980) J Am Chem Soc 102: 6375
133. Hill CL, Hollanderr FJ (1982) J Am Chem Soc 104: 7318
134. Camenzind MJ, Hollander FJ, Hill CL (1982) Inorg Chem 21: 4301
135. Schardt BC, Hollander FJ, Hill CL (1982) J Am Chem Soc 104: 3964
136. Smegal JA, Schedt BC, CL H (1983) J Am Chem Soc 105: 3510
137. Bortoloni O, Meunier B (1983) J Chem Soc Chem Commun 1364
138. Bortolini O, Ricci M, Meunier B (1986) Nouv J Chim 10: 39
139. (a) Ayougou K, Bill E, Charnock JM, Garner CD, Mandon D, Trautwein AX, Weiss R, Winkler H (1995) Angew Chem Int Ed Eng 34: 343 (b) Schappacher M, Weiss R (1987) Inorg Chem 26: 1189
140. Groves JT, Watanabe Y, McMurry TJ (1983) J Am Chem Soc 105: 4489
141. Groves JT, Watanabe Y (1986) Inorg Chem 25: 4808
142. Early works were reviewed by Boucher LJ, Boucher LJ (1972) Coord Chem Rev 7: 289
143. Fabry TL, Simo C, Javaherian K (1986) Biochim Biophys Acta 160: 118
144. Yonetaini T, Asakura T (1969) J Biol Chem 244: 4580
145. Nick RJ, Ray GB, Fish KM, Spiro TG, Groves JT (1991) J Am Chem Soc 113: 1838
146. Nick RJ, Ray GB, Fish KM, Spiro TG, Groves JT (1991) J Am Chem Soc 113: 1838
147. (a) Modi S, Saxena AK, Behere DV, Mitra S (1990) Biochim Biophys Acta 1038: 164 (b) Savena A, Modi S, Behere DV, Mitra S (1990) Biochim Biophys Acta 1041: 83 (c) Khan KK, Mondal MS, Mitra S (1996) J Chem Soc, Dalton Trans 1059 (d) Khan KK, Mondal MS, Mitra S (1998) J Chem Soc Dalton Trans 533
148. Makino R, Uno T, Nishimura Y, Iizuka T, Tsuboi M, Ishimura Y (1986) J Biol Chem 261: 8376
149. Low DW, Abedin S, Yang G, Winkler JR, Gray HB (1998) Inorg Chem 37: 1841

150. (a) Groves JT, Stern MK (1988) J Am Chem Soc 110: 8628 (b) Czernuszewicz RC, Su YO, Stern MK, Macor KA, Kim D, Groves JT, Spiro TG (1988) J Am Chem Soc 110: 4158
151. (a) De Poorter B, Meunier B (1985) J Chem Soc Perkin Trans II 1735 (b) De Poorter B, Ricci M, Meunier B (1985) Tetrahedron Lett 26: 4459 (c) Robert A, Meunier B (1988) New J Chem 12: 885 (d) Meunier B, Guilmet E, De Carvalho M-E, Poilblanc R (1984) J Am Chem Soc 106: 6668 (e) Guilment E, Meunier B (1984) J Mol Cat 23: 115 (f) Bernadou J, Fabiano A-S, Robert A, Meunier B (1994) J Am Chem Soc 116: 9375
152. (a) Razenberg JASJ, Nolte RJM, Drenth W (1986) J Chem Soc Chem Commun 277 (b) van der Made AW, Nolte RJM, Drenth W (1990) Recl Trav Chim Pays-Bas 109: 537
153. (a) Collman JP, Kodadeck T, Brauman JI (1986) J Am Chem Soc 108: 2588 (b) Collman JP, Brauman JI, Meunier B, Hayashi T, Kodadeck T, Raybuck SA (1985) J Am Chem Soc 107: 2000 (c) Collman JP, Brauman JI, Hampton PD, Tanaka H, Bohle DS, Hemnre RT (1990) J Am Chem Soc 112: 7980
154. Groves JT, Lee J, Marla SS (1997) J Am Chem Soc 119: 6269
155. Collins TJ, Gordon-Wylie SW (1989) J Am Chem Soc 111: 4511
156. Collins TJ, Powell RD, Slebodnick C, Uffelman ES (1990) J Am Chem Soc 112: 899
157. Workman JM, Powell RD, Procyk AD, Collins TJ, Bocian DF (1992) Inorg Chem 31: 1548
158. (a) Groves JT, Kruper WJJ (1979) J Am Chem Soc 101: 7613 (b) Groves JT, Haushalter RC (1981) J Chem Soc Chem Commun 165
159. Fujii H, Yoshimura T, Kamada H (1997) Inorg Chem 36: 1122
160. Budge JR, Gatehous BMK, Nesbit MC, West BO (1981) J Chem Soc, Chem Commun 370
161. Groves JT, Kruper WJJ, Haushalter RC, Butler WM (1982) Inorg Chem 21: 1363
162. Buchler JW, Lay KL, Castle L, Ullrich V (1982) Inorg Chem 21: 842
163. Yamaji M, Hama Y, Miyazaki Y, Hoshino M (1992) Inorg Chem 31: 932
164. Greager SE, Murray RW (1985) Inorg Chem 24: 3824
165. (a) Siddall TL, Miyaura N, Huffman JC, Kochi JK (1983) J Chem Soc Chem Commun 1185 (b) Samsel EG, Srinivasan K, Kochi JK (1985) J Am Chem Soc 107: 7606
166. Holm RH (1987) Chem Rev 87: 1401
167. (a) Che C-M, Lau K, Lau T-C, Poon C-K (1990) J Am Chem Soc 112: 5176 (b) Leung W-H, Che C-M (1989) J Am Chem Soc 111: 8812 (c) Che C-M, Wong K-Y, Poon C-K (1985) Inorg Chem 24: 1797 (d) Che C-M, Lai T-F, Wong K-Y (1987) Inorg Chem 26: 2289 (e) Che C-M, Leung W-H (1987) J Chem Soc, Chem Commun 1376
168. (a) Marmion ME, Takeuchi KJ (1988) J AM Chem Soc 110: 1472 (b) Acquaye JH, Muller JG, Takeuchi KJ (1993) Inorg Chem 32: 160
169. (a) Bailey AJ, Griffith WP, White AJP, Williams DJ (1994) J Chem Soc Chem Commun 1833 (b) El-Hendawy AM, Griffith WP, Piggott B, Williams DJ (1988) J Chem Soc Dalton Trans: 1983
170. Masuda H, Taga T, Osaki K, Sugimoto H, Mori M, Ogoshi H (1981) J Am Chem Soc 103: 2199
171. Collman JP, Barnes CE, Brothers PJ, Collins TJ, Ozawa T, Gallucci JC, Ibers JA (1984) J Am Chem Soc 106: 5151
172. Leung T, James BR, Dolphin D (1983) Inorg Chim Acta 79: 180
173. Groves JT, Quinn R (1984) Inorg Chem 23: 3844
174. Groves JT, Quinn R (1985) J Am Chem Soc 107: 5790
175. James BR (1992) Catal Org React 47: 245
176. Leung W-H, Che C-M (1989) J Am Chem Soc 111: 8812
177. Che C-W, Chung W-C, Lai T-F (1988) Inorg Chem 27: 2801
178. Dubourdeaux P, Trvarés M, Grand A, Ramasseul R, Marchon J-C (1995) Inorg Chem Acta 240: 657
179. Leung W-H, Che C-H, Yeung C-H, Poon C-K (1993) Polyhedron 12: 2331
180. Tokita Y, Yamaguchi K, Watanabe Y, Morishima I (1993) Inorg Chem 32: 329
181. Groves JT, Ahn K-H (1987) Inorg Chem 26: 3833
182. Ohtake H, Higuchi T, Hirobe M (1992) J Am Chem Soc 114: 10660

183 (a) Buchler JW, Rohbock K (1972) Inorg Nucl Chem Lett 8: 1073 (b) Buchler JW, Puppe L, Rohbock K, Schneehage HH (1973) Chem Ber 106: 2710
184. Johnson JF, Scheidt WR (1978) Inorg Chem 17: 1280
185. Colin JBC, Cian AD, Weiss R (1983) Angew Chem Int Ed Engl 22: 247
186. Diebold T, Chevrier B, Weiss R (1979) Inorg Chem 18: 1193
187. Tachibana J, Imamura T, Sasaki J (1993) J Chem Soc Chem Commun 1436
188. Mentzen BF, Bonnet MC, Ledon HJ (1980) Inorg Chem 19: 2061
189. Guilard R, Latour J-M, Lecomte C, Marchon J-C, Protas J, Ripoll D (1978) Inorg Chem 17: 1228
190. Richard P, Poncet JL, Barbe JM, Guilard R (1982) J Chem Soc Dalton Trans 1451
191. Czarnecki K, Proniewicz LM, Fujii H, Ji D, Czernuszewicz RS, Kincaid J (1999) Inorg Chem 38: 1543
192. Lemtur A, Chakravorty BK, Dhar TK, Subramanian J (1984) J Phys Chem 88: 5603
193. Oliver Su Y, Czernuszewicz RS, Miller LA, Spiro TG (1988) J Am Chem Soc 110: 4150
194. Macor KA, Czernuszewicz RS, Spiro TG (1990) Inorg Chem 29: 1996
195. Chasteen ND (1981) Biological Magnetic Resonance. In: Berliner LJ, Reuben J (eds) Plenum New York, vol 3, p53
196. Fukui K, Ohya-Nishiguchi H, Kamada H (1993) J Phys Chem 97: 11858
197. Schulz CE, Song H, Lee YJ, Mondal JU, Mohanrao K, Reed CA, Ann Walker F, Scheidt WR (1994) J Am Chem Soc 116: 7196

Characterization, Orbital Description, and Reactivity Patterns of Transition-Metal Oxo Species in the Gas Phase

Detlef Schröder[1], Helmut Schwarz[2], Sason Shaik[3]

[1,2] Institut für Organische Chemie der Technischen Universität Berlin, Straße des 17. Juni 135, D-10623 Berlin, Germany
[3] Department of Organic Chemistry and the Lise-Meitner-Minerva Centre for Computational Quantum Chemistry, The Hebrew University of Jerusalem, 91904 Jerusalem, Israel

[1] *E-mail: df@www.chem.tu-berlin.de*
[2] *E-mail: schw0531@www.chem.tu-berlin.de*
[3] *E-mail: sason@yfaat.ch.huji.ac.il*

Metal-oxo and -peroxo species play key roles in catalytic oxidations using transition metals. This contribution addresses gas-phase studies of diatomic metal-oxo species and their properties which turn out to be of prime importance for the understanding of the observed reactivity patterns. A second topic concerns higher transition-metal oxides such as di- and trioxides. For these species there exists a structural dichotomy with the corresponding dioxygen complexes, e.g., $MO_2 \leftrightarrows M(O_2)$. Most of the metal-based oxidations described involve crossings between surfaces of different spin as crucial steps, and the violation of spin conservation is proposed to determine the reactivity of the late transition-metal oxides. The ability of transition metals to mediate surface crossings via spin-orbit coupling is introduced as a key aspect in oxidation catalysis which has as yet not been fully appreciated. It is further outlined how these fundamental properties may relate to the properties of metal-oxo catalysts in the condensed phase.

Keywords: Gas-phase chemistry, Metal oxides, Oxidation, Spin change

Structure and Bonding, Vol. 97

1 Introduction

An essential aspect in oxidation catalysis by transition metals [M] is their ability to form metal-oxo, -dioxo, and -peroxo species [1]; here [M] stands for a 'bare' or a ligated metal atom. These transition-metal oxides and peroxides exhibit a broad range of reactivity which is determined by: (i) the nature of the metal, (ii) the valence of the metal core, and (iii) the first coordination sphere. In addition, solvent effects, aggregation phenomena, and counter ions influence the activity, substrate specificity and chemoselectivity of transition-metal oxides in catalysis. In metal peroxides [M]-OOR, the additional substituent R (e.g., R = H, alkyl, silyl, or even another metal center) also plays a decisive role.

The manifold of reactivity patterns and properties of metal oxides and peroxides is extremely rich. A tailor-made design of oxidation catalysts requires advanced insight into the bonding schemes, thermochemical parameters, and a detailed knowledge of the kinetics of the elementary steps. These intrinsic properties determine the reactivity of metal oxides and the related peroxides towards oxidizable substrates, thus controlling the reactivity pattern at a molecular level. The associated thermochemical and kinetic parameters define the mechanistic basis for the chemo-, regio-, diastereo-, and even enantioselectivities of metal-based oxidation reactions.

In the most simplified approach, only the metal-oxo species themselves are considered, e.g., diatomic metal oxides MO or small oxo- and peroxo species such as metal dioxides MO_2, metal-oxygen complexes $M(O_2)$, or metal peroxides M(OOR). Gas-phase studies using mass-spectrometric techniques enable to study the reactivity of these metal oxides under well-defined conditions [2]. For technical reasons, these small entities are usually charged in these studies, but the reactivity of neutrals can also be probed by mass spectrometry [3]. A particular advantage of this experimental approach is that the well-defined circumstances allow for direct comparison with accurate theoretical data. In fact, the present stage and level of sophistication of computational chemistry not only provide a reasonable level of credibility, but sometimes also reach the capability of predicting reactivity patterns of small metal-oxo compounds ahead of experimental verification.

The present contribution is restricted to mononuclear metal oxides and is divided into two major sections. The first deals with the chemistry of diatomic transition-metal oxides and serves to illustrate the fundamental reactivity patterns of metal-oxo units. The second describes higher metal oxides, such as di- and trioxides; transition-metal peroxides are only addressed in this context if relevant with respect to the reactivity of metal-oxo species. The substrates to be oxidized are mostly limited to alkanes, as their selective activation is rightly viewed as a major challenge in oxidation catalysis. While most of our considerations deal with cationic species, the fundamental aspects may equally well apply for other charge states if the electronic situations and redox properties are properly taken into account. Notwithstanding, our preference for using cationic metal oxides is not unintentional because neutral and, in particular, anionic species are less likely to activate hydrocarbons. This simply reflects the well-known fact that hydrocarbons are preferentially activated by electrophiles rather than nucleophiles: the electrophiles could be bare metal-oxo cations in the gas phase as well as high-valent metal oxides in acidic solutions or in heterogeneous catalysts with acidic sites.

2
Gas-Phase Studies and Applied Catalysis

Before describing our research on fundamental properties of metal-oxo and -peroxo species in the diluted gas phase (typically conducted in the lower 10^{-6} mbar pressure regime), let us briefly outline how these intrinsic properties are connected to the behavior of transition-metal catalysts in applied processes. Considering a metal-oxo species [M]=O as an example, the intrinsic properties of this unit are determined by: (i) the choice of the metal that affects the nature of the metal-oxygen bond and the net charge of the [M]=O unit, (ii) the ligands other than the oxo unit attached to the metal center, and (iii) the formal valence state of the metal.

Upon proceeding from microscopic to macroscopic systems, these intrinsic properties are then modified by the local environments (e.g., additional ligands, counterions, coordinated solvent molecules) which determine the reactivity patterns as well as the extended environments (e.g., protein backbones in metallo enzymes or the properties of pores in heterogeneous catalysts) affecting substrate specificity and chemoselectivity. In particular, the rate-limiting steps may differ between idealized gas-phase studies and those of real catalytic processes. Among the broad variety of conceivable effects, for the sake of brevity let us just address product desorption which is particularly decisive in the oxidation of hydrocarbons. Thus, upon hydroxylation of a hydrocarbon R—H to the corresponding alkanol R—OH, the latter is often preferentially coordinated to a metal center because of its better donor properties; liberation is usually brought about by thermal activation. In contrast, hydroxylation of an alkane in the highly diluted gas phase is often associated with the release of the alkanol formed according to the Eq. (1) because the exothermicity of the oxidation provides a driving force for product desorption:

$$\text{R—H} + [\text{M}]{=}\text{O} \rightarrow [\text{M}](\text{R—OH}) \rightarrow \text{R—OH} + [\text{M}] \tag{1}$$

In higher pressure regimes, the environment may instead serve as an energy bath, thereby resulting in the formation of long-lived (R—OH)[M] product complexes which can determine the overall rate constants in terms of a product inhibition. In addition, preferential coordination of the alkanol relative to the alkane leads to enhanced residence times for the oxidized product at the active site, thus increasing the risk of overoxidation (see below). These are two of the many factors which give rise to the 'pressure gap' between idealized gas-phase studies and real catalysis under typical operating conditions [4]. Furthermore, minute pathways apparent as side reactions in the gas-phase studies as well as the presence of trace impurities may become major factors for the effectiveness and the turnover numbers in applied catalysis. In turn, understanding these side reactions in terms of elementary steps occurring at the molecular level may help to optimize the performance of real catalysts.

The properties of the model systems described below may thus allow some extrapolation from the gas-phase studies to more dense matter. It needs to be stressed, however, that intrinsic properties cannot be extrapolated to systems which differ in their fundamental aspects. Thus, with appropriate consideration of the boundary conditions, the properties of a mononuclear metal-oxo species [M]=O can be used to propose modifications of real catalysts by ligand effects, additives, co-catalysts, promoters etc. In fortunate cases, even the effects of additional metal atoms close to the active site may be estimated as long as the metal-oxo unit remains intact; e.g., in a mixed binuclear metal oxide of the type [M′]—[M]=O. The central assumption that the reactive entity itself is essentially unchanged usually holds true in homogeneous catalysis and may also be valid for highly dispersed metal catalysts on more or less unreactive supports. If, however, it comes to the association of metal clusters in which oxo-units are μ-bridged to two or more metal centers, the [M]=O unit cannot be regarded as a model system anymore. Instead, the corresponding μ-bridged metal-oxide clusters $[\text{M}]_m\text{O}_n$ of appropriate valence need to be considered in the gas-phase studies. All extrapolations fail, however, if bulk and cooperative effects predominate. The technically important epoxidation of ethene by molecular oxygen on silver contacts appears as such a case [5], and an appropriate gas-phase mnemonic is hardly conceivable.

These considerations shall serve to enable the reader to position properly the role of the gas-phase studies in the context of applied catalysis. Thus, while knowledge of the intrinsic properties of metal-oxo and -peroxo species provides insight into reactivity patterns, activation and passivation mechanisms, etc., a direct translation of gas-phase data to real conditions – not to mention the prediction of better catalysts by engineering the active sites [6] – is impossible and not intended.

3
Properties and Reactivity of Diatomic Transition-Metal Oxides

In 1995, we reviewed the chemistry of cationic transition-metal oxides in the gas phase [2]. Along with updating this review by adding the most salient reports which have appeared lately, we shall concentrate on the fundamental reactivity aspects of transition-metal oxides in the gas phase. In this context, the vast improvement of theoretical methods to describe energetics and reactivity patterns of transition-metal compounds is particularly valuable [7].

4
Bonding Properties of Metal-Oxo Species

A major, concept that emerges from the gas-phase studies of transition-metal oxides concerns the role of spin states, and it is the spin multiplicity which acts as a decisive factor in oxidation reactions (see below). Therefore, it is a prerequisite to consider the nature of the metal-oxo bond in some more detail; here, the notation metal oxo refers to a situation in which an oxygen atom, having no other bonding partners, is directly bound to a metal center. Two possible bonding schemes evolve: (i) a low-spin [M]=O species having a formal double bond between the metal and oxygen, and (ii) a diradicaloid high-spin $[M]^{\bullet}$—$O^{\bullet}$ situation with a covalent *s*-bond and resonating π interactions. While this distinction is somewhat simplistic because intermediate bonding schemes are imposed by the δ-orbitals [8], the gross classification in terms of low- and high-spin situations provides quite an insight as far as understanding of chemical reactivity is concerned.

Bonding patterns are discussed for the metal-oxide cations MO^+ mostly of the 3*d* transition metal series [9–12]. Figure 1 shows the molecular orbital scheme derived for the combination of a 3*d*-block element with atomic oxygen, and it comprises three different types of valence orbitals [11]:

1. Four σ orbitals arise from the combinations of 2*s* and $2p_z$ of oxygen with $3d_{z^2}$ and 4*s* of the metal. Accordingly, 1σ is basically the low-lying 2*s* orbital of oxygen, 2σ forms a polarized, covalent σ-bond between M and O, 3σ is almost a non-bonding orbital with a dominant $3d_{z^2}$ character, and 4σ is the anti-bonding counterpart of 2σ which by and large involves the 4*s* orbital of the metal.
2. The p_x and p_y orbitals of oxygen and the $3d_{xz}$ and $3d_{yz}$ orbitals of the metal give rise to two sets of perpendicular π orbitals, the bonding combinations $1\pi_x$ and $1\pi_y$ and the anti-bonding counterparts $2\pi_x$ and $2\pi_y$.
3. The $3d_{xy}$ and $3d_{x^2-y^2}$ orbitals of the metal do not find symmetry match in the valence orbitals of oxygen and are thus classified as non-bonding δ orbitals.

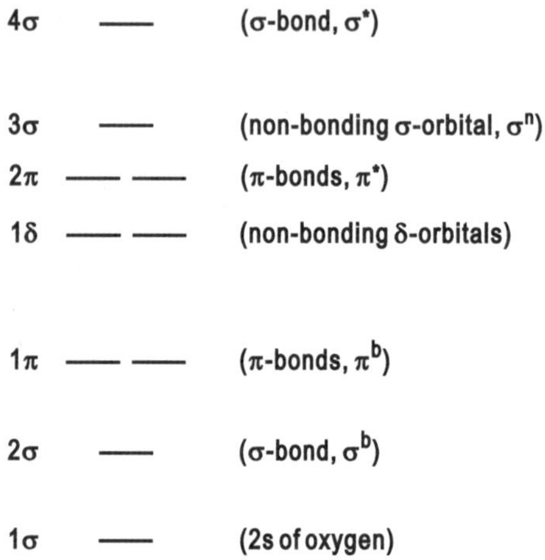

Fig. 1. General molecular orbital scheme of a diatomic transition-metal oxide

Let us now begin to occupy the molecular orbitals in this scheme for MO^+ cations of the 3*d* transition-metal series. In the valence space, ScO^+ has eight electrons giving rise to a $^1\Sigma^+$ ground state with a configuration $1\sigma^2 2\sigma^2 1\pi_x^2 1\pi_y^2$ arising from the perfect pairing of all electrons. Accordingly, all bonding orbitals are doubly occupied while the non-bonding and anti-bonding orbitals are empty (Fig. 2a). This favorable situation gives rise to a strong bond with a dissociation energy at 0 K of $D_0(Sc^+—O)$ = 165 kcal/mol [13] (Table 1). For titanium as the next member of the 3*d* series, one of the non-bonding 1δ orbitals is singly occupied resulting in a $^2\Delta$ ground state. As population of the δ orbital hardly affects the bond between the metal and oxygen, $D_0(Ti^+—O)$ = 159 kcal/mol is large. Following this scheme, VO^+ which has a $^3\Sigma^+$ ground state with a $1\sigma^2 2\sigma^2 1\pi_x^2 1\pi_y^2 1\delta_{xy}^1 1\delta_{x^2-y^2}^1$ configuration and $D_0(V^+—O)$ = 135 kcal/mol. Considering these configurations in terms of the electrons involved in the different bonding blocks (σ, π, and δ), the bonding in ScO^+, TiO^+, and VO^+ finds a simple mnemonic in that of carbon monoxide, for which the molecular orbital scheme is similar to that of ScO^+ (Fig. 2a).

The first ambiguity occurs for the next member of the series, CrO^+, in which the additional electron could either occupy one of the anti-bonding 2π orbitals or the mostly non-bonding 3σ, hence giving rise to either $^4\Pi$ or $^4\Sigma^+$ states. While CrO^+ ($^4\Sigma^+$) has been characterized spectroscopically [14], theory predicts the $^4\Pi$ or $^4\Sigma^+$ states to be very close to each other (ca. 0.1 eV [8, 11, 12]), and a definitive state assignment cannot be made for the time being. This uncertainty notwithstanding, the key aspect in the present context is that Hund's rule wins over the Aufbau principle in that the doublet state CrO^+ ($^2\Delta$) in which the extra electron is spin-coupled into δ-manifold is more than 1 eV less stable than the $^4\Pi$ and $^4\Sigma^+$ quartet states. Among other factors [15], the occupation of anti-bonding orbitals reduces the bond strength in CrO^+ compared to the early transition metals, i.e., $D_0(Cr^+—O)$ = 86 kcal/mol. Much as for chromium, for MnO^+ too the population of the 2π and 3σ manifold is

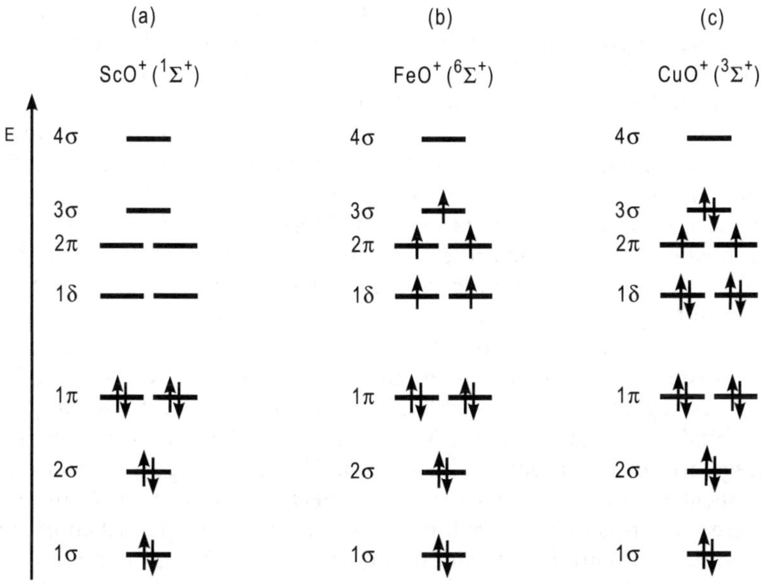

Fig. 2a–c. Molecular orbital schemes for the ground states of: **a** ScO^+; **b** FeO^+; **c** CuO^+

Table 1. Experimental bond dissociation energies, $D_0(M^+—O)$ in kcal/mol, of diatomic transition-metal oxide cations

	D_0			D_0			D_0	
ScO^+	164.6[a]	±1.4	YO^+	167.0[b]	±4.2	LaO^+	206[d]	±4
TiO^+	158.6[a]	±1.6	ZrO^+	178.9[b]	±2.5	HfO^+	173[e]	±5
VO^+	134.9[a]	±3.5	NbO^+	164.4[b]	±2.5	TaO^+	188[e]	±15
CrO^+	85.8[a]	±2.8	MoO^+	116.7[b]	±0.5	WO^+	126[e]	±10
MnO^+	68.0[a]	±3.0				ReO^+	115[f]	±15
FeO^+	80.0[a]	±1.4	RuO^+	87.9[c]	±1.2	OsO^+	100[g]	±12
CoO^+	74.9[a]	±1.2	RhO^+	69.6[c]	±1.4	IrO^+	59[e]	–[h]
NiO^+	63.2[a]	±1.2	PdO^+	33.7[c]	±2.5	PtO^+	77[i]	–[h]
CuO^+	37.4[a]	±3.5	AgO^+	28.4[c]	±1.2	AuO^+	–[j]	
ZnO^+	38.5[a]	±1.2						

[a] [13]
[b] Sievers MR, Chen Y-M, Armentrout PB (1996) J Chem Phys 105: 6322
[c] Chen Y-M, Armentrout PB (1995) J Chem Phys 103: 618
[d] [91]
[e] [29]
[f] [76], p 50
[g] [77]
[h] No error bars given
[i] [25]
[j] For observations of AuO^+, see: Hecq A, Vandy M, Hecq M (1980) J Chem Phys 72: 2876 and Aita CR (1987) J Appl Phys 61: 5182

almost degenerate, and the $^5\Pi$ or $^5\Sigma^+$ states are very close in energy (0.1 eV [8, 16]). The progressive occupation of anti-bonding orbitals inter alia lowers $D_0(\mathrm{Mn^+{-}O})$ to 68 kcal/mol. A simple mnemonic for the bonding situations in CrO^+ ($^4\Pi$) and MnO^+ ($^5\Pi$) is the NO radical [8].

Iron-oxo species are of prime importance in oxidation catalysis. Moreover, the bonding situation of FeO^+ can be regarded as representative for the behavior of the oxo species of the late transition metals [8] and is therefore discussed in more detail. Compared to the ideal situation in ScO^+ ($^1\Sigma^+$), FeO^+ cation must account for the presence of five extra electrons. Perfect pairing of the Fe^+ (^6D) and O (^3P) atomic ground states in terms of a [M]=O double bond would give rise to quartet FeO^+. High-level ab initio calculations, however, predict a sextet ground state [17], and this state assignment has recently been confirmed experimentally [18]. Thus, ground state FeO^+ ($^6\Sigma^+$) has a $1\sigma^2 2\sigma^2 1\pi_x^2 1\pi_y^2 1\delta_{xy}^1 1\delta_{x^2-y^2}^1 2\pi_x^1 2\pi_y^1 3\sigma^1$ configuration with five singly occupied orbitals (Fig. 2b). Quartet states arising from any conceivable spin couplings are higher in energy [2, 11, 17]. The most important consequence with respect to chemical reactivity is that FeO^+ cation can thus not at all be considered in terms of a double bond, but must rather be described in analogy to the bonding situation in triplet dioxygen. In fact, O_2 ($^3\Sigma^-$) and FeO^+ ($^6\Sigma^+$) molecules share $1\sigma^2 2\sigma^2 1\pi_x^2 1\pi_y^2 2\pi_x^1 2\pi_y^1$ configurations in the bonding blocks, i.e. the 2*s* orbital(s) of oxygen, three doubly occupied orbitals due to the bonding σ- and π-combinations, and two singly occupied anti-bonding π-orbitals. The particular interaction between the atoms is best visualized in terms of the corresponding valence-bond picture [8, 10, 11], i.e., the diatomics O_2 ($^3\Sigma^-$) and FeO^+ ($^6\Sigma^+$) exhibit two resonating 3-electron-2-center π bonds perpendicular to each other (Fig. 3). Thus, ground state FeO^+ cation is best described as a high-spin $[M]^\bullet{-}O^\bullet$ diradicaloid species rather than the perfect pairing, low-spin situation [M]=O, which one may anticipate first for a metal-oxo unit.

The bonding scheme in FeO^+ cation leads to a moderate bond strength, $D_0(\mathrm{Fe^+{-}O}) = 80$ kcal/mol, but much more important are the reactivity paradigms which can be derived from the analogy to triplet dioxygen. While

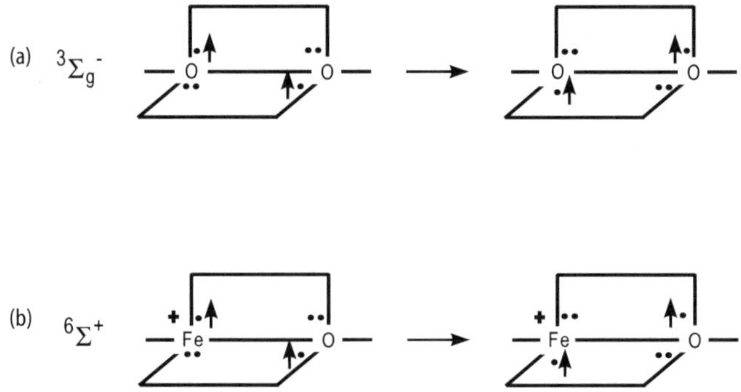

Fig. 3a,b. Valence-bond scheme of the resonating 3-electron-2-center bonds in: **a** triplet dioxygen; **b** FeO^+ cation

the oxidations of almost all organic compounds by O_2 are exothermic, organic matter is metastable in oxygen (and air) because kinetic barriers are significant. Moreover, the combustion chemistry occurring after ignition is usually non-specific, and selective, partial oxidation is often difficult to achieve without catalysis. This behavior can precisely be attributed to the triplet ground state of dioxygen. For example, the initial step in the oxidation of methane by O_2 to yield methyl hydroperoxide is exothermic only if a concerted process occurs, e.g., at Eq. (2). In a step-wise process, e.g., at Eqs. (3a,b), however, the hydrogen abstraction in the first step is highly endothermic:

$$CH_4 + O_2 \rightarrow H_3C{-}OOH, \quad \Delta_r H = -13.5\,\text{kcal/mol} \tag{2}$$

$$CH_4 + O_2 \rightarrow CH_3^\bullet + {}^\bullet OOH, \quad \Delta_r H = +55.1\,\text{kcal/mol} \tag{3a}$$

$$CH_3^\bullet + {}^\bullet OOH \rightarrow H_3C{-}OOH, \quad \Delta_r H = -68.6\,\text{kcal/mol} \tag{3b}$$

The concerted insertion at Eq. (2) is, however, spin-forbidden because methane and methyl hydroperoxide are singlets while O_2 has a triplet ground state. Consequently, non-radicaloid oxidations with molecular oxygen experience a spin-inversion bottleneck. In contrast, the stepwise sequence via reactions at Eqs. (3a,b) circumvents the bottleneck, but requires large activation energies which often coincides with the onset of combustion. In fact, many catalysts employed for the partial oxidation of alkanes induce initial homolytic R—H bond cleavages to the corresponding radicals [19]; thus the often observed poor selectivities have their origin in the very first step. As outlined further below, this reactivity scheme applies also to FeO^+ cation in that its concerted insertion into an R—H bond to yield the intermediate $R{-}Fe^+{-}OH$ cannot occur unless spin inversion has taken place.

Bonding in the remaining MO^+ cations of the 3*d* metals follows the scheme outlined for FeO^+ cation. Thus, CoO^+ exhibits a $^5\Delta$ state in which one of the δ orbitals is doubly occupied. Then, the δ-manifold is completely filled in NiO^+ ($^4\Sigma^-$), and even in CuO^+ the analogy to triplet oxygen persists having a $^3\Sigma^-$ ground state with $1\sigma^2 2\sigma^2 1\pi_x^2 1\pi_y^2 1\delta_{xy}^2 1\delta_{x^2-y^2}^2 2\pi_x^1 2\pi_y^1 3\sigma^2$ configuration (Fig. 2c). Only with ZnO^+ does the 2π manifold begin to be filled further. The increased occupation of non- and anti-bonding orbitals is associated with decreasing bond strengths, i.e., $D_0(Fe^+{-}O)$ = 80 kcal/mol, $D_0(Co^+{-}O)$ = 75 kcal/mol, $D_0(Ni^+{-}O)$ = 63 kcal/mol, $D_0(Cu^+{-}O)$ = 37 kcal/mol, and $D_0(Zn^+{-}O)$ = 39 kcal/mol. The slight increase with ZnO^+ can be attributed to a more pronounced contribution of ionic configurations to the strongly polarized metal-oxygen bond [20].

This bonding scheme of transition-metal oxides is by no means confined to the cationic species, and analogous arguments apply for neutral and anionic species as well [9, 21]. For example, the O_2 mnemonic of the bonding block holds true not only in FeO^+ ($^6\Sigma^+$), but also in the $^5\Delta$ and $^4\Delta$ ground states of the neutral and the anionic counterparts [22]. Due to the lower oxidation state of the metal, e.g., Fe(II) in neutral FeO and Fe(I) in FeO^-, the ability of these oxides as oxidizing agents is significantly lowered, however, as might be inferred from the increasing bond strengths, i.e., $D_0(Fe^+—O)$ = 80 kcal/mol, $D_0(Fe—O)$ = 101 kcal/mol, and $D_0(Fe^-—O)$ = 132 kcal/mol [23]. This ordering provides a further rationale for the fact that most of the oxidation reactions described here involve cationic species, because the neutral and anionic counterparts are of lower valence state, and thus weaker oxidizers. For simple non-functionalized hydrocarbons as the substrates of major interest, the lack of reactivity towards nucleophilic reagents further disfavors oxidations with metal-oxide anions. In fact, we are not aware of any C—H bond activation of an alkane by transition-metal oxide anions.

Except for comprehensive computational studies of the 4*d*-metal monoxides [24], much less is known about the properties of 4*d*- and 5-transition-metal oxides. Nevertheless, none of the findings reported so far conflicts with the bonding schemes outlined above. For example, even though the atomic properties of platinum largely differ from its 3*d*-congener nickel, NiO^+ and PtO^+ share $^4\Sigma^-$ ground states with high-spin situations [8, 11, 12, 25]. Note, however, that spin is no longer a good quantum number for 5*d* elements because relativistic effects are significant [26]. In this context, even the few complete mechanistic schemes which have recently been deduced from high-level calculations of 5*d*-metal oxides [25, 27] have to be viewed as first-order approximations because present theoretical methods still do not enable accurate incorporation of spin-orbit coupling in the algorithms used to locate and optimize minima and transition structures in systems of moderate size.

5
Reactivity of Metal-Oxo Species

Before addressing the course of alkane oxidation by transition-metal oxo species, some fundamental conditions are to be defined which must be met in order to afford an effective catalyst having a metal-oxo unit as active site: (i) the metal-oxo species must be capable of transferring an oxygen atom to a given substrate, and (ii) re-oxidation of the reduced form of the metal needs to be feasible.

5.1
Thermochemical and Mechanistic Considerations

Let us specify these requirements for the metal-mediated hydroxylation of an alkane R—H according to the catalytic sequence given at Eqs. (1) and (4); in the latter, ⟨O⟩ stands for any putative O-atom donor, e.g., peroxides, ozone, dioxygen etc.:

$$\text{R—H} + [\text{M}]{=}\text{O} \rightarrow \text{R—OH} + [\text{M}] \tag{1}$$

$$[\text{M}] + \langle\text{O}\rangle \rightarrow [\text{M}]{=}\text{O} \tag{4}$$

The C—H bond dissociation energies of alkanes range from 103 kcal/mol for methane to about 90 kcal/mol for tertiary positions, while the related C—O bonds of the alkanols formed in the reaction at Eq. (1) amount to 90–95 kcal/mol. For example, $D_0(H_3C—H) = 103.3$ kcal/mol [28] vs $D_0(H_3C—OH) = 90.3$ kcal/mol [28] in the conversion methane → methanol compared to $D_0((H_3C)_3C—H) = 95.0$ kcal/mol [28] vs $D_0((H_3C)_3C—OH) = 94.6$ kcal/mol [29] for the oxidation *iso*-butane → *tert*-butanol. Accordingly, O-atom transfer from a metal oxide to an alkane is exothermic, if $D_0(M—O) < [101.4$ kcal/mol $+ D_0(R—OH) - D_0(R—H)]$, that is $D_0(M—O) < 88.4$ kcal/mol for the hydroxylation of methane ($R = CH_3$) and $D_0(M—O) < 101.0$ kcal/mol for the tertiary C—H bond in *iso*-butane ($R =$ *tert*-C_4H_9). From a thermochemical point of view, reactivity is therefore expected to increase with decreasing $D_0(M—O)$. However, if the M—O bond strengths get too low, re-oxidation of the reduced form to the metal-oxo species according to the reaction at Eq. (4) may become difficult. For example, CuO^+ is expected to ba a potent reagent for C—H bond activation because $D_0(Cu^+—O)$ amounts to only 37 kcal/mol and the binding pattern suggests a substantial radical-type character on oxygen [9, 30]. However, precisely because of the low affinity for oxygen, generation of CuO^+ from Cu^+ in the gas phase under conditions which would permit further reactivity studies with CuO^+ has not been achieved so far [31]. From a thermochemical point of view, the bond strengths of those metal-oxo species that are potentially attractive for catalysis thus fall in the range of 70–110 kcal/mol, thereby allowing for hydroxylation of hydrocarbons as well as for facile re-oxidation.

Thus, oxidation catalysis meets inter alia two conceptual concerns: (i) the discovery of highly reactive metal-oxo species, and (ii) the development of methods for the re-oxidation of the reduced counterparts. Various concepts are possible to fulfill these requirements, e.g., separation of substrate oxidation and re-oxidation of the catalyst in time or space. An ideal scenario would, however, involve catalysts which exhibit high specificities in either states, i.e., the oxidized forms only react with the substrate and the reduced forms only react with the terminal oxidant. Such procedures would allow for the continuous feed of substrate/oxidant mixtures in a simple manner. Of course, handling these mixtures without hazard poses some technical problems, yet it has been managed in several industrial processes.

Another general concern in oxidation catalysis is the risk of overoxidation which results in undesired consumption of material and the production of excessive heat. It is obvious that a metal-oxo species capable of activating alkanes can also afford oxidation of the hydroxylated products. For example, the C—H bonds of methanol are much weaker than those in methane, i.e., $D_0(H—CH_2OH) = 94.5$ compared to $D_0(H_3C—H) = 103.3$ kcal/mol [28]. Moreover, the polar oxidation products are likely to bind more strongly to the active site(s) compared to the hydrocarbon substrate, thereby increasing the residence time of the product on the catalyst and hence favoring overoxidation even further. Two obvious strategies to minimize overoxidation are operation at low conversions and/or addition of moderators. For example, the desorption of a hydroxylation product may be facilitated by adding steam to the feed because H_2O may replace ROH ligands at the active site(s).

While overoxidation appears as a general problem in oxidation catalysis, let us address one specific system in some more detail. Thus, the time-honored Gif-systems developed by the late Sir Derek Barton [32] bring about the oxidation of cyclohexane to cyclohexanone while oxidation of cyclohexanol is less efficient compared to that of the cycloalkane precursor [33]. While the precise mechanistic details of the Gif-oxidation are not yet entirely clear, some speculation about a rationalization of this conceptually interesting effect are indicated as they may help to exploit other strategies to prevent overoxidation. Based on the arguments raised above, it is quite obvious that thermochemical parameters cannot account for the preferred alkane oxidation by the Gif-systems, because the C—H bonds of cyclohexane are stronger than the α-C—H bond in cyclohexanol. For the same reason, it appears puzzling that the activation of the weaker C—H bond of the alcohol is kinetically hampered compared with alkane activation, unless there exists a mechanistic switch.

There is consensus that the Gif-systems seem to proceed via the attack of the alkane by a metal-oxo species leading to the formation of alkyl radicals as intermediates [34, 35] which are then trapped by molecular oxygen present in the mixture to yield hydroperoxides [32]. Subsequently, the peroxides rearrange to the ketones in terms of the Hock reaction. Whatever the precise mechanistic details of the complex reaction sequence are, let us propose the presence of a second ligand, here a hydroxyl group, at the active site in order to rationalize the surprising preference for oxidation of alkanes vs alkanols. Approach of the hydrocarbon substrate can only yield products if it occurs towards the metal-oxo unit (path a in Scheme 1). The alkanol could, however, either approach the oxo-unit (resulting in oxidation) or react with the additional ligand via substitution (path b in Scheme 1); for a related scenario see [36]. Given the low polarizability of C—H bonds in alkanes, the R—OH dipole may in fact enforce path b for the alkanol. Covalent attachment at the active site may further disfavor C—H bond activation via the metal-oxo unit and thereby can account for the observed selectivity. Similar arguments can be raised for the moderate overoxidation of the keto products in the Gif-systems, e.g., formation of hemiketals with the metal bound hydroxyl ligand. While this scenario is purely speculative, it needs to be stressed that neither thermochemical nor kinetic arguments can account for the disfavored oxidation of alkanols by the Gif-systems. It is the competition with another process in terms of a mechanistic switch (here coordination at a neighboring position)

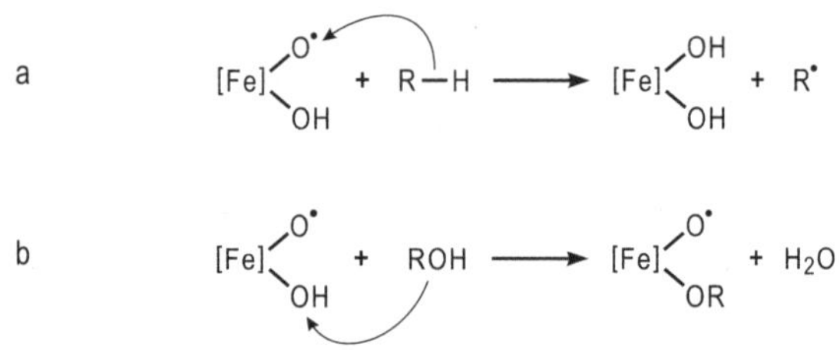

Scheme 1.

which may serve to provide a conceptual guidance. So far, the Gif-systems appear to be quite unique in this particular respect, and the mechanistic proposal made here may thus stimulate further research on catalytic systems in which overoxidation experiences a bottleneck.

5.2 General Considerations for Alkane Hydroxylation

Based upon the above consideration of thermochemical criteria, the following generalizations can be drawn for the transition-metal mediated oxidations; note that this comparison is confined to metal-oxo units as reactive sites and particularly the reactivity of metal peroxides (e.g., the Sharpless epoxidation) is not covered here.

1. Unless hypervalent, metal-oxo species of early transition metals are unlikely to bring about alkane oxidation because of thermochemical restrictions arising from the large oxophilicities of these elements. In fact, the oxo species of the early transition metals are poorer C—H- and C—C-bond activators in comparison with the bare metals themselves [37]. This inertness of the early transition-metal oxides fits nicely with the CO bonding mnemonic mentioned above. If hypervalent, however, the oxides of these metals are expected to behave as oxygen-centered radicals. These features are reflected in the gas-phase properties of the corresponding oxide cations. For example, TiO^+ as a formal titanium(III) compound is incapable of oxygenating hydrocarbons; in fact, the opposite can occur, e.g., bare Ti^+ reduces water to afford $TiO^+ + H_2$ [38, 39]. In marked contrast, the formally hypervalent TiO_2^+ cation behaves as a typical oxygen-centered radical and even abstracts a hydrogen atom from water, $D_0(HO—H) = 118.1$ kcal/mol [28], according to the reaction at Eq. (5) [40]:

$$TiO_2^+ + H_2O \rightarrow Ti(O)OH^+ + HO^\bullet \tag{5}$$

   Radical-type reactions of this kind often have low selectivities and are thus of limited value for the selective partial oxidations; complete oxidation may be useful, however, and the rutile-based photo-oxidative treatment of waste waters falls into this category.
2. Some transition-metal oxides in their highest oxidation states (i.e., d^0 compounds) can bring about the oxidation of various organic compounds, e.g., CrO_2Cl_2, MnO_4^-, RuO_4, and OsO_4. However, the organic substrates to be oxidized are either confined to reactive ones, e.g., bearing allylic or benzylic C—H bonds, olefins etc., or the reactions proceed via radical-type mechanisms [41, 42].
3. Oxides of the late transition metals form the basis of various powerful oxidation catalysts which are capable of activating alkanes. Moreover, some of these oxidants circumvent mere radical-type pathways and thereby allow for regio- and stereoselective functionalizations of alkanes. The most prominent examples in this respect are cytochrome P-450 and methane

monooxygenase [43]; both enzymes share iron cores and afford the oxidation of a broad variety of substrates including methane. Let us therefore focus on the iron based systems which are most relevant and can further be regarded as being representative for the behavior of other oxo species of late transition metals; for a recent survey of the gas-phase chemistry of iron in general, see [44].

5.3
Spin Inversion as a Key Aspect in Alkane Hydroxylation by Transition-Metal Oxo Species

The conventional view of potential-energy surfaces regards species of different spin multiplicities as being entirely separated from each other; let us term this behavior single-state reactivity (SSR). Here, we will show that the thermal reactivity of metal-oxo species is dominated by interaction of surfaces having different spin. We term this behavior as two-state reactivity (TSR), i.e., the interference of two spin surfaces along the reaction path of a thermally activated process with the crucial feature that the rate-determining step is associated with the spin inversion [45].

For didactic purposes, let us discuss the most simple oxidation brought about by MO^+ cations in some more detail. Dating back from 1994, several in-depth gas-phase studies dealt with Eq. (6) in the FeO^+/H_2 system [46]. Subsequent theoretical studies [9, 10, 47] have demonstrated that this simple reaction may serve to illustrate central features of TSR as well as its role as a key aspect in the reactivity of transition-metal oxides:

$$FeO^+ + H_2 \rightarrow Fe^+ + H_2O \tag{6}$$

Considering the ground state of reactants and products, the reaction at Eq. (6) is formally spin-allowed as Fe^+ and FeO^+ are sextets and H_2 and H_2O are singlets. Initiated by some counterintuitive experimental observations, a number of detailed theoretical studies have been performed. To cut a long story short, the lowest-energy path of the reaction at Eq. (6) involves a double spin flip from the sextet to the quartet surface close to the entrance and back to the high-spin surface in the exit channel (Fig. 4).

As the potential-energy surface shown in Fig. 4 has been discussed quite comprehensively [9, 10, 47], here we would like to focus on two particular questions:

1. Why is the rate-determining transition structure (TS) of the quartet surface lower in energy than the sextet TS?
2. How is the crossover between different spin multiplicities brought about?

The answer to the first question is inherent to the orbital description of the metal-oxo species made above. Thus, the high-spin, sextet ground state of FeO^+ has a diradicaloid bonding scheme and cannot react in a concerted manner with the substrate. Bond activation must therefore proceed stepwise and hence involves considerable activation barriers as is the case with

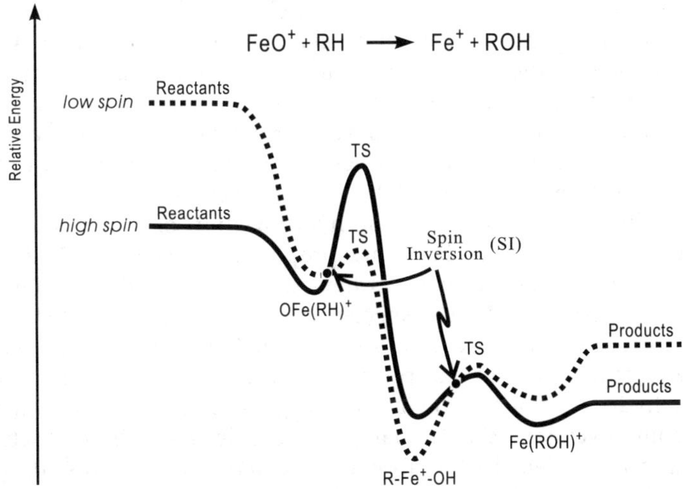

Fig. 4. Schematic potential-energy surfaces for alkane hydroxylation by FeO^+; see [11, 47] for details. Note that product formation involves two spin inversions to occur along the lowest lying reaction path

oxidations using molecular oxygen (see above). In the excited quartet states, however, bond cleavage and bond formation can occur in concert which in turn lowers the energy demand of the TS. The same arguments apply for the insertion intermediate H—Fe^+—OH in that iron can form two covalent bonds on the quartet surface, whereas an anti-bonding *s*-orbital is occupied in the sextet electromer.

The fundamentally different reactivity patterns of the high- and low-spin states of metal oxo species can be sketched in terms of a simple spin counting (Scheme 2). Thus, R—H bond insertion of a high-spin state would inevitably lead to a partially anti-bonding interaction, whereas the low-spin state can afford two covalent bonds via perfect pairing. Accordingly, the low-spin insertion intermediates are lower in energy than their high-spin electromers and for precisely the same reasons the insertion barriers are lowered on the low-spin surface. In even a more general sense, the high-spin metal-oxo species can only undergo single-bond reactions, i.e., atom abstractions, radical-type processes, or electron transfer, whereas in addition to these, the low-spin states can also promote two-bond reactions, such as concerted bond insertions, anion transfers, etc. [10]. Note that this classification of the

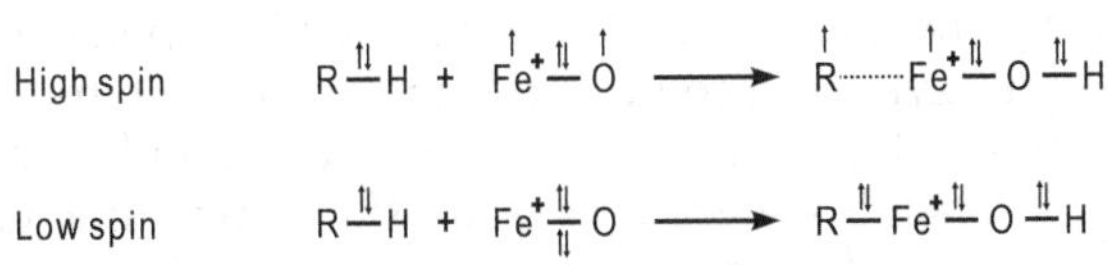

Scheme 2.

reactivity of the metal-oxo species agrees well with the O_2 bonding analogy. Thus, the triplet ground state of molecular oxygen is kinetically stable and mostly involved in radical-chain reactions, while concerted processes having low barriers are common for the excited singlet state of dioxygen.

TSR is not at all restricted to the particular activation of dihydrogen in the reaction at Eq. (6). Thus, the hydroxylations of methane by FeO^+, CoO^+, and NiO^+ cations [2] involve TSR [47b, 48], the FeO^+-mediated oxidation of benzene to phenol [49], and even the hydroxylation of a complex substrate such as norbornane [50] by FeO^+ are proposed to occur via TSR [51, 52]. Moreover, Yoshizawa and coworkers have demonstrated that in the hydroxylations of methane by FeO^{n+} species ($n = 0$–2) the concerted, low-spin pathways are preferred irrespective of the actual charge of the systems [53]. Accordingly, TSR is not an exception but appears as a general mechanistic pattern for hydroxylations involving oxo species of late transition metals.

While spin inversion is a crucial parameter in TSR, these reactions are formally spin-forbidden. In the absence of external fields, crossing between surfaces of different multiplicities can be mediated by spin-orbit coupling of the surfaces involved. Even though relativistic effects do not affect the energetics of 3*d* metals very much, communication between surfaces of different multiplicities via spin-orbit coupling is a significant factor in the reactivity of these metals. Detailed theoretical studies of the reaction at Eq. (6) have demonstrated that the experimentally observed reaction efficiency coincides with the probability of spin-inversion between the sextet and quartet surfaces [47a]. Occurrence of TSR has even been proposed in the reactions of early transition metals [38, 54] for which spin-orbit coupling is lowest in the 3*d* series.

Mediation of TSR by spin-orbit coupling has some important implications for the reaction kinetics if surface crossing is rate limiting. In particular, spin inversion via spin-orbit coupling must not obey Arrhenius-type kinetics, because the 'residence time' in the crossing region and hence the probability to invert spin is inversely proportional to temperature. Thus, the rate of TSR may decrease at higher temperatures [55], while at the same time the spin-allowed high-spin channels, namely single-state reactivity, can compete effectively [47b].

$$FeO^+ + CH_4 \rightarrow Fe^+ + CH_3OH \tag{7a}$$

$$FeO^+ + CH_4 \rightarrow FeOH^+ + CH_3^\bullet \tag{7b}$$

The experimental results obtained for the FeO^+/CH_4 system [46d] may serve as an example for the unusual energy dependencies of both reaction rates and branching ratios. Thus, the rates of the reactions at Eqs. (7a,b) exhibit rather different energy dependencies (Fig. 5). As outlined in Scheme 2, formation of the closed-shell hydroxylation product methanol, concomitant with generation of Fe^+ in the reaction at Eq. (7a), is a two-bond process and thus requires TSR. As the crossing probability decreases with shortening the lifetime of the reactant encounter complex, the Fe^+ product channel drops rapidly with increasing collision energy. The reaction at Eq. (7b), however, leads to the

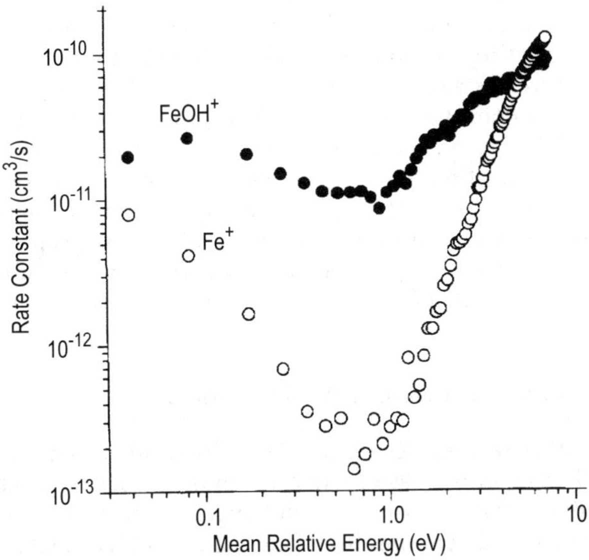

Fig. 5. Rate constants of the reactions at Eqs. (7a,b) in the FeO^+/CH_4 couple as a function of energy; adopted from [46d] to which the reader is referred for further details

release of a $CH_3^\bullet$ radical and can thus occur via TSR *as well as* SSR. The latter obeys a regular energy dependence such that the rate constant of this process remains almost constant over quite a large energy range. The competition between both channels inter alia changes in the $Fe^+/FeOH^+$ branching ratios from about 30:70 at lowest energies to a minimum of about 2:98 in the range 0.5–1.0 eV and then reaches 50:50 at higher energies. Similarly, pronounced variations in rate constants due to SSR/TSR competition occur in the FeO^+/H_2 [46] and V^+/CS_2 [55] systems. Thus, while TSR in itself provides an intriguing mechanistic scenario [45], an even more intriguing deviation from classical behavior evolves in cases in which TSR competes with SSR. Based upon the insight gained in the gas-phase studies of these extremely simplified oxidation 'catalysts', we have proposed the competition between single- and two-state reactivity to act as a mechanistic distributor in alkane hydroxylations by cytochrome P-450 [56]. In 1998, Newcomb and coworkers [57] argued that the interplay of SSR and TSR may indeed resolve some of the controversial explanations put forward in order to explain the severe experimental anomalies observed in P-450-mediated oxdiations. Very recently, further evidence for the crucial role of a spin-state crossing effect has been reported by Jin and Groves [58] for oxomanganese(V) prophyrins.

$$Ti^+ + H_2O \rightarrow TiO^+ + H_2 \tag{8}$$

Interestingly, surface crossing is also observed in the reactions of early transition-metals, but here it occurs in an opposite sense. For example, the exothermic reaction at Eq. (8) of ground state Ti^+ (^4F) with water yields TiO^+ ($^2\Delta$) concomitant with neutral dihydrogen, and thus a net change results from crossing the quartet to the doublet surface [38, 59]. Although the high-spin transition structure has not been located [39], it is safely assumed to be more energy demanding than the low-spin TS. Thus, the reaction at Eq. (8) fits the TSR scheme nicely, i.e., a surface crossing en route from reactant complex to the lowest-lying transition structure. Substantial variations of the energy-dependent cross sections [38] suggest that in the reaction at Eq. (8) spin inversion is also a rate limiting factor. Due to the high oxophilicity of low-valent titanium, however, oxygen transfer occurs from the substrate to the metal, i.e., a reduction of the substrate rather than an oxidation takes place.

5.4 Miscellaneous Reactions of Transition-Metal Monoxides

Before closing this section, let us briefly address other achievements made since the publication of the 1995 review [2]. In the 3*d* series, reactivity studies mostly focused on FeO^+ by extending the range of substrates to alkanols [60], olefins [61], substituted benzenes [62], heterocycles [63], as well as main group hydrides [64]. While we refer the reader to the original sources for further details, some aspects are noteworthy in the present context. As expected from the cursory discussion above, the oxidation of alkanols by FeO^+ proceeds more rapidly than that of the parent hydrocarbons. Interestingly, the reaction mechanisms differ fundamentally from those observed with the alkanes in that initial coordination of the metal-oxide cation to the hydroxy group determines the fate of the reactions. For example, a major pathway in the reactions of some alkanols with FeO^+ constitutes a direct hydroxide ion abstraction to afford the corresponding carbocations and neutral Fe(O)(OH); nevertheless, C—H bond activation can still take place to some extent [60]. More extreme, phenol avoids C—H bond activation as an initial step [62b] even though FeO^+ is capable of hydroxylating benzene [49]. Mass spectrometric studies in conjunction with extensive ^2H and ^{13}C labeling reveal that O—H bond activation of phenol predominates to afford $C_6H_5O{-}Fe^+{-}OH$ as central intermediate from which the subsequent products evolve. Initial occurrence of ring hydroxylation by FeO^+ as a major path is rigorously excluded by ^{18}O-labeling [62b]. Finally, the reactions of FeO^+ with aromatic amines are initiated by electron transfer (ET) from the amine to the metal oxide [62a]; this result is in close analogy to the ET driven oxidative dealkylation of *N*-alkylanilines by cytochrome P-450.

A few other reactivity studies have been performed with 3*d*-metal monoxides and organic substrates since 1995 [65]. Significant progress has been achieved for a series of small inorganic reagents. Thus, Sievers and Armentrout examined the reduction of $CO_2 \rightarrow CO$ by various early transition metals in quite some detail [66]; a highlight concerns the study of the V^+/CO_2 system in which it was possible to map out crucial parts of the potential-energy surface experimentally [67]. Also noteworthy is the reaction of FeO^+ with NO_2 [46c] as one of the few examples in which the metal is further oxidized, yielding NO^+ together with neutral FeO_2, i.e., a formal iron(IV) compound (see below). In this context, the ongoing studies of Plane and coworkers on the role of iron oxides in atmospheric chemistry are worth mentioning [68].

Several anionic and neutral transition-metal oxides have been examined recently by different means of optical spectroscopy [22, 69]. Andrews and coworkers have undertaken systematic investigations of the metal-oxide chemistry in doped rare gas matrices; for example, neutral and cationic $OM(CO)^{+/o}$ complexes (M = Ti, V [70]) as well as $OM(CO)^{o/-}$ species (M = Cr—Cu [71]) were examined by matrix-isolation spectroscopy.

Among the 4*d* and 5*d* series, few reactivity studies with oxidizable substrates have been reported. The reactivities of neutral metal oxides towards alkanes have been examined, but in these studies only the depletion of the reactants, namely MO, has been monitored while the product structures remain undetermined [72]. Oxygen-atom transfer to alkanes does not occur with MoO^+ [73], HfO^+ [74], TaO^+ [75], WO^+ [74], ReO^+ [74, 76], and OsO^+ [77] which is in accordance with their large bond-dissociation energies (Table 1). MoO^+ can, however, bring about oxidation of methanol to formaldehyde [78]. Interestingly, ReO^+ and OsO^+ are even capable of activating hydrocarbons including methane [74, 77], but instead of O-atom transfer, dehydrogenation takes place presumably to yield bisligated oxometal carbenes according to the reaction at Eq. (9):

$$MO^+ + CH_4 \rightarrow M(O)(CH_2)^+ + H_2 \quad (M = Re,\ Os) \tag{9}$$

Particularly noteworthy is a rather extensive theoretical study of the reaction paths in the $Pt^+/CH_4/O_2$ system [25, 79] which allows for a catalytic oxidation of methane by dioxygen [80]. In this context, the recently reported catalytic sequences for gas-phase oxidations with $Pt_nO_n^-$ cluster anions are of interest [81]; note, however, that the oxidizable substrate chosen was again not an alkane but carbon monoxide, which is more likely to be attacked by anionic species.

Finally, an interesting observation in the series of lanthanide monoxides should be mentioned [82]. Although oxygen-atom transfer cannot be achieved with lanthanides due to their large oxophilicities, the LnO^+ cations are much more capable of inducing olefin oligomerizations than their formally isovalent MX_2^+ homologs (X = F, OH, Cl, OCH_3). This difference demonstrates the decisive role of the metal-oxo unit and has been attributed to the particular reactivity of this unit in terms of intermediate electron transfer as a rate-determining step [82]. More recently, Gibson used an elegant device to extend these studies to several cationic actinide monoxides and obtained similar results [83].

6 High-Valent Transition-Metal Oxides

So far, the discussion has, by and large, been confined to metal-monoxide cations, i.e., compounds with formal metal(III) oxidation states. Most applied transition-metal based oxidants exhibit higher oxidation states in their active forms; often these are termed high-valent metal-oxo species. Typical examples include chromium(VI) compounds such as CrO_2Cl_2, CrO_3, and CrO_4^{2-}, permanganate, Mo, and Ru oxides, and the time honored osmium(VIII) tetroxide. Here we review the present status of the gas-phase experiments conducted with high-valent metal oxides. Note that metal peroxides are only considered if directly relevant to the chemistry of metal-oxo species; in particular, neither the Sharpless epoxidation nor the reactions mediated by methyltrioxorhenium are discussed because they do not involve metal-oxo units in the decisive oxidation steps.

7
Structural Dichotomy

If more than one oxygen atom is attached to a metal center, a structural dichotomy results. Thus, for the elemental composition $[M,O_2]$ three fundamentally different bonding situations exist: (i) an end-on metal dioxygen complex **I**, (ii) a side-on complex **II**, and (iii) an inserted metal dioxide **III** (Scheme 3). Depending on the spin coupling, these structures possess different bonding mnemonics. For example, high-spin coupled **II** can be regarded as a mere complex of the metal with triplet dioxygen, whereas its low-spin congener is best described as a covalently bound metal peroxide. Similar arguments apply to $[M,O_n]$ species with $n > 2$.

These different types of bonding have been distinguished in a comparative study of oxide cations of chromium, iron, and rhenium [84]. For $[Cr,O_2]^+$ structures **I** and **III** coexist as well-separated minima on different spin surfaces, i.e., the end-on dioxygen complex $Cr(O_2)^+$ ($^6A''$) and the dioxide cation CrO_2^+ ($^2A_1$), respectively [85]. In the case of $[Fe,O_2]^+$, sextet states of structures **II** and **III** are very close in energy and separated by a rather low barrier [86]. Ionization of the high-valent rhenium peroxide $CH_3Re(O_2)_2O$ [87] leaves the peroxo units intact; thus $CH_3Re(O_2)_2O^+$ cation may serve as a representative for the low-spin, peroxo variant of structure **II**. As a rough but by no means general guide, structure **III** is energetically favored for low oxidation states as well as for those metals which have a preference for the formation of high-valent oxides, while structures **I** and **II** prevail for the late transition metals which do not support high oxidation states. This structural dichotomy needs to be kept in mind in the evaluation of the gas-phase properties of $[M,O_n]^{+/o/-}$ because more than a single isomer and/or state may be generated under the experimental conditions. In fact, coexistence of structural isomers has been demonstrated for anionic [69, 88], neutral [89], cationic [84–86], and dicationic transition-metal oxides [85].

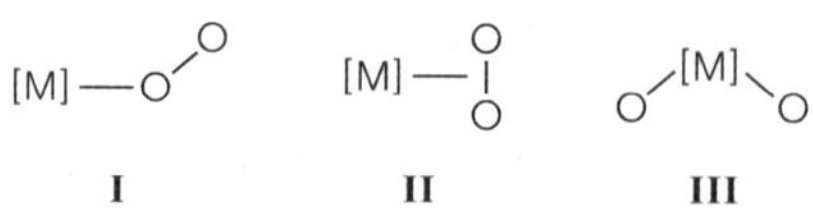

Scheme 3.

8 Transition-Metal Dioxides

Analogous thermochemical criteria as outlined above for alkane oxidation by metal monoxides apply to transition-metal dioxides, except that D_0(M—O) is to be replaced by D_0(OM—O). If not prevented by a spin-inversion bottleneck, loss of triplet dioxygen needs to be considered as an additional reaction path in evaluating thermochemical stabilities. Thus, a metal dioxide is metastable with respect to the dissociation $MO_2 \rightarrow M + O_2$, if D_0(OM—O)+D_0(M—O) < D_0(O—O) = 118 kcal/mol. For ionic $[M,O_2]^{+/-}$-species, the additional electrostatic interaction in the $M(O_2)^{+/-}$-complexes provides roughly 10–30 kcal/mol extra stabilization of structures **I** and **II**.

Unlike the monoxides, the electronic structures of transition-metal dioxides are more difficult to categorize because of bending, symmetry breaking, and the structural dichotomy mentioned above. Let us therefore elucidate the bonding patterns as well as selective aspects of the reactivity of transition-metal dioxides by progressing through the 3*d* series and only briefly addressing the 4*d* and 5*d* homologs; for a comprehensive theoretical study of neutral 4*d* dioxides, see [90].

8.1 ScO_2^+

While $D_0(Sc^+$—O) = 165 kcal/mol is largest for all 3*d* monoxide cations, $D_0(OSc^+$—O) = 40 kcal/mol [91] is rather low (Table 2). This result is not surprising because scandium cation has only two valence electrons to bind the first oxygen atom and thus no bonding capabilities are left for a second one. Even though no reactivity studies have so far been performed with ScO_2^+, the weakness of the OSc^+—O bond implies that ScO_2^+ can act as an efficient O-atom donor. For the very same reason re-oxidation $ScO^+ + \langle O \rangle \rightarrow ScO_2^+$ will be problematic, therefore rendering this ion unattractive as far as oxidation catalysis is concerned. Similar considerations apply for YO_2^+ with $D_0(OY^+$—O) = 41 kcal/mol and LaO_2^+ with $D_0(OLa^+$—O) = 23 kcal/mol [91]. It would be interesting to know whether these dioxide cations are low- or high-spin coupled and whether the oxygen atoms are equivalent, e.g., symmetrical ScO_2^+ vs the asymmetric structures $OSc^+\cdots O$ and $ScOO^+$.

8.2 $TiO_2^{+/0/-}$

Neutral titanium has four valence electrons and can thus precisely saturate the demands of two oxo ligands. Indeed, neutral TiO_2 undergoes perfect pairing resulting in a $^1A_1$ singlet ground state with D_0(OTi—O) = 144 kcal/mol; the latter value is much too high in the context of alkane oxidation [40, 92]. Due to the electron-withdrawing oxo ligands, TiO_2 has a sizable electron affinity as well as ionization energy, i.e., EA(TiO_2) = 1.6 eV [69f] and IE(TiO_2) = 9.5 eV

Table 2. Bond dissociation energies, $D_0(OM^+—O)$ in kcal/mol, of transition-metal dioxide cations

	D_0		D_0		D_0
ScO_2^+	40[a]	YO_2^+	41[a]	LaO_2^+	23[a]
TiO_2^+	81[b]	ZrO_2^+	89[b]	HfO_2^+	
VO_2^+	90[b]	NbO_2^+	132[b]	TaO_2^+	140[h]
CrO_2^+	66[c]	MoO_2^+	128[f]	WO_2^+	132[h]
				ReO_2^+	65[i]
FeO_2^+	66[d]	RuO_2^+	79[g]	OsO_2^+	105[j]
		RhO_2^+	78[h]	IrO_2^+	125[k]
				PtO_2^+	75[h]
CuO_2^+	99[e]			AuO_2^+	–[l]

[a] [91]
[b] [40]
[c] [85]
[d] [86]
[e] According to ab initio studies, the ion structure is $Cu(O_2)^+$, see [104]
[f] [66b]
[g] Derived from data given in: Norman JH, Staley HG, Bell WE (1968) Adv Chem Ser 72, Gould RF (ed), ACS Washington, p 101
[h] [29]
[i] [76]
[j] [77]
[k] Norman JH, Staley HG, Bell WE (1965) J Chem Phys 42: 1123
[l] For an observation of AuO_2^+, see: Aita CR (1987) J Appl Phys 61: 5182

[40]. Ground state TiO_2^- ($^2A_1$) anion [69f, 92, 93] has one extra electron and is best described as a titanium(III) dioxide, i.e., $O{=}Ti^{\cdot}{-}O^- \leftrightarrow {}^-O{-}Ti^{\cdot}{=}O$. Accordingly, $D_0(OTi^-—O)$ = 150 kcal/mol exceeds the bond strength in neutral titanium(IV) dioxide; once again this bond energy is much too high to render this oxo species useful in oxidation processes.

In marked contrast, TiO_2^+ ($^2B_2$) cation has a moderate bond strength, i.e., $D_0(OTi^+—O)$ = 81 kcal/mol and behaves as a highly reactive oxidant which not only activates alkanes but even abstracts a hydrogen atom from water – see the reaction at Eq. (5). This behavior is in accordance with the description of TiO_2^+ as a hypervalent compound with radical character centered on oxygen [40]; the latter is due to the fact that the three valence electrons of titanium cannot saturate the needs of two oxo ligands. Even though TiO_2^+ is a potent reagent for bond activation, as an oxygen-centered radical, unselective hydrogen-atom abstraction is facile. This dioxide ion is thus not considered as a valuable reagent for selective oxidation processes. Likewise, the 4*d* congener ZrO_2^+ reacts in a radical-like manner [40], and the reactivity patterns of hafnium oxide cations suggest a similar behavior of HfO_2^+ [74]. Further, neutral and anionic zirconium and hafnium dioxides have been characterized by matrix spectroscopy [93]. Note that most dioxide cations of lanthanide and actinide metals are either of low reactivity or show radical-type behavior [82b, 94].

8.3 VO_2^+

The vanadium(III) species VO_2^- and the neutral vanadium(IV) dioxide VO_2 ($^2A_1$) are unfavorable candidates for metal-oxo based oxidations, i.e., $D_0(OV^-—O) = 150$ kcal/mol and $D_0(OV—O) = 132$ kcal/mol [40, 95]. In marked contrast, the vanadyl cation VO_2^+ is capable of activating hydrocarbons concomitant with a reduction from formal vandium(V) to vanadium (III). In VO_2^+, the four valence electrons of the metal precisely fulfill the needs of the oxo units giving rise to a $^2A_1$ ground state, sketched as $O{=}V^+{=}O$. This perfect pairing situation notwithstanding, the net positive charge and the electron withdrawing oxo ligands enhance the reactivity of VO_2^+ and result in a moderate bond strength in a range desired for catalysis. The absolute value of $D_0(OV^+—O)$ is a matter of debate though. Sievers and Armentrout [67] derived $D_0(OV^+—O) = 71$ kcal/mol from the onset of highly endothermic oxygen-atom transfer from CO to VO^+ cation, whereas recent theoretical and experimental results suggest $D_0(OV^+—O) = 90$ kcal/mol [40]. As the endothermic reaction studied by Sievers and Armentrout may only provide a lower bound for $D_0(OV^+—O)$, due to kinetic hindrance of O-atom transfer from the strongly bound CO molecule, we list the larger value in Table 2.

Vanadyl cation is capable of oxidizing a variety of substrates [40] beginning with ethane activation according to the reaction at Eq. (10). We restrict the discussion of the oxidation properties of VO_2^+ to this particular example:

$$VO_2^+ + C_2H_6 \rightarrow V(OH)_2^+ + C_2H_4 \qquad (10)$$

The $V(OH)_2^+$ cation formed in the reaction at Eq. (10) is a vanadium(III) compound and exhibits a triplet ground state, whereas the other reagents are singlets. Thus, again the oxidation occurs via spin-inversion. However, because ground state VO_2^+ is a singlet, the course of the reaction fundamentally differs from the TSR scenario outlined above. According to theoretical studies [40], the lowest-lying TS **1/2** for C—H bond activation evolve from the singlet surface of the reactants via the sequence **1** → **2** → **3** (Fig. 6). This is obvious from the arguments raised above in the discussion of TSR because the concerted reaction path evolves from the low-spin surface, i.e., the singlet reactants. Hence, bond activation does not require spin inversion and the subsequent crossing to the triplet surface occurs in the product complex and is thus just a bottleneck in product release. Any eventual restrictions in spin inversion occurring after TS **1/2** as the rate-determining step, i.e., in structures **2** and **3**, can thus by and large be expressed in terms of the pre-exponential factor in an Arrhenius formalism, while the actual bond activation is expected to follow classical kinetics. Therefore, we may rather classify the reaction at Eq. (10) as SSR followed by spin inversion rather than TSR in which spin inversion is part of the rate-determining step. In fact, there are not many arguments in favor of a situation which we may term as inverse TSR, i.e., rate-limiting crossing from a low- to a high-spin surface along the reaction coordinate, because in most but unusual cases the low-spin TS is expected to

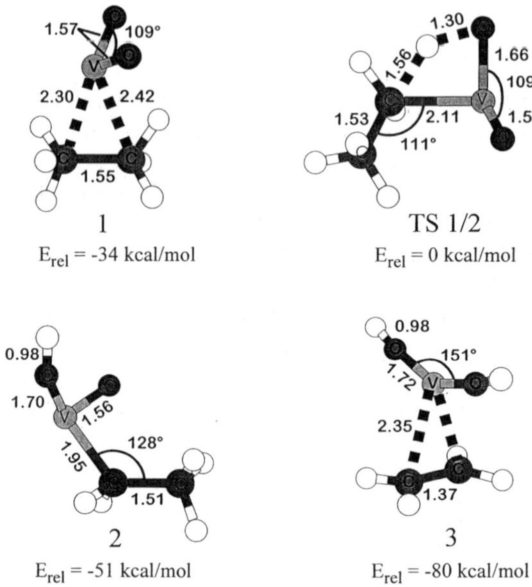

Fig. 6. Geometries of stationary points relevant in the reaction at Eq. (10); adopted from [40]

be lower in energy than its high-spin congener if the reactants already have low-spin ground states.

Like VO_2^+ ($^1A_1$), the 4*d* congener NbO_2^+ ($^1A_1$) also exhibits a perfect pairing ground state, the enhanced $D_0(ONb^+—O)$ = 132 kcal/mol disfavors its application in catalytic oxidation of alkanes [40]. Similarly, the oxophilicity of tantalum does not allow for O-atom transfer with TaO_2^+ with $D_0(OTa^+—O)$ = 140 kcal/mol [29]. Indeed, TaO_2^+ is formed as the terminal product in the stoichiometric coupling of carbon dioxide and methane to ketene [27, 75].

8.4 CrO_2^+

Chromium-dioxide cation has a $^2A_1$ ground state, and the moderate bond strength $D_0(OCr^+—O)$ = 66 kcal/mol allows for the oxidation of various hydrocarbons [85]. In addition to the ground state dioxide CrO_2^+, the existence of the dioxygen complex $Cr(O_2)^+$ as an isomeric species arising from the ground states of Cr^+ (^6S) and O_2 ($^3\Sigma_g$) has been demonstrated in this case; circumstantial evidence for a quartet species has also been obtained.

Chromium dioxide is one of the most powerful gas-phase oxidants examined so far. Thus, CrO_2^+ activates various substrates including dihydrogen and methane. Moreover, complete reduction from CrO_2^+ to Cr^+ is observed to some extent, i.e., transfer of two oxygen atoms to the substrate. An example is the reaction at Eq. (11) occurring with methane; the nature of the neutral molecule(s) formed is unknown, but water and formaldehyde are a likely product combination which finds support in the inter alia observed formation of $Cr(CH_2O)^+$ from the O_2^+/CH_4 couple:

$$CrO_2^+ + CH_4 \rightarrow Cr^+ + [C, H_4, O_2] \tag{11}$$

$$Cr^+ + O_2 \rightarrow Cr(O_2)^+ \rightarrow CrO_2^+ \tag{12}$$

In this respect, the possible sequence depicted in the reaction at Eq. (12) is also of interest. Though it involves a net change from the sextet to the doublet surface, spin inversion in the reaction at Eq. (12) appears to occur with a finite probability in the gas phase, because under multiple collision conditions, $Cr(O_2)^+$ was found to end up as CrO_2^+ [85]. Thus, combination of the reactions at Eqs. (11) and (12) is a possible strategy for catalytic oxidations applying dioxygen as a terminal oxidant. While it may appear counterintuitive that a low valent metal, i.e., Cr^+ cation, can promote oxidations, this is precisely what has been observed by Bakac and Espenson in the chromium(II)-mediated oxidations of alkanols by dioxygen in protic solutions [96].

Unfortunately, CrO_2^+ is too reactive in that its reactions with higher alkanes show little selectivities giving rise to a manifold of products. Selectivity poses quite a dilemma in designing reactive oxidants, because those species which are capable of activating the reasonably strong C—H bonds in non-polar substrates such as alkanes are often capable of activating *any* bond in the substrate as well as in the oxidation products. In fact, high reactivity is often associated with reduced selectivity, and for this very reason intrinsic reactivity needs to be modified by the local environment in order to improve selectivity while keeping reactivity high.

For the description of the bonding pattern of CrO_2^+, let us refer to the 4*d* homolog MoO_2^+ which has been analyzed in some detail [73a]. In the MoO_2^+ ($^2A_1$) ground state, the combination of Mo with two O atoms gives a set of doubly occupied *s*- and π orbitals which are all bonding except for a doubly occupied orbital of b_1 symmetry which, by and large, comprises the $2p_x$ orbitals of oxygen and is thus referred to as non-bonding. The uncoupled electron resides in a σ orbital of a_1 symmetry with slightly anti-bonding character. Thus, the bonding situation of MoO_2^+ can be described in terms of two metal-oxygen double bonds resulting in a formal Mo(V). The balance of the bonding orbitals increases the stability of MoO_2^+ and leads to a significantly enhanced bond strength compared with chromium, i.e., $D_0(OCr^+—O) = 66$ kcal/mol vs $D_0(OMo^+—O) = 128$ kcal/mol. In fact, MoO^+ can reduce CO_2 to CO concomitant with formation of MoO_2^+ [66b, 73a]. Accordingly, O-atom transfer is less likely to occur with MoO_2^+, and along the series of MoO_n^+ cations (n = 1–3), the dioxide shows the lowest reactivity towards hydrocarbon substrates [73, 97]. Similarly, WO_2^+ with $D_0(OW^+—O) = 132$ kcal/mol is not attractive as far as catalytic oxidations are concerned [74].

8.5 MnO_2^+

While MnO_2 is a well-known oxidant in the condensed phase, gas-phase oxidations with neutral and cationic MnO_2 have not been reported so far. $[Mn,O_2]^-$ anions have been made by reacting $Mn(CO)_n^-$ anions with dioxygen [98] and laser desorption of bulk manganese oxides [99], respectively. Interestingly, depending on the mode of ion generation different reactivities towards methanol were found, i.e., the ions formed from the $Mn(CO)_n^-/O_2$ couple are unreactive, whereas efficient oxidation to formaldehyde occurs with the ions desorbed from solid manganese oxide. These opposing findings indicate the role of the structural dichotomy mentioned above even for anionic species, i.e., formation of $Mn(O_2)^-$ in one case and MnO_2^- in the other. Due to the relevance of manganese in oxidation catalysis, further studies of group 7 metal oxides are very much indicated. For example, the 5*d* congener ReO_2^+ activates methane according to Eq. (13); note, that the structure of the ionic product is uncertain [74, 76]:

$$ReO_2^+ + CH_4 \rightarrow [Re, C, H_2, O_2]^+ + H_2 \tag{13}$$

8.6 $FeO_2^{+/o/-}$

Experimental studies of transient FeO_2 date back to Addison and coworkers in 1965 who examined the thermolysis of volatile iron nitrates [100], and various other reports on this species have appeared more recently [68, 69a, 86, 89b, 99, 101]. Interestingly, $[FeO_2^{+/o/-}]$ appears as a system for which the dichotomy between dioxygen complex and dioxide structure is particularly pronounced and, in addition, several low-lying states exist. In fact, the variations with respect to electronic structures as well as the richness of accessible structural isomers render iron chemistry one of the challenges for contemporary experimental and theoretical methods. For example, the computational predictions for neutral $[Fe,O_2]$ suggest the dioxide structure **III** is more stable, but the ground-state assignments vary from singlet $^1A_1$ to triplet $^3B_1$, quintet $^5B_2$, and even septet $^7A_1$ [68b, 89b, 101]. Based on their theoretical results, Cao et al. [101a] even doubted that the otherwise accurate and reliable photodetachment experiments with FeO_2^- [69a] involved ground state ions. Thus, we refrain from a more detailed discussion here, and the interested reader should consult the original references.

Two aspects which appear to be settled reasonably well concern the $[Fe,O_2]^+$ cation which shows an almost degenerate bonding situation between structures **II** and **III** [86]:

1. The dioxygen complex $Fe(O_2)^+$ ($^6A_1$) and the dioxide FeO_2^+ ($^6A_1$) are very close in energy, of the same multiplicity and symmetry, and separated by a rather low barrier (Fig. 7). The facile interconversions of structures **II** and **III** is of conceptual interest as it provides a spin-allowed route for the activation of dioxygen. In fact, in the 3*d* series oxidation of organic ligands

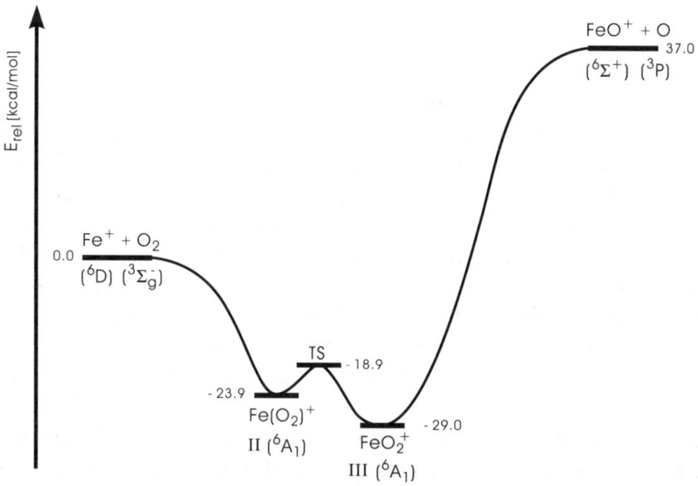

Fig. 7. Schematic potential-energy surface of the $[Fe,O_2]^+$ system; adopted from [86] to which the reader is referred for further details

L is a unique feature of iron in the reactions of $M(L)^+$ species with dioxygen [102]. Due to the energetic proximity, structures **II** and **III**, however, $[Fe,O_2]^+$ show an ambivalent behavior in that substrates weakly coordinating to iron undergo oxidation, i.e., a reaction typical for a metal dioxide, while for substrates which favorably bind to bare Fe^+, simple ligand displacement occurs to yield $Fe(L)^+$ and O_2; this is precisely the behavior expected for a dioxygen complex.

2. The pronounced effects of dynamic electron correlation observed for $[Fe,O_2]^+$ is remarkable with respect to the theoretical description of transition-metal oxides. CASSCF calculations with reasonably large basis sets predict $Fe(O_2)^+$ $(^6A_1)$ to be 32 kcal/mol more stable than FeO_2^+ $(^6A_1)$, while inclusion of perturbation theory at the CASPT2D level with the same basis sets disfavors $Fe(O_2)^+$ $(^6A_1)$ over FeO_2^+ $(^6A_1)$ by 5 kcal/mol. Thus, dynamic correlation changes the relative stabilities by as much as 37 kcal/mol. Quite obviously, the theoretical treatment of $[Fe,O_2]^+$ is far from being complete, and further studies of this fundamental, rather challenging, problem are advised. As a note of caution, we add that the pronounced effect of correlation energy may also affect the accuracy of thermochemical predictions made for transition-metal oxides using the nowadays popular hybrid methods [7].

Unlike RuO_2^+, the 5*d* congener OsO_2^+ has been studied comprehensively [77], and OsO_2^+ is capable of oxidizing a broad range of substrates. In particular, a potential route for the OsO_2^+-catalyzed oxidation of methane to formaldehyde according to Eq. (14) has been proposed [77]:

$$CH_4 + O_2 \rightarrow CH_2O + H_2O \tag{14}$$

8.7
CoO_2-ZnO_2

So far, gas-phase studies of cobalt and nickel dioxides are limited to anions and neutrals and mostly dealt with their spectroscopic features [69e, 99, 103]. In the case of nickel, isomeric $Ni(O_2)^-$ and NiO_2^- ions were distinguished by their photoelectron detachment spectra [69e]. Oxidations of organic molecules have to date only been reported for the anionic species with methanol as a substrate and with moderate to low efficiencies [99]. Except for some thermochemical information (Table 2), none of the 4*d* and 5*d* congeners has been subjected to detailed structural studies nor have their reactivities towards alkanes been reported. Note however, that cationic iridium and platinum complexes react with dioxygen [74, 80], thus suggesting the intermediate $[M,O_2]^+$ species.

Copper deserves particular attention because nature has chosen this metal as an alternative to iron for the transport and activation of oxygen in living systems [43]. Interestingly, many of these metalloenzymes have mononuclear copper centers, even though one would not expect efficient binding of oxygen with a single copper center due to the filled 3*d* shell. Theoretical studies of neutral $Cu(O_2)$ indicate that electron transfer plays an important role in this respect [104], i.e., the binding has a significant contribution of a $[Cu^+ \cdot O_2^{-\bullet}]$ configuration and thus compares to a superoxide. The potential-energy surface of neutral $Cu(O_2)$ is extremely flat, however, and a definitive assignment as either structure **I** or **II** cannot be made with the information available [103]. For the cationic species, experimental evidence for structure **III** has been provided [105], and the co-existence of isomeric $Cu(O_2)^-$ and CuO_2^- anions has been demonstrated recently [88]. Considering the biochemical importance of copper in dioxygen activation, further studies of neutral and ionic copper oxides are anticipated.

Gas-phase studies of zinc, cadmium, and mercury dioxides have not been reported, but the example of copper may tell us that chemical intuition, which would exclude dioxides with these elements, is not safe against surprises, and one should thus not generally dismiss these elements as oxidation catalysts.

9
Miscellaneous Higher Transition-Metal Oxides

Among the cationic trioxides, gas-phase activations of hydrocarbons have only been studied for MoO_3^+ [73, 97], ReO_3^+ [76], and OsO_3^+ [77] in some detail. Of course the structural variety mentioned above for $[M,O_2]$ is even more pronounced for $[M,O_3]$; for M = Mo, Re, and Os these three species can be considered as transition-metal trioxides, however. For example, computational studies predict a C_{3v}-symmetrical structure for ground state MoO_3^+ ($^2A_2$) [73]. As Mo^+ has only five valence electrons, it cannot satisfy the demands of three oxo units and can thus be described as hypervalent with high radical character on the oxygen atoms. Therefore, it is not surprising that typical radical-type reactions occur. For example, in analogy to TiO_2^+ and ZrO_2^+ cations, MoO_3^+

abstracts a hydrogen atom from water. Accordingly, MoO_3^+ is rather reactive and can even activate methane. In line with its description as an oxygen centered radical, the reactions are not very specific and H-atom abstraction as well as electron transfer (ET) are major routes observed for MoO_3^+. Occurrence of ET is obvious considering IE(MoO_3) = 11.7 eV, and similarly one could not expect much specifity for the 5*d* congener WO_3^+ because IE(WO_3) = 12.5 eV is even larger [29]. In contrast, atom abstractions and ET are much less pronounced for ReO_3^+ and OsO_3^+, while both ions bring about the activation of various hydrocarbons [76, 77]. Interestingly, OsO_3^+ does not activate methane, whereas ReO_3^+ dehydrogenates methane according to Eq. (15):

$$ReO_3^+ + CH_4 \rightarrow [Re, C, H_2, O_3]^+ + H_2 \tag{15a}$$

$$ReO_3^+ + CH_4 \rightarrow [Re, H_4, O_2]^+ + CO \tag{15b}$$

Recently, reactivities of some $[M,O_3]^-$ anions (M = Mn, Fe [99], and Nb [106]) have been examined, but these anions do not even bring about the oxidation of methanol to formaldehyde [107]. The only metal tetroxide cation studied so far is – of course – OsO_4^+; not quite unexpectedly, it behaves very much as an oxygen-centered radical and atom abstractions as well as ET prevail [77]. $[Fe,O_4]^-$ made by chemical ionization of $Fe(CO)_5/O_2$ mixtures [86], neutral $[V,O_4]$ observed in O_2-doped matrices [108], and $[Re,O_n]^+$ ions (n = 4–8) formed in O_2-seeded supersonic expansions appear to have peroxide structures [76, 109].

Finally, some substituted metal-oxo species have been examined in the gas phase, ranging from the iron(IV) compound $Fe(O)(OH)^+$ [110] and the chromium(VI) species $CrOF_3^+$ [111] as well as CrO_2Cl^+ [112] to the molecular cations of the rhenium(VII) compounds CH_3ReO_3 [113] and $CH_3Re(O_2)_2O$ [84]. Except for the reactions of $Fe(O)(OH)^+$ already described in the earlier review [2], no particular reactivity patterns have been observed which are relevant in the context of alkane hydroxylation.

10 Conclusions

Gas-phase studies of metal-oxo species provide a wealth of mechanistic insight into fundamental aspects of oxidation catalysis. In particular, it is the combination of experimental studies with theoretical methods that brings about major progress. In this context, the two-state reactivity paradigm for the reaction of metal-oxo species is of paramount importance. Due to their inherent reactivity, the proximity of various electronic states, vast correlation effects, the role of relativistic effects etc., both experimental and theoretical studies have to apply advanced methods in order to tackle the challenging properties of transition-metal oxides.

In a more general sense, we consider close cooperation between various branches of research as being essential to make substantial contributions to this area. This demand for interdisciplinarity ranges from theoretical physics (e.g., surface crossing) via chemical physics (e.g., spectroscopy), physical chemistry (e.g., kinetics and thermochemistry), inorganic chemistry (e.g., bonding properties), physical organic chemistry (e.g., reaction mechanisms) to synthetic inorganic and organic synthesis (e.g., volatile metal compounds, isotopically labeled substrates). Last but not least, direct contact with catalyst research in academia and industry is required in order to define the questions to be answered in the hope of improving the understanding of oxidation catalysis at a molecular level.

Acknowledgements. Financial support by the Deutsche Forschungsgemeinschaft, the Volkswagen Stiftung, the Fonds der Chemischen Industrie, and the Gesellschaft der Freunde der Technischen Universität Berlin is gratefully acknowledged. Numerous colleagues from the Bayer AG, BASF AG, and DEGUSSA-HÜLS AG are acknowledged for inspiring discussions and directing our attention to industrial oxidation processes. Furthermore, we appreciate the ongoing and fruitful cooperation with Professor P.B. Armentrout, Salt Lake City, and thank Dr. M. Beyer for sending us a copy of [76].

11
References

1. Bernadou J, Meunier B (1998) Chem Commun 2167
2. Schröder D, Schwarz H (1995) Angew Chem 107: 2023; Angew Chem Int Ed Engl 34: 1973
3. (a) Schalley CA, Hornung G, Schröder D, Schwarz H (1998) Chem Soc Rev 27: 91; (b) Schalley CA, Hornung G, Schröder D, Schwarz H (1998) Int J Mass Spectrom Ion Processes 172: 181
4. (a) Ceyer ST (1990) Science 249: 133; (b) Ceyer ST (1994) In: Proceedings of the Robert A. Welch Foundation, 38th Conference on Chemical Research, Chemical Dynamics of Transient Species, Houston, p 157
5. (a) Jorgensen KA (1989) Chem Rev 89: 431; (b) Jorgensen KA, Hoffmann R (1990) J Phys Chem 94: 3046
6. Thomas JM (1997) Chem Eur J 3: 1557
7. (a) Koch W, Hertwig RH (1998) DFT Applications to transition metal problems. In: Schleyer PvR, Allinger NL, Clark T, Gasteiger J, Kollmann PA, Schaefer HF, Schreiner PR (eds) The encyclopedia of computational chemistry. Wiley, Chichester; (b) Truhlar DG, Morokuma K (eds) (1999) Transition state modeling for catalysis. ACS Symp Ser 721, ACS, Washington
8. Shaik S, Danovich D, Fiedler A, Schröder D, Schwarz H. (1995) Helv Chim Acta 78: 1393
9. Yamaguchi K, Takahara Y, Fueno T (1984) In: Smith VH, Schaefer HF, Morokuma K (eds) Applied quantum chemistry. Reidel, Boston, p 155
10. Carter EA, Goddard WA (1988) J Phys Chem 92: 2109
11. Fiedler A, Schröder D, Shaik S, Schwarz H (1994) J Am Chem Soc 116: 10,734
12. Sodupe M, Branchadell V, Rosi M, Bauschlicher CW (1997) J Phys Chem A 101: 7854
13. If not mentioned otherwise, for the sake of consistency all metal-ligand bond strengths are taken from the following compilation: Freiser BS (ed) (1996) Organometallic ion chemistry. Kluwer, Dordrecht
14. Dyke JM, Gravenor BWJ, Hastings MP, Morris A (1985) J Chem Phys 89: 4613
15. Schilling JB, Beauchamp JL (1988) Organometallics 7: 194
16. Ryan MF, Fiedler A, Schröder D, Schwarz H (1995) J Am Chem Soc 117: 2033

17. Fiedler A, Hrušák J, Koch W, Schwarz H (1993) Chem Phys Lett 211: 242
18. Husband J, Aguirre F, Ferguson P, Metz R (1999) J Chem Phys 111: 1433
19. (a) Lunsford J (1995) Angew Chem 107: 1059; Angew Chem 34: 970; (b) Weldon MK, Friend CM (1996) Chem Rev 96: 1391
20. Bauschlicher CW, Partridge H (1998) J Chem Phys 109: 8430
21. Piechota J, Suffczynski M (1997) Z Phys Chem 200: 39
22. Drechsler G, Boesl U, Bäßmann C, Schlag EW (1997) J Chem Phys 107: 2284
23. Harvey JN, Heinemann C, Fiedler A, Schröder D, Schwarz H (1996) Chem Eur J 2: 1230
24. (a) Siegbahn PEM (1993) Chem Phys Lett 201: 15; (b) Siegbahn PEM, Blomberg MRA, Svensson M (1993) J Phys Chem 97: 2564
25. Pavlov M, Blomberg MRA, Siegbahn PEM, Wesendrup R, Heinemann C, Schwarz H (1997) J Phys Chem A 101: 1567
26. Pyykkö P (1988) Chem Rev 88: 563
27. Sändig N, Koch W (1998) Organometallics 17: 2344
28. Berkowitz J, Ellison GB, Gutman D (1994) J Chem Phys 98: 2744
29. Lias SG, Bartmess JE, Liebman JF, Holmes JL, Levin RD, Mallard WG (1988) J Phys Chem Ref Data, 17: Suppl 1
30. Hippe D, Peyerimhoff SD (1992) Mol Phys 76: 293
31. Rodgers MT, Walker B, Armentrout PB (1999) Int J Mass Spectrom 182/183: 99
32. Barton DHR (1998) Tetrahedron 54: 5805
33. Barton DHR, Boivin J, Gastiger M, Morzycki J, Hay-Motherwell RS, Motherwell WB, Obzalik N, Schwarzentruber KM (1986) J Chem Soc Perkin Trans 1: 947
34. (a) Minisci F, Fontana F, Araneo S, Recupero F, Zhao L (1996) Synlett 119; (b) MacFaul PA, Ingold KU, Wayner DDM, Que L (1997) J Am Chem Soc 119: 10,594
35. For a general debate of this aspect, see: Walling C (1998) Acc Chem Res 31: 155; MacFaul PA, Wayner DDM, Ingold KU (1998) Acc Chem Res 31: 159
36. Kim J, Harrison RG, Kim C, Que L (1996) J Am Chem Soc 118: 4373
37. Jackson TC, Carlin TJ, Freiser BS (1986) J Am Chem Soc 108: 1120
38. Chen Y-M, Clemmer DE, Armentrout PB (1994) J Phys Chem 98: 11,490
39. Irigoas A, Fowler JE, Ugalde JM (1998) J Phys Chem 102: 293
40. Harvey JN, Diefenbach M, Schröder D, Schwarz H (1999) Int J Mass Spectrom 182/183: 85
41. (a) Ziegler T, Li J (1995) Organometallics 14: 214; (b) Limberg C, Köppe R, Schnöckel H (1995) Angew Chem 110: 512; Angew Chem Int Ed Engl 37: 496; (c) Deng L, Ziegler T (1996) Organometallics 15: 3011
42. Mayer JM (1998) Acc Chem Res 31: 441
43. Kaim W, Schwederski B (1991) Bioanorganische Chemie, Teubner, Stuttgart
44. Capron L, Mestdagh H, Rolando C (1998) Coord Chem Rev 178–180: 269
45. Plattner DA (1999) Angew Chem 111: 86; Angew Chem Int Ed 38: 82
46. (a) Schröder D, Fiedler A, Ryan MF, Schwarz H (1994) J Phys Chem 98: 68; (b) Clemmer DE, Chen Y-M, Khan FA, Armentrout PB (1994) J Phys Chem 98: 6522; (c) Baranov V, Javahery G, Hopkinson AC, Bohme DK (1995) J Am Chem Soc 117: 12801; (d) Schröder D, Schwarz H, Clemmer DE, Chen Y, Armentrout PB, Baranov VI, Bohme DK (1997) Int J Mass Spectrom Ion Processes 161: 175
47. (a) Danovich D, Shaik S (1997) J Am Chem Soc 119: 1773; (b) Filatov M, Shaik S (1998) J Phys Chem A 102: 3835
48. (a) Yoshizawa K, Shiota Y, Yamabe T (1997) Chem Eur J 3: 1160; (b) Yoshizawa K, Shiota Y, Yamabe T (1998) J Am Chem Soc 120: 564; (c) Yoshizawa K, Shiota Y, Yamabe T (1999) J Chem Phys 111: 538
49. (a) Schröder D, Schwarz H (1992) Helv Chim Acta 75: 1281; (b) Becker H, Schröder D, Zummack W, Schwarz H (1994) J Am Chem Soc 116: 1096
50. (a) Schwarz J, Schwarz H (1995) Helv Chim Acta 78: 1013; (b) Schwarz J (1996) PhD Thesis, TU Berlin D83, Shaker, Aachen
51. Yoshizawa K, Shiota Y, Yamabe T (1999) J Am Chem Soc 121: 147
52. Harris N, Shaik S, Schröder D, Schwarz H (1999) Helv Chim Acta 82: 1784

53. Yoshizawa K, Shiota Y, Yamabe T (1998) Organometallics 17: 2825
54. Irigoras A, Fowler JE, Ugalde JM (1999) J Am Chem Soc 121: 574 and references cited therein
55. (a) Schröder D, Heinemann C, Schwarz H, Harvey JN, Dua S, Blanskby SJ, Bowie JH (1998) Chem Eur J 4: 2550; (b) Rue C, Armentrout PB, Kretzschmar I, Schröder D, Harvey JN, Schwarz H (1999) J Chem Phys 110: 7858
56. (a) Shaik S, Filatov M, Schröder D, Schwarz H (1998) Chem Eur J 4: 193; (b) Filatov M, Harris N, Shaik S (1999) J Chem Soc Perkin Trans 2: 391
57. Toy PH, Newcomb M Hollenberg PF (1998) J Am Chem Soc 120: 7719
58. Jin N, Groves JT (1999) J Am Chem Soc 121: 2923
59. Clemmer DE, Aristov N, Armentrout PB (1993) J Phys Chem 97: 544
60. Schröder D, Wesendrup R, Schalley CA, Zummack W, Schwarz H (1996) Helv Chim Acta 79: 123
61. Schwarz J, Wesendrup R, Schröder D, Schwarz H (1996) Chem Ber 129: 1463
62. (a) Brönstrup M, Schröder D, Schwarz H (1999) Can J Chem 77: 774; (b) Brönstrup M, Schröder D, Schwarz H (2000) Chem Eur J 6: 91
63. Bakhtiar R, Jacobson DB (1996) J Am Soc Mass Spectrom 7: 938
64. (a) Brönstrup M, Kretzschmar I, Schröder D, Schwarz H (1998) Helv Chim Acta 81: 2348; (b) Brönstrup M, Schröder D, Schwarz H (1999) Chem Eur J 5: 1176
65. Seemeyer K, Schröder D, Kempf M, Lettau O, Müller J, Schwarz H (1995) Organometallics 14: 4465
66. (a) Sievers MR, Armentrout PB (1998) Int J Mass Spectrom 179/180: 103; (b) Sievers MR, Armentrout PB (1998) J Phys Chem A 102: 10,574; (c) Sievers MR, Armentrout PB (1998) Int J Mass Spectrom 185/187: 117; (d) Sievers MR, Armentrout PB (1999) Inorg Chem 38: 397
67. Sievers MR, Armentrout PB (1995) J Chem Phys 102: 754
68. (a) Helmer M, Plane JMC (1994) J Chem Soc Farady Trans 90: 31; (b) Helmer M, Plane JMC (1994) J Chem Soc Farady Trans 90: 395; (c) Plane JMC, Rollason RJ (1996) J Chem Soc Farady Trans 92: 4371; (d) Rollason RJ, Plane JMC (1998) J Chem Soc Farady Trans 94: 3067
69. (a) Fan J, Wang L-S (1995) J Chem Phys 102: 8714; (b) Allen MD, Ziurys LM, Browm JM (1996) Chem Phys Lett 257: 130; (c) Wu H, Desai SR, Wang L-S (1996) J Am Chem Soc 118: 5296; (d) Knight LB, Babb R, Ray M, Banisaukas TJ, Russon R, Dailey RS, Davidson ER (1996) J Chem Phys 105: 10,237; (e) Wu H, Wang L-S (1997) J Chem Phys 107: 16; (f) Wu H, Wang L-S (1997) J Chem Phys 107: 8221; (g) Loock H-P, Simard B, Wallin S, Linton C (1998) J Chem Phys 109: 8980; (h) Li X, Wang LS (1998) J Chem Phys 109: 5264
70. Zhou M, Andrews L (1999) J Phys Chem 103: 2066
71. Zhou, M, Liang B, Andrews L (1999) J Phys Chem A 103: 2013
72. (a) Parnis JM, Lafleur RD, Rayner DM (1994) Chem Phys Lett 218: 544; (b) Parnis JM, Lafleur RD, Rayner DM (1995) J Phys Chem 99: 673; (c) Campbell ML (1998) J Chem Soc Faraday Trans 94: 353
73. (a) Kretzschmar I, Fiedler A, Harvey JN, Schröder D, Schwarz H (1997) J Phys Chem A 101: 6252; (b) Kretzschmar I, Schröder D, Schwarz H (1997) Int J Mass Spectrom Ion Processes 167/168: 103
74. Irikura KK, Beauchamp JL (1991) J Phys Chem 95: 8344
75. Wesendrup R, Schwarz H (1995) Angew Chem 107: 2176; Angew Chem Int Ed Engl 34: 2033
76. Beyer M. (1996) Dissertation, Technische Universität München
77. Irikura KK, Beauchamp JL (1989) J Am Chem Soc 111: 75
78. Fialko EF, Kikhtenko AV, Goncharov VB, Zamaraev KI (1997) J Phys Chem 101: 5772
79. Achatz U, Beyer M, Joos S, Fox BS, Niedner-Schatteburg G, Bondybey VE (1999) J Phys Chem (submitted)
80. Wesendrup R, Schröder D, Schwarz H (1994) Angew Chem 106: 1232; Angew Chem Int Ed Engl 33: 1174
81. Shi Y, Ervin KM (1998) J Chem Phys 108: 1757

82. (a) Cornehl HH, Wesendrup J, Harvey JN, Schwarz H (1997) J Chem Soc, Perkin Trans 2: 2283; (b) Cornehl HH, Wesendrup R, Diefenbach M, Schwarz H (1997) Chem Eur J 3: 1083
83. (a) Gibson JK (1997) Organometallics 16: 4214; (b) Gibson JK (1998) J Am Chem Soc 120: 2683; (c) Gibson JK, Haire RG (1998) J Phys Chem A 102: 10746; (d) Gibson JK (1999) Inorg Chem 38: 165
84. Schröder D, Fiedler A, Herrmann WA, Schwarz H (1995) Angew Chem 107: 2714; Angew Chem Int Ed Engl 34: 2517
85. Fiedler A, Kretzschmar I, Schröder D, Schwarz H (1996) J Am Chem Soc 118: 9941
86. Schröder D, Fiedler A, Schwarz J, Schwarz H (1994) Inorg Chem 33: 5094
87. Herrmann WA, Fischer RW, Scherer W, Rauch MU (1993) Angew Chem 105: 1209; Angew Chem Int Ed Engl 32: 1157
88. Wu H, Desai SR, Wang L-S (1995) J Chem Phys 103: 4363
89. (a) Downs AJ, Greene TM, Gordon CM (1995) Inorg Chem 34: 6191; (b) Andrews L, Chertihin GV, Ricca A, Bauschlicher CW (1996) J Am Chem Soc 118: 467; (c) Cao Z, Duran M, Solá (1997) Chem Phys Lett 274: 411
90. Siegbahn PEM (1993) J Phys Chem 97: 9096
91. Clemmer DE, Dalleska NF, Armentrout PB (1992) Chem Phys Lett 190: 259
92. Walsh MB, King RA Schaefer HF (1999) J Chem Phys 110: 5224
93. Chertihin GV, Andrews L (1995) J Phys Chem 99: 6356
94. Heinemann C, Cornehl HH, Schröder D, Dolg M, Schwarz H (1996) Inorg Chem 35: 2463
95. Wu H, Wang LS (1998) J Chem Phys 108: 3310
96. Bakac A, Espenson JH (1993) Acc Chem Res 26: 519
97. Cassady CJ, McElvany SW (1992) Organometallics 11: 2367
98. (a) Fokkens H, Gregor IK, Nibbering NMM (1991) Rapid Commun Mass Spectrom 5: 368; (b) van den Berg KJ, Ingemann S, Nibbering NMM, Gregor IK (1994) Rapid Commun Mass Spectrom 8: 895
99. Oliveira MC, Marcalo J, Vieira MC, Almoster Ferreira MA (1999) Int J Mass Spectrom 185–187: 825
100. Addison CC, Johnson BFG, Logan N (1965) J Chem Soc 4490
101. (a) Cao S, Duran M, Solà M (1997) Chem Phys Lett 274: 411; (b) Kellogg CB, Irikura KK (1999) J Phys Chem A 103: 1150
102. (a) Schröder D, Schwarz H (1993) Angew Chem 105: 1493; Angew Chem Int Ed Engl 32: 1420; (b) Boissel P, Marty P, Klotz A, de Parseval P, Chaudret B, Serra G (1995) Chem Phys Lett 242: 157
103. Bauschlicher CW, Langhoff SR, Partridge H, Sodupe M (1993) J Phys Chem 97: 856
104. Hrušák J, Koch W, Schwarz H (1994) J Phys Chem 101: 3898
105. Sülzle D, Schwarz H, Moock KH, Terlouw JK (1991) Int J Mass Spectrom Ion Processes 108: 269
106. Jackson P, Fisher KJ, Willett GD (1999) Int J Mass Spectrom (submitted)
107. Keese RG, Chen B, Harms AC, Castleman AW (1993) Int J Mass Spectrom Ion Processes 123: 225
108. (a) Almond MJ, Atkins RW (1994) J Chem Soc Dalton Trans 835; (b) Chertihin GV, Bare WD, Andrews L (1997) J Phys Chem A 101: 5090
109. Beyer M, Berg C, Albert G, Achatz U, Joos S, Niedner-Schatteburg G, Bondybey VE (1997) J Am Chem Soc 119: 1466
110. Schröder D, Schwarz H (1991) Angew Chem 103: 987; Angew Chem Int Ed Engl 30: 991
111. Mazurek U, Schröder D, Schwarz H (1998) Coll Czech Chem Comm 63: 1498
112. Walba DM, DePuy CH, Grabowski JJ, Bierbaum VM (1984) Organometallics 3: 495
113. Schröder D, Herrmann WA, Fischer RW, Schwarz H (1992) Int J Mass Spectrom Ion Processes 122: 99

Quantum Chemical Studies on Metal-Oxo Species Related to the Mechanisms of Methane Monooxygenase and Photosynthetic Oxygen Evolution

P.E.M. Siegbahn[1], Robert H. Crabtree[2]

[1] Department of Physics, Stockholm University, Box 6730, S-113 85 Stockholm, Sweden
[2] Yale Chemistry Department, PO Box 208107, 225 Prospect St., New Haven, CT 06520-8107, USA
[1] *E-mail: ps@physto.se*
[2] *E-mail: robert.crabtree@yale.edu*

During the last few years it has become possible to study enzymatic mechanisms by means of high accuracy quantum chemistry methods. In the present review, examples are given of applications of density functional theory to two mechanisms involving redox active centers. In methane monooxygenase an iron dimer complex oxidizes methane to methanol. A hydrogen abstraction mechanism is suggested involving a high-valent Fe(IV,IV) intermediate, termed **Q**. In Photosystem II an oxyl radical mechanism is suggested for the formation of O_2. Different models of the oxygen evolving cluster are discussed. Spin-state considerations are emphasized.

Keywords: Density functional theory, Methane monooxygenase, Photosystem II, Spin-state crossings, Oxyl radicals

1 Introduction

Oxidation chemistry is one of the most difficult problems within the general area of inorganic and bioinorganic reaction mechanisms. Even in model systems, experimental evidence is often hard both to obtain and to interpret – the difficulties only increase on moving to enzymes. At least in methane monooxygenase (MMO), X-ray crystal structures of the enzyme are available as a guide but in Photosystem II (PSII) the structure of the key oxygen-evolving tetramaganese cluster (OEC) is still uncertain.

These factors make quantum chemical studies an indispensable component of any serious attack on these mechanistic problems. Thanks to the large improvement of quantum chemical methods in recent years, chemically relevant, reliable results can now be obtained even on paramagnetic clusters of the types found in these enzymes. At the simplest level, such studies can suggest which proposed intermediate structures have energies either so high that they can be eliminated as possibilities or so low that they can best be considered as particularly likely intermediates in the enzyme mechanism. In many cases, a good estimate of the kinetic barriers separating plausible intermediates can also be obtained, so that the plausibility of alternate kinetic pathways can also be compared.

2
The Spin State Problem

In addition, quantum chemical studies can sometimes be interpreted to suggest general principles or trends that must apply to any mechanism proposed. As an example of a general point of this sort, it has only been very recently that the problem of alternate spin states has received close attention [1]. Experimental studies of the spectroscopic and magnetic properties of a model complex or enzyme intermediate can normally only give information on the ground spin state of a system. It may be, however, that an unobserved excited spin state of the cluster may prove to have sufficiently higher reactivity than the ground state that the entire reaction pathway may pass via the excited spin state. Quantum chemistry is one of the few ways of probing this issue, which is discussed below in relation to both MMO and PSII.

Experiments and calculations on small unsaturated transition metal complexes have shown that spin-crossings are very common in typical redox reactions like oxidative addition. In fact, a spin-crossing is the rule rather than the exception. In a systematic theoretical study of oxidative addition of H_2 to second transition row MH_x complexes, 19 of the 24 reactions studied had to go through a spin-crossing [2]. Experimentally, these spin-transitions are known often to be quite efficient [3, 4]. A prerequisite for a fast transition is normally that at least one of the crossing potential energy surfaces is rather flat [5]. The rate of spin state changes may well be much faster in 'real' condensed phase complexes than in the gas phase small molecules considered in this work. One case from the condensed phase where spin-state changes were shown to have a decisive effect was for oxidative addition of linear alkanes to MCp(CO) for cobalt, rhodium and iridium [6, 7]. The difference in relative positioning of the singlet and triplet states was shown to lead to a high reactivity for Rh and Ir, while Co is entirely unreactive.

A situation which is less common than a single spin-state crossing is one where two crossings occur. Such a situation would be harder to detect experimentally since the spin of the reactants and products could be the same. One of the first times a reaction of this type was demonstrated was for the case of the insertion of a nickel atom into the O—H bond of water [3]. In this case both the reactant and the product are triplets but the barrier on the triplet surface is exceedingly high. In the region of the molecular complex there is therefore an initial transition to a singlet surface which has a low barrier for the insertion. The reaction is completed in the region of the products where the complex returns to the triplet state. Similar situations have more recently been demonstrated by Shaik et al. [8], who emphasized the potential role of differential reactivity of different spin states. In the reaction of gas-phase FeO^+ with a CH bond, Shaik, Schwartz and coworkers suggest that the ground high spin state, which correlates with the ground high spin state of the products, has a much higher barrier than does the excited low spin state. They argue that the reaction is slowed by the necessity to undergo a double spin state conversion from high to low and back. Extrapolating to the case of CH activation by P-450, they also suggest that the low spin state can undergo oxene-like reactions in which the FeO group inserts into the C—H bond to give R—Fe—O—H directly. This would open up several possible pathways to ROH and could in principle account for some of the apparently contradictory experimental data, where the rebound rate exceeds the maximum values required to consider the radical as having a separate existence. Yoshizawa et al. [9] have suggested a similar explanation for the case of MMO. However, there are also entirely different, more probable, explanations for the contradictory experiments as will be discussed further below.

3 Quantum Chemical Studies

It is only in the last 4–5 years that quantum chemical methods have advanced to the point that reliable calculations can begin to be carried out on systems as complicated as enzyme active sites. The most important reason for this development is that Density Functional Theory (DFT) has developed into a much more accurate tool than before. In particular, the introduction of terms depending on the gradient of the density to describe the exchange interaction has proven to improve substantially the accuracy [10, 11]. This improvement, together with the improvement obtained by introducing a few semi-empirical parameters and a part of the Hartree-Fock exchange, led to the B3LYP method which has an accuracy that is not far away from that obtained by the most accurate ab initio methods at a small fraction of the cost [12]. The B3LYP method has been used for most of the DFT studies discussed in the present review. This development has meant that much less effort in the quantum chemical studies has to be spent on reaching sufficient accuracy. Instead, for an enzyme problem most time is often spent on selecting the appropriate model system for study. For testing of different pathways, model ligands (H_2O, NH_3, OH, etc.) are normally used, but once the general patterns have been established, it is often possible to move to larger ligands much closer to the experimental ones (e.g., imidazole for His, HCOO for Asp or Glu) to check that the energies remain reasonable. The results of typical DFT calculations consist of an optimized structure and the energy corresponding to that structure. The structure can sometimes be compared with physical data on model compounds or enzyme intermediates. The energy is used to decide if the structure in question can be fitted into a sensible catalytic cycle. Normally a plausible series of intermediates for an exothermic enzyme reaction must have energies in a relatively smoothly descending pattern, so that no one step has an excessive endo-or exothermicity. One also has to attempt to show that no other rearrangement product has a much lower energy that would give rise to a thermodynamic sink for the system. In such a case the catalytic reaction would be brought to a halt by the build-up of the 'sink' species. Ideally, the barriers to the key conversion steps should be estimated to check feasibility of the pathway. Recent reviews have been written on the study of enzymes by DFT methods and further details of this type of studies can be found there [13, 14].

3.1 Methane Monooxygenase

Monoxygenases use O_2 as primary oxidant but activate it by reduction by two electrons, so the net conversion introduces one oxygen of O_2 into the substrate while the other is reduced to H_2O. MMO brings about the selective hydroxylation of a variety of unactivated CH bonds, such as in the conversion of methane to methanol [15–17]. The product methanol is much more sensitive to further oxidation, however, so a chemical system normally gives unselective oxidation to CO_2. The hydrophobic pocket of MMO expels the product methanol before the iron cluster is activated for a new catalytic cycle. The enzyme active site contains a dimeric non-heme iron cluster not unlike those found in the oxygen binding protein hemerythrin (Hr) and ribonucleotide reductase (RNR), a protein involved in DNA synthesis [18, 19].

Where a C—H bond to be attacked is sufficiently weak, as in camphor, a cytochrome P-450 dependent monooxygenase enzyme can be involved [20], as in P-540cam from *Pseudomonas putida*. In this case, the active site contains a heme – an iron in a porphyrin ring – and the mechanism is believed to go via an iron oxo formally considered as Fe(V) but better described as having an Fe(IV)=O structure with one oxidizing equivalent present as an electron 'hole' in the heme organic π-system. The reaction is believed to proceed by abstraction of an H atom from the substrate by the oxo group to give an iron hydroxo species and a carbon radical. The radical subsequently abstracts the OH group from iron to give the product. Since the overall reaction in the case of MMO is similar, except that the C—H bond of the substrate is unusually strong (104 kcal/mol), it has generally been assumed that a high valent iron oxo was also a key intermediate in MMO, but that its non-heme structure made it more reactive and able to attack methane. In RNR a similar oxo species has been invoked [19] while in Hr [21] the O—O bond of O_2 is never broken because this is a transport protein.

The MMO from *Methylococcus capsulatus* (Bath) consists of three proteins, a hydroxylase (protein A), a reductase (protein C), and a regulatory protein, (protein B). The C component uses NADH as primary reductant that provides the electrons necessary to reduce O_2, the B component allows C to feed these electrons to A, and the A component interacts with O_2 and carries out the substrate oxidation step. Fe_2(II,II), Fe_2(II,III), and Fe_2(III,III) forms of the enzyme are accessible, the first of which interacts with O_2 to give an Fe_2(III,III) peroxo species. Indeed the Fe_2(II,II) form of the A protein is able to carry out one turnover of substrate hydroxylation without the need for proteins B or C. The active site iron cluster has an Fe···Fe distance of 3.4–3.5 Å in its isolated forms. A crystal structure [22] shows three bridges between the iron atoms: a glutamate, a hydroxo bridge, and an acetate that is believed to be an artefact of the isolation procedure.

By analogy with P-450 an Fe_2(IV,IV) oxo species might be involved. Consistent with this proposal, the Fe_2(III,III) form is weakly active with H_2O_2 as substrate [23]. Strong evidence for non-heme iron oxo species have been obtained in relevant model systems [24].

Rather unexpected results were obtained for *M. trichosporium* MMO with chiral ethane (CH_3CHDT) as substrate, where 35% inversion of configuration occurred and a deuterium kinetic isotope effect (KIE) of 4.2 was observed [25]. The quantitative interpretation of the data was complicated by the apparent requirement for an unreasonably short radical lifetime [26, 27].

Rapid freeze-quench techniques allowed a series of intermediates to be identified, including the key species, denoted **Q**, that appears to be the active oxidant that attacks methane. Mössbauer spectroscopy on **Q** suggests it is an Fe_2(IV,IV) species. As well as discussing the mechanistic data, Feig and Lippard list essentially all the mechanistic variants currently proposed [15]. In 1996 Wallar and Lipscomb [16] and in 1997 Valentine and Lippard [28] reviewed the most recent developments in the area.

Theoretically, Yoshizawa et al. [29] used the extended Hückel method to look at binuclear iron clusters relevant to MMO. They considered CH activation both by bridging peroxide and terminal oxo intermediates and found the latter to be preferred. A C_{3v} distortion of CH_4, earlier suggested by Shestakov and Shilov, [30] was assigned a role in helping to activate CH_4. An interaction between a filled a_1 orbital of a C_{3v}-distorted methane and empty d-orbitals on Fe was considered to be particularly important. In addition, 5-coordination of at least one of the Fe atoms was considered best able to provide a suitable site for the reaction. Yoshizawa et al. [9] later amplified these studies, proposing that the activation of methane C—H bond by the high valent $Fe_2(\mu\text{-}O)_2$ diamond core of compound **Q** involves formation of a **Q**(CH_4) complex with both an Fe—C bond and an Fe—H bond, followed by a concerted H atom abstraction via a four-centered transition state.

In the first DFT study of MMO a quite simple model of the iron dimer complex was used [31]. Instead of the actual glutamates and histidines, hydroxyl and water ligands were used. The number of hydroxyl ligands was chosen to give an overall neutral complex with the desired oxidation states. For MMO there are very good experimental indications that the iron dimer complexes involved are all neutral. For example, normal charge counting on the reduced Fe_2(II,II) complex with four carboxylates and two imidazoles leads to a neutral complex. The starting structure for a possible compound **Q** iron dimer structure was furthermore chosen with two six-coordinated irons and two μ-oxo bridges. At the end of the B3LYP optimization this led to two five-coordinated irons instead. One of the resulting OH···H_2O bridges was then replaced by a carboxylate bridge in analogy to the protein complex. This leads to a suggested structure **1** for compound **Q**.

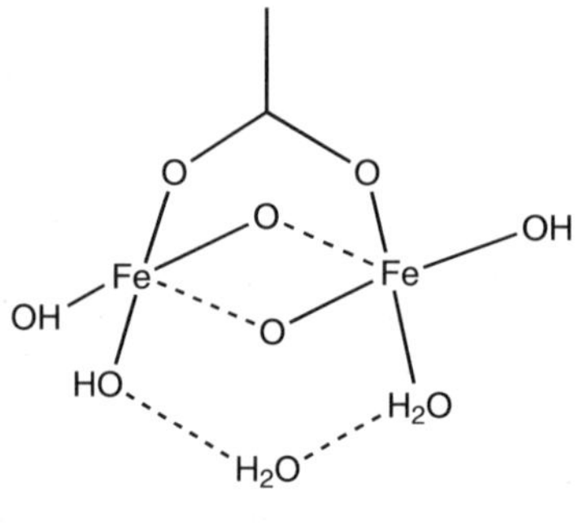

1

The optimized structure using ferromagnetic spin coupling has at least one striking feature and this is the quite different Fe—O distances of the diamond core obtained, with two short bonds of 1.74 Å and 1.77 Å and two longer ones of 2.00 Å and 2.05 Å. This was explained by the presence of Jahn-Teller (JT) distortions with the weak JT axis going through the long Fe—O bond on one side of the iron atoms and with an empty site on the other side, leading to 5-coordination for the irons. Independently an EXAFS measurement suggested a very similar structure [32]. In addition, EXAFS gave an unusually short Fe—Fe distance for compound **Q** of only 2.46 Å. The Fe—Fe distance in the optimized geometry is 2.79 Å which is quite short, but not as short as the one measured by EXAFS, which suggests that some feature of the bonding may have been missing in this model. Very similar models to study MMO by DFT methods have also been used recently by Basch et al. [33], yielding similar structures.

Still using a simple iron dimer model as in **1**, several new possible structures for compound **Q**, not considered in the previous study, were investigated in a more recent study [34]. In the most interesting of these structures there are two, rather than one, bridging carboxylate. When ferromagnetic coupling was used this led to a shortening of the Fe—Fe distance from 2.79 Å to 2.74 Å. When d-functions were added on the oxygens the bond distance decreased to 2.67 Å. In the final attempt to make the Fe—Fe bond still shorter, antiferromagnetic spin-coupling was used. This is very much harder to converge but after some effort the proper spin-coupling was achieved. With d-functions on oxygen, the optimization converged to a structure with a short Fe—Fe distance of 2.47 Å, fortuitously close to the EXAFS result. The effect of using antiferromagnetic spin-coupling is thus a shortening by as much as 0.20 Å. Further analysis of this effect shows that the spins on the irons have not simply been antiferromagnetically coupled but they have also been substantially reduced in size. It turns out that it is this reduction, leading to low-spin coupling of the individual irons, rather than the antiferromagnetic coupling itself, that causes the reduction in the Fe—Fe distance. In fact, ferromagnetic coupling of two low-spin irons give a nearly identical structure and a very similar total energy as the antiferromagnetically coupled singlet.

At the next higher level of modeling, the actual functional groups of the glutamates and the histidines were included in the calculation. Using antiferromagnetic coupling this model converged to the structure in Fig. 1. The Fe—Fe distance of 2.54 Å is reasonably close to the experimental EXAFS distance of 2.46 Å, considering the model used and the fact that, for example, relativistic effects are not included. It is important to note that the structures of the simple model, **1**, and the more advanced one in Fig. 1 have many similarities, which supports the use of the simple models in the previous study. Even though the symmetry is lower for the larger model and that both irons are 6-coordinated and some of the ligands are nitrogen derived, the JT distortions are, for example, quite similar. The spin distribution is also quite similar, but the asymmetry of the complex causes the two μ-oxo oxygens to be different with the oxygen *trans* to the histidines having a somewhat higher spin, +0.57 compared to −0.37, which should make it more reactive. Since the oxygen with the higher spin is the one pointing towards the substrate pocket in MMO, this could be an important effect.

Once a model is decided upon for compound **Q**, the MMO substrate hydroxylation reaction can also be studied. This reaction has been studied extensively experimentally and several possible mechanisms have been suggested [15, 16]. These mechanisms fall essentially into two categories, radical and non-radical mechanisms. In the radical mechanisms the first step is a hydrogen abstraction from the hydrocarbon, while the non-radical mechanisms suggest a concerted insertion pathway. Starting the theoretical investigation of this reaction with the simple model, **1**, a methane molecule was moved towards the empty coordination site of one of the irons. A four-

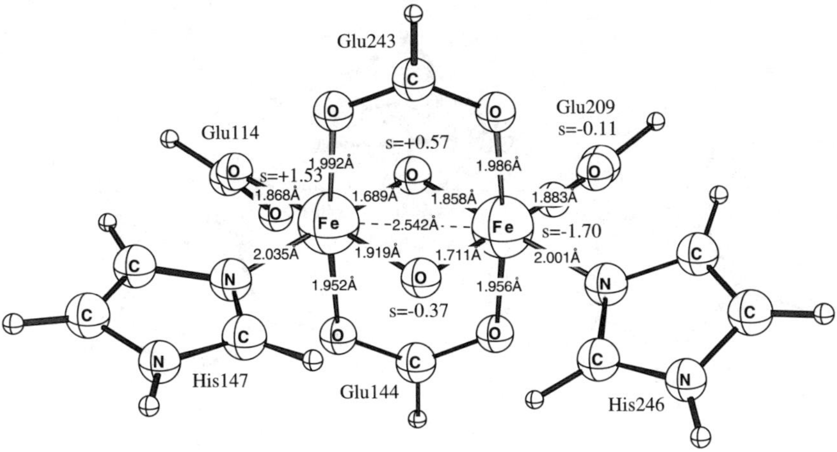

Fig. 1. Suggested model for compound **Q** of MMO with two bridging carboxylates. The structure was obtained using antiferromagnetic spin coupling (^1A)

centered starting structure for the transition state was the initial guess, with one Fe—CH_3 bond and one FeO—H bond forming in a concerted manner. As the optimization proceeded the structure changed and eventually converged to a transition state of a pure hydrogen abstraction type. However, a rearrangement is necessary, where one of the bridging μ-oxo groups move out to form a terminal oxo ligand which then abtracts a hydrogen atom from methane.

The larger model for compound **Q** in Fig. 1 is much less flexibly coordinated than the previous smaller one. If the doubly bridging carboxylates are kept during the reaction, there are not many places on the complex where a methane molecule could be activated. By far the most plausible place for activation is at the μ-oxo bridge *trans* to the histidines. This is close to the hydrophobic pocket in the protein to which methane can diffuse and where it will fit. Several spin-states (^1A, ^3A, etc.) were investigated but no low energy activation of methane was found until the ferromagnetic ^9A state was tried. On this surface there is a smooth activation of methane and the fully optimized transition state is shown in Fig. 2. The calculated barrier height for activating methane including promotion and zero-point effects is about 10 kcal/mol.

The transition state shown in Fig. 2 is clearly one of an almost pure hydrogen abstraction. This is quite different from the results of other B3LYP studies using different types of models [9]. A major feature of these other

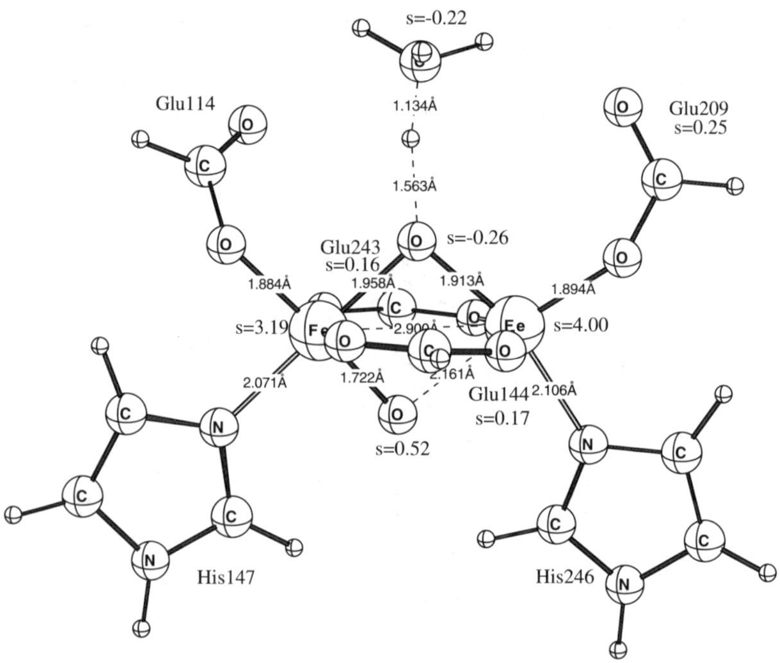

Fig. 2. Optimized transition state for CH_4 activation in MMO. The structure was obtained using ferromagnetic spin coupling (^9A)

models is that they have a net positive charge and are coordinatively unsaturated. One of these models is FeO^+, while the other one is more realistic with one four-coordinated and one five-coordinated iron atom and a total charge of +3. With these models a four-centered transition state is found. The reason for this is the large electrostatic attraction obtained between methane and the positive metal complex, which leads to deep minima for molecular methane of 19.5 kcal/mol for FeO^+ and of 27.3 kcal/mol for the larger model. No such minimum was obtained for the model in Fig. 2.

From the above studies it can thus be concluded that the transition state for activating methane in MMO is most likely one of a pure hydrogen abstraction. This is in line with the conventional rebound mechanism [16] but appears inconsistent with the very short lifetimes measured by radical clocks [26, 27]. As described above, an explanation for this contradiction involving an insertion transition state is considered very unlikely since all higher level models point in the same direction of a pure hydrogen abstraction. Instead, the most likely explanation for the contradictory experiments is the following. Very shortly after the pure hydrogen abstraction transition state is passed, there could be a crossing with the potential surface leading to the methanol product, which is extremely attractive and lacks any additional barrier. In fact, if this crossing occurs before the minimum of the methyl radical is reached, no free radical is likely to be found by the radical clock experiments. In line with this scenario, the methanol surface is found to be quite low in the region after the hydrogen abstraction transition state. No exact crossing point has as yet been determined for MMO since this is a quite difficult problem in many dimensions. However, in the case of P-450 Shaik and coworkers have recently suggested a similar reaction path and actually succeeded in determining a spin-crossing point which does explain the contradictory experimental data [35].

3.2 The Oxygen Evolving Center of Photosystem II

The mechanism of dioxygen evolution by the Oxygen Evolving Center of Photosystem II (OEC) in green plants, the mechanistic steps involved and, in the absence of a crystal structure, even the structure of the OEC itself are still uncertain [36]. The OEC itself consists of one calcium and four manganese ions per PSII complex, all essential for oxygen evolution, and one chloride cofactor, removal of which does not completely suppress O_2 evolution [37]. An EXAFS peak assigned to an Mn···Mn separation of 2.7 Å [36a], suggests that the manganese ions are arranged in two pairs of $Mn(\mu\text{-}O)_2Mn$ clusters, a very common structure in inorganic model compounds [36f–h]. The two Mn dimers, the Cl and the Ca are believed to be close to and possibly even linked together in the OEC. Water is believed to bind and undergo oxidation at Mn, but the roles of chloride and calcium are much less well defined [36d], chloride usually being considered as a probable ligand for Mn, or Ca, or both.

In photosynthetic oxygen evolution, each photon absorbtion event leads to an electron transfer which oxidizes the P680 chromophore of PSII to the $P680^+$ cation radical. This radical in turn oxidizes the OEC by one electron via the intermediate oxidation of a tyrosine residue, denoted Tyr_z. From a saturating flash experiment, which showed maximal O_2 evolution on the third, seventh, and eleventh flashes, Kok et al. [38] proposed a generally accepted scheme in which the OEC accumulates oxidizing equivalents by advancing it in one-electron steps from an S_0 oxidation level through S_1, S_2, and S_3 until at the most oxidized S_4 oxidation level, O_2 is released and the cluster returns to S_0. EPR spectroscopy suggests that alternative states of the cluster are possible at the S_2 state because two different EPR signals (the multiline signal at $g = 2$ and the $g = 4.1$ signal) are seen [39] depending on the temperature and sample preparation. The invariance of the EXAFS Mn K-edge on going from S_2 to S_3 suggests the Mn atoms are not oxidized in this step [40a]; contrary conclusions have been reported [40b,c].

In a critical experiment, Messinger and coworkers [41] have shown by $H_2{}^{18}O$ labeling that, of the two O atoms in S_2 and S_3 that become O_2 after S_4, one is in a slowly exchanging site and the other in a very rapidly-exchanging site. The slow exchange site has been considered to be a terminal oxo, and the fast exchange site a water or hydroxo group. In any case a symmetrical Mn=O + O=Mn mechanism seems to be unlikely.

Ca-depletion abolishes O_2-evolution but reconstitution with Sr partially restores activity and modifies the EPR multiline spectrum of the S_2 state of the Mn cluster, suggesting that Ca may be close to the Mn cluster. Strontium EXAFS experiments on PSII samples with Sr exchanged for Ca have shown two Mn—Sr distances of 3.5 Å [42]. It is believed that strontium has taken the position normally occupied by calcium. Ca-depleted samples are inactive because S_2Tyr_z is the highest state that can be attained. Replacement of chloride by bromide [43] does not affect activity but iodide partially and fluoride completely inhibits O_2 evolution. Fluoride binding appears to elongate the 2.7 Å EXAFS Mn···Mn vector to 2.8 Å, suggesting binding at or near the Mn cluster.

Important information for making Mn oxidation state assignments for the S states comes from NMR proton relaxation studies (NMR-PRE) [44], where Mn(II) and Mn(IV) are strongly relaxing and Mn(III) weakly relaxing ions. Microwave power saturation studies [44b] of the S_2 EPR signal of PSII can also probe the oxidation states involved. In conjunction with XANES data [40], both NMR and EPR results suggest that an Mn(II) exists in S_0, that one Mn(III) is oxidized to Mn(IV) in the S_1 to S_2 conversion, and that no change in Mn oxidation state occurs on going from S_2 to S_3. An extra paramagnet produced in S_3 and coupled to the Mn cluster is thought [44b] to prevent observation of the S_2 multiline spectrum in S_3.

In the process when Tyr_z is oxidized by P680 it loses a proton to become a neutral tyrosyl radical. In some way Tyr_z is recreated by obtaining an electron and a proton in each step of water oxidation. Even though the details of this process are still under debate, this is one of the most important experimental findings on which the models discussed below are built. Independently of the mechanism of this process, it means that the energy available to the water oxidizing complex in each step is approximately equal to the bond strength of the Tyr_z O—H bond, which is equal to 86.5 kcal/mol. This quantity can be modified, but only slightly, by changes of the charge of the cluster and changes in hydrogen bonding occurring during the *S*-state transitions. Two leading models for recreation of Tyr_z exist. In the first, termed the hydrogen abstraction mechanism, [45], see Fig. 3a, the tyrosyl radical obtains the proton and electron from a water molecule coordinated to manganese in a hydrogen atom transfer step. In the second model, the electron transfer model, see Fig. 3b, the tyrosyl radical obtains the electron from the manganese complex and the proton from a nearby protonated base. In that model, the water molecules which will eventually form water will lose their protons to a different base. This model has recently been elaborated further and some new aspects and modifications of the proton translocation mechanism have been introduced [46]. The models for water oxidation, discussed below, are independent of which of these schemes is the correct one.

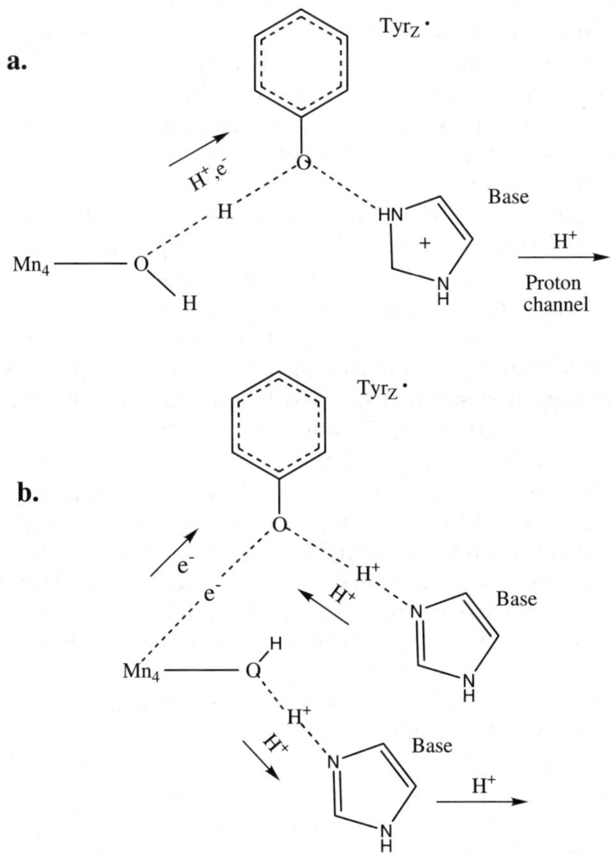

Fig. 3a,b. Schematic picture of: **a** the hydrogen abstraction scheme; **b** the electron transfer scheme

Not only did Babcock et al. [45a–e] propose H atom abstraction from bound water as the key step in O—H bond breaking but they also suggested a role for metalloradicals in O_2 formation via an Mn(IV)=O/(HO)Mn(IV) coupling induced by H transfer to Tyr_z. Baldwin and Pecoraro [45f] have experimental evidence from model compounds that μ—O—H bonds are relatively weak in Mn_2(III,IV) dimers, consistent with H abstraction being important. Yachandra et al. have considered a pair of bridging oxyl radicals as precursors to O—O bond formation [36a]. Both Pecoraro et al. [45g] and Brudvig et al. [47b] have discussed O—O bond formation via nucleophilic attack by a hydroxo group on an Mn(V)=O oxo species. The proposed overall requirement for charge neutrality has also been emphasized by several authors as most appropriate for a cluster buried deep in the low dielectric medium of the protein [45, 48].

Krishtalik [49] applied a thermodynamic analysis to the problem and came to several conclusions about the O_2-evolution mechanism. He proposed that the barrier produced by the repulsion of unbound O-atoms during their mutual approach may be partly surmounted using the binding energy of water molecules to the manganese ions. The concerted transfer of electrons and protons [45, 48] involving the participation of bases stronger than water was considered to improve the process energetics. The most likely reaction path was considered to be rate-determining two-electron water oxidation to the hydrogen peroxide oxidation level, followed by two fast oxidation steps of H_2O_2 to HO_2 and on to O_2.

In 1992, Proserpio et al. [50] looked at possible O_2 evolution mechanisms with the extended Hückel method. The oxygen-oxygen coupling process was examined starting from some known Mn cluster types and assuming oxo-oxo coupling as the route to O_2. Tetranuclear models were also considered. An in-plane approach of two oxo ligands was proposed to have the lowest barrier for peroxide bond formation. A slight energetic advantage was suggested in the peroxide to dioxygen step when the oxo ligands are coordinated to at least three Mn ions compared to coordination to only two Mn ions. No clear cut preferred mechanism could be devised, however. In other early work, Proserpio et al. [51] also used the extended Hückel method to study the problem. The energies of two complexes containing peroxo groups bound at a dinuclear fragment within a tetranuclear cluster ('dimer-of-dimers' model) were minimized using the universal force field technique. The equilibrium geometries of the peroxo bound tetranuclear models indicated asymmetric coordination of the peroxo oxygen with Mn—O—O—Mn torsion angles of ca. 60°, consistent with ground state (triplet) oxygen release by PSII.

A recent series of papers [48, 52–54] developed a quantum chemical picture of the photosynthetic O_2 evolution problem. One paper [48a] examined the energetic requirements for breaking O—H bonds of water and hydroxide ligands coordinated to manganese. Another reported initial proposals on how the O—O bond might be formed [52]. In the third and fourth papers [53] – the most relevant ones to this review – the properties of the key oxo species that promotes O—O bond formation were discussed. In addition, a number of general principles emerged which strongly constrain possible pathways.

In the first paper [48a] the H abstraction model was tested via quantum chemical calculations. The B3LYP method was used, and both monomeric and dimeric manganese model systems were studied. It was found that, by coordination to a manganese center, the first O—H bond strength of water is lowered to a value 0.2 kcal/mol lower than that in tyrosine – a critical point as described above. The second hydrogen abstraction energy was quite similar. Since thermoneutrality in the reaction (or a weak exothermicity) is a requirement for the hydrogen abstraction model, these calculations are in accord with this model.

In the second paper [52], several pathways for O—O bond formation were examined. For example, the approach of two Mn=O oxos to form O_2 were investigated, as well as the approach of one Mn—OH hydroxo towards an Mn=O oxo to form a peroxide. All of these attempts led to exceedingly high barriers for O—O bond formation. The reason for this could be traced to the fact that in all cases an oxygen radical, oxyl, is formed early along the reaction pathways, which for the systems investigated required too much energy. Further investigations of O—O bond formation pathways therefore focused on this particular problem (see below).

The third paper [53] proposed a full mechanism based on a few quite general results and principles that were thought to rule out a very large number of otherwise viable pathways. One of these refers to the question of spin states. In redox reactions of a paramagnetic metal cluster, the starting and final species may be on different potential energy surfaces and therefore have different ground spin states. In such a case there must be a transfer from the starting surface to the product surface at some point as one goes from starting material to products. In the case of the OEC Mn cluster, a reduction of the Mn ions occurs as the O—O bond is formed in one of the later *S*-states, which means that the local spin on manganese increases in this step. This leads directly to a spin state problem of the type mentioned. To take a specific case, if a 5-coordinate complex $(H_2O)(HO)_3$Mn(V)=O were to convert to $(H_2O)(HO)_2$Mn(III)—O—O—H in one step of O_2 formation, the spin would rise from triplet in the starting material to quintet in the product; similar changes happen whatever the precise structures involved.

In the O—O formation step, the reaction could occur on the potential energy surface of the reactants or on that of the products. In the former case, an excited state of the products will be formed; in the latter, the reaction has to start in an excited state for the reactants. There is in practice no third alternative as has been discussed, for example, for the simple reaction between an Ni atom and water [3] and also by Shaik et al. [1] who recently considered this point in the case of P-450. In the present case of O—O bond formation a large number of model calculations have shown that it is the excited state of the reactants that needs to be low-lying for obtaining a low barrier.

Mn(V) oxo species with biologically relevant ligands have at least two spin states, a triplet with an Mn(V)=O structure and a quintet (or antiferromagnetic triplet) with an Mn(IV)—$O^{\bullet}$ oxyl structure. The most stable structures of Mn(V) oxo are 5-coordinate while those of Mn(IV) oxyl are six coordinate. Six coordination stabilizes the Mn(IV) oxyl form relative to the Mn(V) oxo because Mn(IV), as a d^3 ion, has a greater preference for an octahedral environment than does d^2 Mn(V). As might be anticipated, the Mn(IV) oxyl, having an open shell oxo and an Mn—O single bond, is far more reactive and interest centered on factors that favor formation of this state. Five coordinate Mn(V) oxo species like those that have been isolated and characterized tend to have quintet states at very high energy and so they are weakly reactive, consistent with their isolation as stable species. On the other hand, six-coordinate species have accessible quintet states, consistent with the fact that none has ever been isolated. One experimental system [55] that is thought to involve a six-coordinate Mn(V) oxo species is found to evolve O_2 catalytically, in line with the theoretical expectation. It should furthermore be added that, for complexes where the quintet state is low lying, the triplet state is also found to have oxyl radical character.

A serious problem with this approach is that, given the OH bond strength of TyrO—H, only unreactive oxo complexes can be directly obtained via H abstraction from Mn—OH by $\text{TyrO}^\bullet$, the H abstractor present in the OEC, via the route of Eqs. (1) and (2):

$$\text{Mn(II)(HO—H)} + \text{TyrO} = \text{Mn(III)O—H} + \text{TyrO—H} \quad (1)$$

$$\text{Mn(III)O—H} + \text{TyrO} = \text{Mn(IV)=O} + \text{TyrO—H} \quad (2)$$

Reactive oxo complexes have very strong O—H bonds in the precursors, typically 20 kcal/mol larger than that in tyrosine. No other hydrogen abstraction pathways leading to the same 6-coordinated oxyl complex proved to be possible.

The solution finally proposed is represented by Fig. 4. Only if one of the four Mn ions alone changes oxidation state can a feasible route to the desired oxyl be constructed. The initially 5-coordinate MnA is Mn(II) and bears an H_2O ligand. Two H abstractions give an unreactive 5-coordinate Mn=O at the S_2 state. At this point conversion to 6-coordination was proposed via chelation of the Ca cofactor via a bridging water. A third H abstraction is now possible from a bridging OH_2 to give an oxo or oxyl at S_3. Whether the oxo or oxyl is the ground state is not important – it is only necessary that the oxyl state be accessible. In the key step, $S_3'' \rightarrow S_3'''$, O—O bond formation takes place by a reaction between the oxyl and an outer sphere OH_2 molecule (Fig. 5). In this step an H atom is transferred from the H_2O to a μ-oxo group to form an incipient outer sphere OH radical that can readily form an O—O bond. The resulting Mn(III)—O—O—H species readily loses a fourth H atom to give O_2.

The assignment of the S_0, S_1, and S_2 states are consistent with the main experimental information available, as confirmed by the similarity of the oxidation states suggested with models previously proposed by Klein et al. [36, 40] and by Hoganson and Babcock [45e] who both relied purely on experimental information. In particular, the NMR-PRE [44a], EPR power saturation [44b], and the XANES [40] data indicate Mn cluster oxidation state changes with S state, entirely in accord with all three proposals.

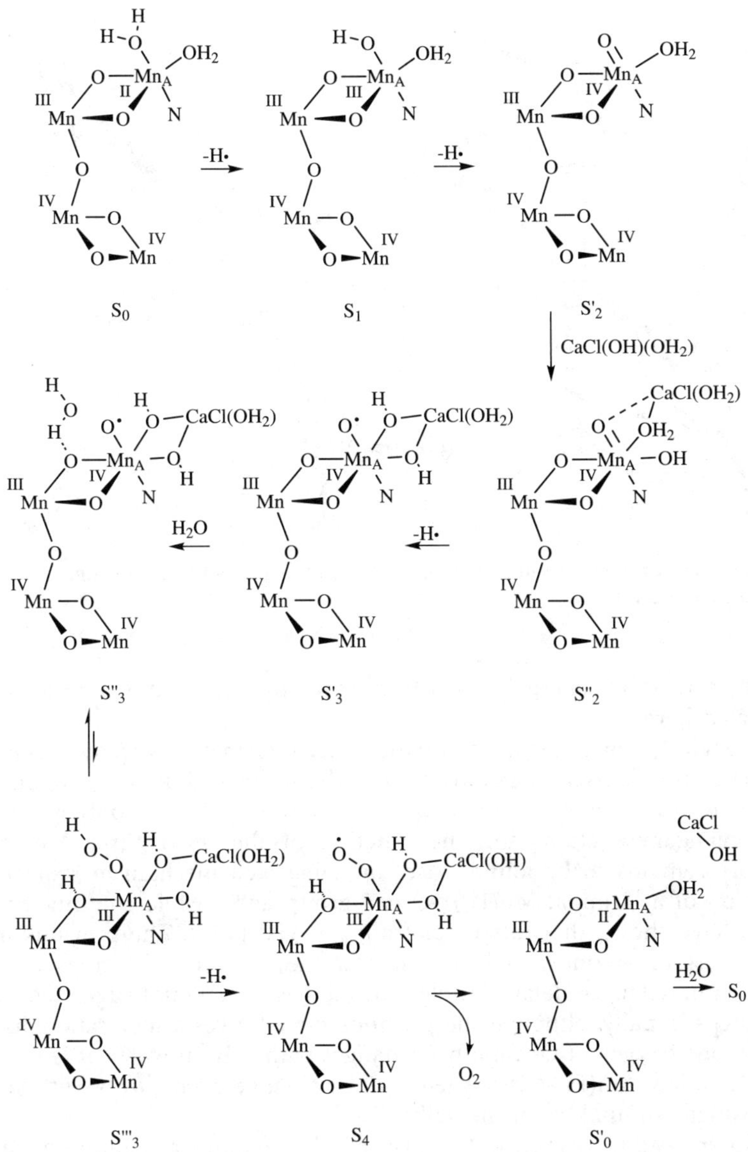

Fig. 4. Proposed sequence of *S*-states using the dimer-of-dimers model for the OEC

The model is also in reasonable accord with the Messinger ^{18}O labeling studies [41] because one O of O_2 comes from the slow-exchanging oxo and the other from the fast-exchanging second coordination sphere water. The paramagnet interacting with the Mn cluster in S_3, proposed in early work [44], is identified here as the oxyl center. Finally, the S_2 to S_3 step has the

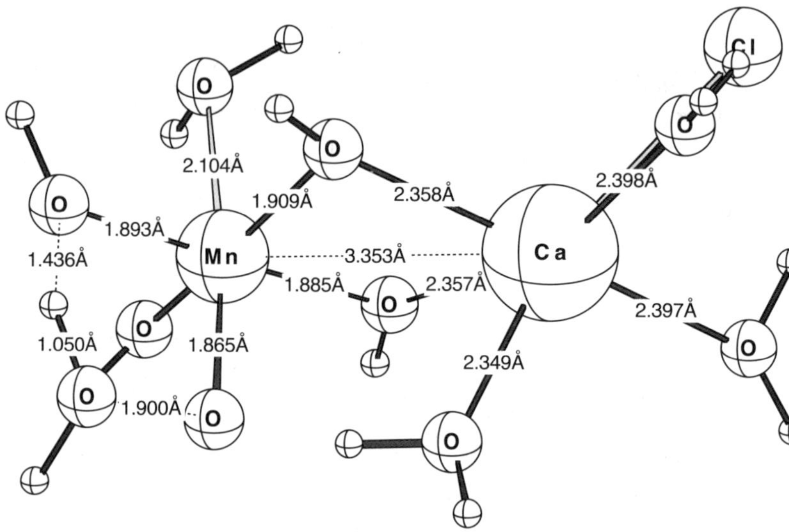

Fig. 5. Structure of the transition state for O—O bond formation using a simple model with one manganese center

highest activation energy, consistent with the structural rearrangement postulated here.

Although the suggested oxygen radical mechanism solves many of the more significant problems in water oxidation, there are still some questions that need more investigation. For example, in the model there is only one redox-active manganese atom, and the function of the other three Mn-centers therefore remains to be found. Also, the suggested mechanism requires the formation of a terminal Mn(IV)=O oxo-group at S_2, while continued model studies have shown that this is far from a trivial point. The previous model calculations furthermore only considered the most critical steps of water oxidation in detail, so detailed molecular models are still not suggested for the other steps. Finally, although the previous model gives a nice rationalization for why one oxygen of O_2 rapidly exchanges with solvent water, it is not clear how a terminal Mn(IV)=O oxygen could exchange even slowly with solvent water which is found experimentally.

In order to address some of these issues a large number of additional model calculations using a larger manganese cluster has recently been performed [54]. In most of these models there are three Mn-centers – see Fig. 6 – so the role of the fourth Mn has still not been addressed. The main conclusions from the Mn_3-model calculations are as follows. The most essential features of the previous model remain the same, including the surprising fact that only one manganese is redox-active and that an oxyl-radical appears in S_3. However, in the Mn_3-model the oxyl does not appear as a terminal oxyl but as a μ-oxo bridging oxyl. This means that the problem of creating a terminal oxo ligand, which is energetically very costly, is removed. The roles of the non-redox

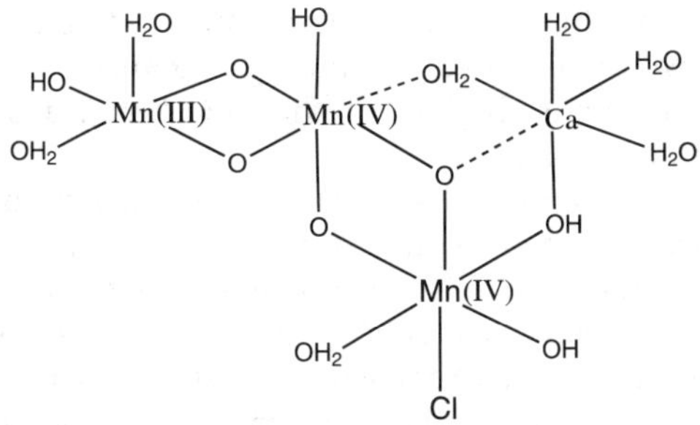

Fig. 6. The Mn_3 model for the S_2-state of the OEC

active Mn-centers are to tune the quite important and sensitive *trans*-effects that affect the O—H bond strengths. A list of six different important *trans*-effects were found, including one which suggests an explanation for the surprising EXAFS-finding that two of the Mn-Mn distances increase in the S_2 to S_3 transition. To achieve this structural change a cluster of the type shown in Fig. 6 is required, where the two μ-oxo-bridging Mn-dimers are directly linked by the central Mn-atom in a bent configuration. The role of the chloride cofactor is also to tune the *trans*-effects, while the function of the Ca-cofactor is to provide chelation energy, just as in the original oxyl-radical mechanism. To obtain maximum energetic effect by the chelation, Ca should bridge two of the three Mn-centers, leading to two short Mn-Ca distances in agreement with conclusions drawn from experiments for Sr-substituted clusters [42]. The water-oxidation is thought to occur by removing protons from water positioned in the corners of an incomplete cube where two Mn, one Ca, two μ-oxo, and two waters form seven corners of the cube. The external water enters in the eighth, originally empty, corner. As in the previous model, this external water should easily exchange with solvent water. A mechanism for the slowly exchanging oxygen was also proposed, in which the bridging oxyl radical first obtains a proton from an Mn-Ca bridging hydroxyl in a low-barrier, water assisted process.

4 Conclusions

The application of theoretical methods to the problem of enzyme mechanisms is in a very early stage of development and it is not yet certain how well our simple quantum models represent the enzyme. Benchmarking of the quantum chemical methods employed, however, suggests that inadequacies in these methods are not the main factors limiting the quality of the final results. The procedure adopted in the work described above is to choose sufficiently simplified quantum model compounds to make the problem tractable and see whether the results obtained have promise. The results indicate that there are indeed sufficient points of contact between the theoretical results obtained and the known biophysical and bioinorganic data to provide a solid basis for further work along these lines. In future work, more refined model calculations will be done to prove or disprove the findings obtained so far. For example, once better structural data is obtained on the OEC, better quantum models will automatically become available.

As described above, theoretical studies have already been very valuable in providing very clear energetic constraints to rule out otherwise plausible mechanisms that invoke energetically unreasonable structures or pathways. Indeed, few mechanisms survive this test. The one or ones that remain seem to be in good accord with the current experimental data and they also make predictions that have yet to be tested experimentally. As in all mechanistic work, other mechanisms cannot be ruled out. Nevertheless, both for MMO and PSII, significant progress has already been made and we expect quantum chemical studies to be a growing presence in bioinorganic chemistry in the years ahead.

5 References

1. Shaik S, Filatov M, Schroeder D, Schwartz H (1998) Chem Eur J 4: 193
2. Siegbahn PEM, Blomberg MRA, Svensson M (1993) J Am Chem Soc 115: 4191–4200
3. Mitchell SA, Blitz MA, Siegbahn PEM, Svensson M (1994) J Chem Phys 100: 423–433
4. Carroll JJ, Weisshaar JC, Siegbahn PEM, Wittborn CAM, Blomberg MRA (1995) J Phys Chem 99: 14,388–14,396
5. Mitchell SA (1992) In: Fontijn A (ed) Gas phase metal reactions. Elsevier, Amsterdam
6. Bengali AA, Bergman RG, Moore CB (1995) J Am Chem Soc 117: 3879
7. Siegbahn PEM (1996) J Am Chem Soc 118: 1487–1496
8. (a) Shaik S, Danovich D, Fiedler A, Schröder D, Schwarz H (1996) Helvetica Chim Acta 78: 1393–1407; (b) Filatov M, Shaik S (1998) J Phys Chem A 102: 3835–3846
9. (a) Yoshizawa K, Shiota Y, Yamabe T (1998) Organometallics 17: 2825–2831; (b) Yoshizawa K, Ohta T, Shiota Y, Yamabe T (1997) Chem Lett 1213–1214; (c) Yoshizawa K, Ohta T, Shiota Y, Yamabe T (1998) Bull Chem Soc Jpn 71: 1899
10. Becke AD (1988) Phys Rev A 38: 3098
11. Becke AD (1992) J Chem Phys 96: 2155–2160
12. Becke AD (1993) J Chem Phys 98: 5648–5652
13. Siegbahn PEM, Blomberg MRA (1999) Ann Rev Phys Chem 50: 221–249
14. Siegbahn PEM, Blomberg MRA (1999) Chem Rev (in press)
15. Feig AL, Lippard SJ (1994) Chem Rev 94: 759
16. Wallar BJ, Lipscomb JD (1996) Chem Rev 96: 2625
17. Dalton H (1980) Adv Appl Microbiol 6: 1
18. (a) Sjöberg BM (1994) Structure 2: 793–796; (b) Sjöberg BM (1997) Structure and Bonding 88: 139–173
19. (a) Stubbe J (1990) Biol Chem 265: 5330; (b) Mao SS, Holler TP, Yu GX, Bollinger JM, Booker S, Johnston MI, Stubbe J (1992) Biochemistry 31: 9733–9743
20. Ortiz de Montellano PR (1986) Cytochrome P-450: Structure, mechanism, biochemistry. Plenum Press, NY
21. Wilkins PC, Wilkins RG (1987) Coord Chem Rev 79: 195
22. Rosenzweig AC, Frederick CA, Lippard SJ, Nordlund P (1993) Nature 366: 537
23. Andersson KK, Froland A, Lee SK, Lipscomb JD (1991) New J Chem 15: 411
24. (a) Stassinopoulos A, Caradonna J (1990) J Am Chem Soc 112: 7071; (b) Liesing RA, Brennan BA, Que L Jr (1991) J Am Chem Soc 113: 3988
25. Priestley ND, Floss HG, Frohland WA, Lipscomb JD, Williams PG, Morimoto H (1992) J Am Chem Soc 114: 7561–7562

26. Liu KE, Johnson CC, Newcomb M, Lippard SJ(1993) J Am Chem Soc 115: 939–947
27. Choi SY, Eaton PE, Hollenberg PF, Liu KE, Lippard SJ, CC, Newcomb M, Putt DA, Upadhyaya SP, Xiong Y (1996) J Am Chem Soc 118: 6547–6555
28. Valentine AM, Lippard SJ (1997) J Chem Soc Dalton Trans 3925
29. Yoshizawa K, Yamabe T, Hoffmann R (1997) New J Chem 21: 151
30. Shestakov AF, Shilov AE (1996) J Mol Catal A 105: 1
31. Siegbahn PEM, Crabtree RH (1997) J Am Chem Soc 119: 3103
32. Shu L, Nesheim JC, Kauffmann K, Munck E, Lipscomb JD, Que L Jr (1997) Science 275: 515
33. Basch H, Mogi K, Musaev DG, Morokuma K (1999) J Am Chem Soc 121: 7249–7256
34. Siegbahn PEM (1999) J Inorg Chem 38: 2880–2889
35. Shaik S, private communication
36. (a) Yachandra VK, Sauer K, Klein MP (1996) Chem Rev 96: 2927–2950; (b) Ruettinger W, Dismukes GC (1997) Chem Rev 97; (c) Brudvig GW (1990) In: Hoff AJ (ed) Advanced EPR: applications in biology and biochemistry. Elsevier, Amsterdam, pp 839–865; (d) Yocum CF (1992) In: Pecoraro VL (ed) Manganese redox enzymes. VCH, NY, pp 71–83; (e) Kambara T, Govindjee (1985) Proc Nat Acad Sci USA, 82: 6119; (f) Pecoraro VL (1988) Photochem Photobiol 48: 249; (g) Christou G (1989) Acc Chem Res 22: 328; (h) Armstrong WH (1992) In: Pecoraro VL (ed) Manganese redox enzymes. VCH, NY, pp 261–286; (i) Yocum CF (1991) Biochim Biophys Acta 1059: 1
37 Lindberg K, Andreasson L (1992) Biochemistry 35: 14,259
38. Kok B, Forbush B, McGloin M (1970) Photochem Photobiol 11: 457
39. (a) Hansson O, Aasa R, Vanngard T (1987) Biophys J 51: 825; (b) de Paula JC, Brudvig GW (1985) J Am Chem Soc 107: 2643
40. (a) Roeloffs TA, Liang W, Latimer MJ, Cinco RM, Rompel A, Andrews JC, Sauer K, Yachandra VK, Klein MP (1996) Proc Natl Acad Sci USA 93: 3335; (b) Ono T, Noguchi T, Inoue Y, Kusinoki M, Matsushita T, Onayagi H (1992) Science 258: 1335; (c) Iuzzolini L, Dittmer J, Dörner W, Meyer-Klaucke W, Dau H (1998) Biochemistry 37: 17,112–17,119
41. (a) Messinger J, Badger M, Wydrzinski T (1995) Proc Natl Acad Sci USA 92: 3209–3213; (b) Messinger J, Hillier M, Badger M, Wydrzinski T (1995) In: Mathis P (ed) Photosynthesis: from light to biosphere. Kluwer, Dordrecht, vol 2, pp 283
42. Cinco RM, Robblee JH, Rompel A, Fernandez C, Yachandra VK, Sauer K, Klein MP (1998) J Phys Chem B102: 8248–8256
43. Beck WF, Brudvig GW (1988) Chem Scr 28A: 93
44. (a) Sharp RR (1992) In: Pecoraro VL (ed) Manganese redox enzymes. VCH, NY, pp 177–196; (b) Styring SA, Rutherford AW(1988) Biochemistry 27: 4915; (c) Evelo RG, Styring SA, Rutherford AW, Hoff AJ (1988) Biochim Biophys Acta 973: 428
45. (a) Hoganson CW, Lydakis-Simantiris N, Tang XS, Tommos C, Warncke K, Babcock GT, Diner BA, McCracken J, Styring S (1988) Photosynth Res 46: 177; (b) Babcock GT (1995) In: Mathis P (ed) Photosynthesis from light to biosphere. Kluwer, Dordrecht, vol 2, pp 209; (c) Tommos C, Tang XS, Warncke K, Hoganson CW, Styring S, McCracken J, Diner BA, Babcock GT (1995) J Am Chem Soc 117: 10,325; (d) Tommos C, Babcock GT (1998) Acc Chem Res 31: 18; (e) Hoganson CW, Babcock GT (1997) Science 277: 1953; (f) Baldwin MJ, Pecoraro VL (1996) J Am Chem Soc 118: 11,325; (g) Pecoraro VL, Baldwin MJ, Caudle MT, Hsieh WY, Law NA (1998) Pure Appl Chem 70: 925; (h) Caudle MT, Pecoraro VL (1997) J Am Chem Soc 119: 3415
46. (a) Haumann M, Bögershausen O, Cherepanov D, Ahlbrink R, Junge W (1997) Photosynth Res 51: 193–208; (b) Ahlbrink R, Haumann M, Cherepanov D, Bögershausen O, Mulkidjanian A, Junge W (1998) Biochemistry 37: 1131–1142; (c) Haumann M, Junge W (1999) Biochim Biophys Acta 1411: 86–91
47. (a) Limburg J, Crabtree RH, Brudvig (1998) In: Meunier B (ed) Biomimetic oxidations catalyzed by transition metal complexes. ICP Press, Singapore; (b) Limburg J, Szalai VA, Brudvig GW (1999) J Chem Soc Dalton 1353

48. (a) Blomberg MRA, Siegbahn PEM, Styring S, Babcock GT, Åkermark B, Korall P (1997) J Am Chem Soc 119: 8285–8292; (b) Blomberg MRA, Siegbahn PEM (1997) Theor Chem Acc 97: 72–80
49. (a) Krishtalik LI (1986) Biochim Biophys Acta 849: 162; (b) Krishtalik LI (1990) Bioelectrochem Bioenerg 23: 249
50. Proserpio DM, Hoffmann R, Dismukes GC (1992) J Am Chem Soc 114: 4374
51. Proserpio DM, Rappe AK, Gorun SM (1993) Inorg Chim Acta 213: 319
52. Siegbahn PEM (1997) In: Comba P, Banci L (eds) Molecular modeling and dynamics of bioinorganic systems. Kluwer, pp 233–253
53. Siegbahn PEM, Crabtree RH (1999) J Am Chem Soc 121: 117–127
54. Siegbahn PEM Inorg Chem (submitted)
55. Limburg J, Vrettos JS, Liable-Sands LM, Rheingold AL, Crabtree RH, Brudvig GW (1999) Science 283: 1524

Characterization and Properties of Non-Heme Iron Peroxo Complexes

Jean-Jacques Girerd[1], Frédéric Banse[2], Ariane Jalila Simaan[3]

Laboratoire de Chimie Inorganique, UMR CNRS 8613, Université Paris-Sud, 91405 Orsay cedex, France

[1] *E-mail: jjgirerd@icmo.u-psud.fr*
[2] *E-mail: fredbanse@icmo.u-psud.fr*
[3] *E-mail: jasimaan@icmo.u-psud.fr*

Iron-peroxo Fe(III)O_2 and hydroperoxo Fe(III)OOH systems are important intermediates between the initial Fe(II)-dioxygen adduct and the "activated" form of the catalytic site in many mono-iron biomolecules. To the same peroxidic level correspond, in diiron enzymes, bridged peroxo Fe(III)—O—O—Fe(III) intermediates. This review is concerned with the preparation and spectroscopic characterization of such intermediates in non-heme chemical systems, the properties of the natural systems being quoted as references. Although none have been crystallized, it seems very likely that Fe(III)OOH systems present a η^1-coordination mode for the hydroperoxo group. These Fe(III)OOH units have given clear signatures in UV-vis, resonance Raman and mass spectrometry. By EPR it was found that in Fe(III)OOH, the Fe(III) is low-spin ($S = 1/2$) and we propose here a simple rationalization of the characteristics of the EPR g-tensor. The electronic properties of the Fe(III)(η^1-OOH) known so far, point toward a strong Fe—O bond and a weak O—O bond, in total agreement with the reactivity scheme implying a cleavage of the O—O bond to lead formally to a Fe(V)O unit. Alkylperoxo systems are also included in this review. Fe(III)-peroxo systems Fe(III)O_2 have been prepared and described. They contain high-spin Fe(III) and those identified seem to be of the η^2 type. The Fe—O bond is weaker and the O—O one is stronger than in the Fe(III)OOH systems. The implication of these Fe(III)O_2 units in catalysis is unclear. "Complementary" systems, such as Fe(III)(η^1-OO) or Fe(III)(η^2-OOH) have been evoked in publications but not identified spectroscopically. These systems certainly deserve to be actively looked for. Finally, bridged peroxo Fe(III)—O—O—Fe(III) systems have been characterized and most remarkably even crystallized and studied by X-ray diffraction in three cases. Their configuration was found *cis*-planar or *cis-gauche*. Here the main question is how enzymes and possibly models can go from the Fe(III)—O—O—Fe(III) state to the di-μ-oxo Fe(IV)O_2Fe(IV) active state?

Keywords: Peroxo complexes, Iron, Non-Heme, EPR, Resonance Raman, UV-vis

1 Introduction

Dioxygen interaction with a heme-iron is of fundamental importance in several proteins implied in aerobic life processes. Considering for instance human heme proteins, the following examples can be quoted:

1. Dioxygen transport proteins like myoglobin or hemoglobin [1]
2. The cytochrome P450 which is so important in the metabolism of exogenous molecules, among them medicaments [2]
3. The cytochrome c oxidase, the terminal respiratory enzyme able to reduce dioxygen into water [3].

In myoglobin or hemoglobin [1] the dioxygen molecule ligates to the Fe(II) ion in a reversible way: when the dioxygen pressure decreases, the Fe—O bond is broken and the dioxygen molecule leaves the iron ion.

In cytochrome P450, the fixation of dioxygen is followed by reduction by one external electron which weakens the O—O bond. Double protonation induces the breakage of the O—O bond and one oxygen atom stays on the iron ion and the other is eliminated as a water molecule. The Fe(V)O unit or its equivalent P·Fe(IV)O inserts its oxygen into a C—H bond to create a C—OH alcohol group (Scheme 1).

In the cytochrome c oxidase, a dioxygen molecule is inserted in between an Fe(II) ion and a Cu(I) ion and is likely to form an Fe(III)—O—O—Cu(II) unit. This unit appears to undergo a reduction by two electrons and protonation to give two water molecules (Scheme 2). However, the real mechanism is more sophisticated [3].

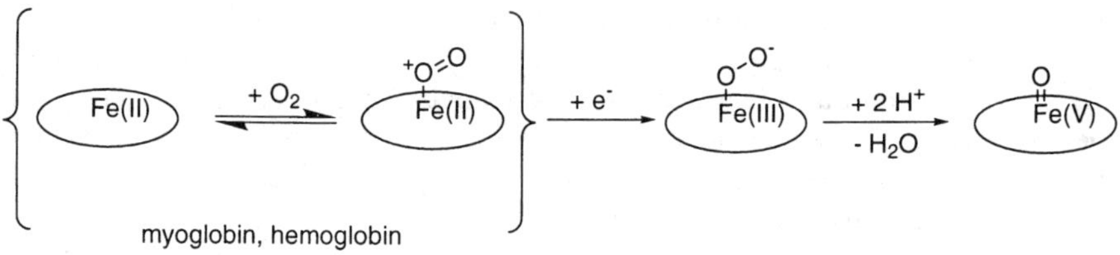

Scheme 1.

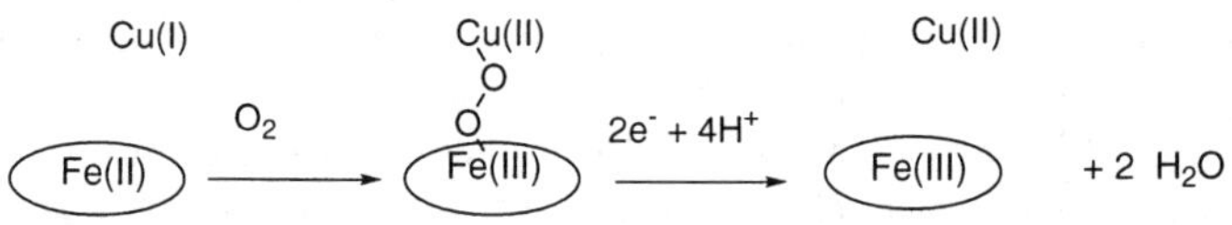

Scheme 2.

Several other heme proteins interacting with dioxygen are found in nature. This fascinating type of reaction has been reviewed in other chapters of this book.

In recent years, several biomolecules implying reaction of dioxygen with non-heme iron have been discovered. These systems are in fact numerous and it is likely that many remain to be discovered. Nature often uses different means to reach the same chemical result: some of these non-heme enzymes achieve, by a different route, biochemical functions that are otherwise performed by a heme system.

For instance, dioxygen transport is effected in sea worms not by a myoglobin type protein but by hemerythrin, a diiron non-heme system [1]. One of the two Fe(II) ions is pentacoordinated and thus has an open site for the (reversible) coordination of the dioxygen. Dioxygen links to this Fe(II) ion to form an Fe(III)Fe(III)OOH adduct (Scheme 3).

Methane monooxygenase [4] (MMO), found in some bacteria, has some analogy with cytochrome P450 in its ability to insert an oxygen atom into a C—H bond. It is so powerful that it can do so even on the poorly reactive methane molecule CH_4, giving methanol CH_3OH. Unlike cytochrome P450,

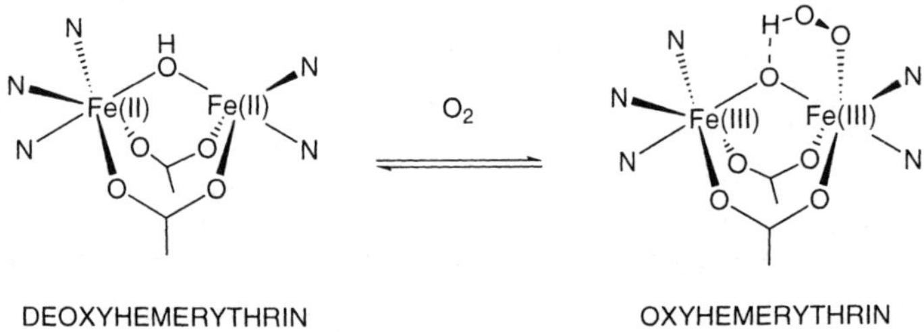

Scheme 3.

MMO contains two iron ions at the active site, each one in a non-heme environment. The mechanism of the catalytic reaction is not entirely understood but the main features are now known (Scheme 4). Dioxygen gives a peroxo bridge between the two Fe(III) ions. One possibility is that this peroxo unit rearranges to give a highly oxidizing Fe(IV)Fe(IV) unit (see the chapter by Siegbahn and Crabtree in this book).

Some analogy may also exist between Cytochrome P450 and Bleomycin (BLM) [5] which is a natural antibiotic synthesized by fungi to protect themselves against bacteria. It contains a non-heme iron ion which interacts with dioxygen to lead to "activated bleomycin" which has been shown to contain an Fe(III)OOH unit [6] which is thought to decompose to an Fe(V)O group (Scheme 5) able to initiate DNA cleavage. It has been discovered that bleomycin can efficiently cleave the DNA molecules of some tumor cells and it is thus used as a drug in chemotherapy of some human cancers (see the chapter by Burger in this book).

Important research efforts have been directed toward the comprehension of the interaction between dioxygen and non-heme iron ions. Structural motifs implied are, as mentioned above:

1. The Fe(III)—O—O—Fe(III) one formed upon insertion of dioxygen between two Fe(II) ions as in MMO
2. The Fe(III)OOH one formed when dioxygen reacts with an Fe(II) ion to give an adduct which is reduced by one electron and protonated, as in bleomycin and hemerythrin

It is clear that in these Fe(III)—O—O—Fe(III) and Fe(III)OOH units the dioxygen molecule is reduced to the peroxo level. These entities can indeed in some cases be prepared by reaction of Fe(III) (or Fe(II)) with hydrogen peroxide. Moreover, if an alkyl peroxide ROOH is used in place of HOOH, an Fe(III)-alkylperoxo can be prepared. It has also been shown that FeO_2 systems can be prepared by deprotonation of Fe(III)OOH.

Overall, the four types of entities known today with non-heme iron are depicted in Scheme 6.

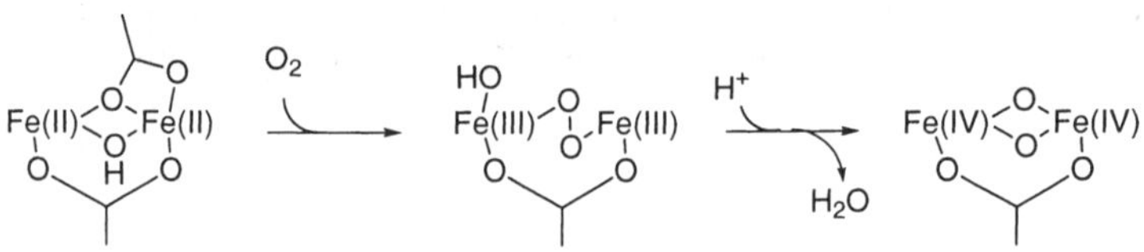

Scheme 4.

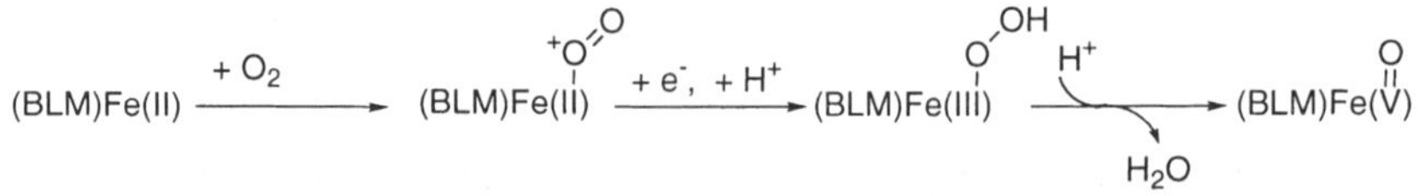

Scheme 5.

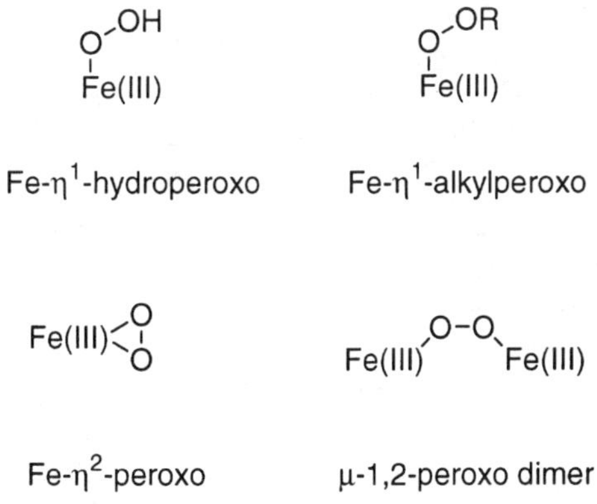

Scheme 6.

This review is devoted to the chemical and structural description of these species and their spectroscopic signatures together with an elementary discussion of their electronic properties.

2 Fe-Hydroperoxo Units Fe(III)OOH

Except for the ligand Ph_4DBA^{2-} which gives dinuclear iron complexes (see below) [7], all the ligands represented in Fig. 1 have allowed the preparation of mononuclear Fe-hydroperoxo Fe(III)OOH complexes.

In general these complexes have been obtained by reaction of the Fe(II) complex with an excess of H_2O_2, following the reaction:

$$\text{Fe(II)} \xrightarrow[\text{H}_2\text{O}_2]{\text{excess}} \text{Fe(III)OOH} \tag{1}$$

2.1 UV-Visible Spectroscopy

The mononuclear Fe(III)OOH complexes are characterized by an absorption in the visible region attributed to a LMCT from the hydroperoxo group to the Fe(III) ion (Table 1). This absorption gives these complexes a purple color.

Apparently, in MeOH, for the series trispicen [8], trispicMeen [9, 10], TPEN [8], ettpen [11], N4py [12–14], and Py5 [15], the higher the number of pyridines coordinated to Fe(III) the lower the band is in energy:

1. With 3 pyridines (trispicen, trispicMeen, TPEN, ettpen) 531–541 nm
2. With 4 pyridines (N4py) 548 nm
3. With 5 pyridines (Py5) 592 nm.

This can be fortuitous or related to the π-accepting power of pyridine which will increase that of Fe(III) and thus displace the LMCT toward low energies.

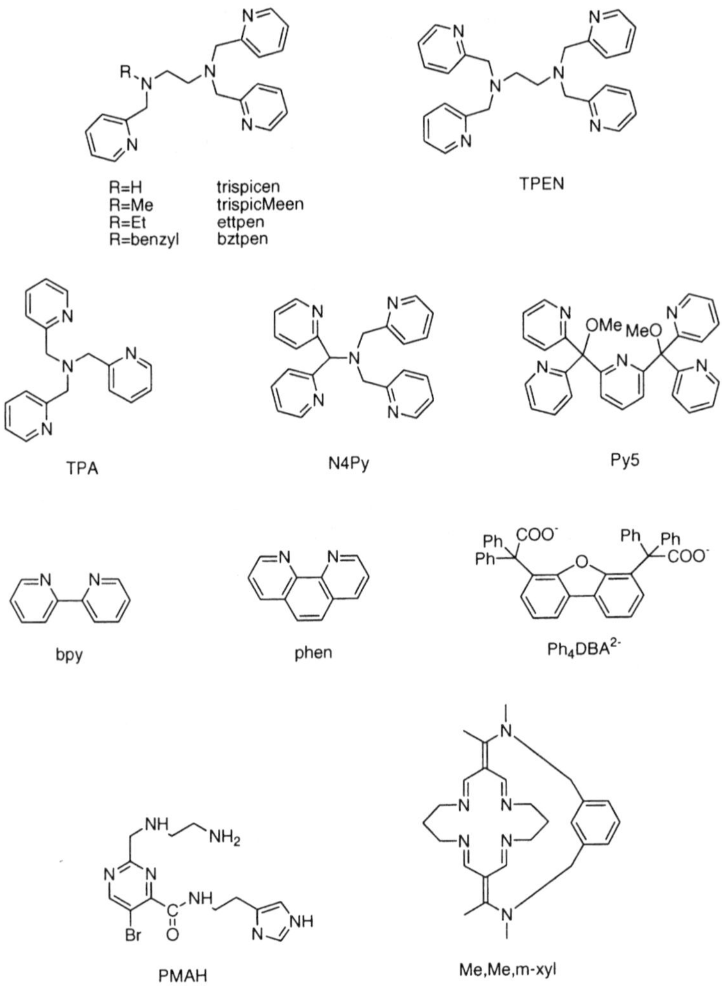

Fig. 1. Ligands which have allowed the characterization of Fe(III)OOH complexes, named according to literature

Complexes of PMAH [16] and Me,Me,*m*-xyl ligands [17] with Fe(II) have been reacted with O_2 and a reducing equivalent to give the Fe(III)OOH unit:

$$\text{Fe(II)} \xrightarrow{O_2,H^+,e^-} \text{Fe(III)OOH} \tag{2}$$

UV-vis informations for the LMCT hydroperoxo-Fe(III) band have not been reported with these ligands.

With Ph_4DBA^{2-} a model of hemerythrin has been built by Mizoguchi and Lippard [7]. The Fe(II)Fe(II) complex reacts to give an Fe(III)Fe(III)OOH

Table 1. Wavelength (molar extinction coefficient) of the hydroperoxo → Fe(III) LMCT for Fe(III)OOH units

Complex	$\lambda_{max}(\varepsilon/M^{-1} \cdot cm^{-1})$	Refs.
$[(trispicMeen)Fe(OOH)]^{2+}$	537 nm (1000) in methanol	[9, 10]
$[(trispicen)Fe(OOH)]^{2+}$	531 nm (950) in methanol	[8]
$[(TPEN)Fe(OOH)]^{2+}$	541 nm (900) in methanol	[8]
$[(ettpen)Fe(OOH)]^{2+}$	536 nm (1000) in methanol	[11]
$[(TPA)Fe(OOH)]^{2+}$	538 nm (1000) in acetonitrile	[18]
$[(N4Py)Fe(OOH)]^{2+}$	532 nm (1100) in acetone	[12–14]
	548 nm (1100) in methanol	
$[(Py5)Fe(OOH)]^{2+}$	592 nm (?) in methanol	[15]
"$[Fe_2(\mu\text{-}Ph_4DBA)(OOH)]$"	470 nm (2600) in CH_2Cl_2	[7]

hydroperoxo adduct which is a good mimic of the corresponding group in oxyhemerythrin. The LMCT hydroperoxo-Fe(III) band was observed at 470 nm. The higher energy is in agreement with a decrease in the electron affinity of Fe(III) due to ligation to negative carboxylate groups.

2.2 Mass Spectrometry

Electrospray Ionization Mass Spectrometry (ESI-MS) appeared to be very useful in the identification of the Fe(III)OOH intermediates. First, Sam et al. [6] demonstrated using ESI-MS, that activated bleomycin contains an Fe(III)OOH unit. The attribution was corroborated with ^{18}O substitution. Several aminopyridine type model complexes have been similarly studied and have been proven to contain the same Fe(III)OOH unit [9–12, 15, 18]. Information on equilibria in solution has also been obtained as well as evidence for ligand degradation [9, 10] and Fe(IV)O species [11, 18].

The demonstration of the existence of the unit Fe(III)OOH in these species, added to considerations from the structure of the pentadentate ligands such as N4py or Py5, make very likely the attribution of an η^1 coordination mode of the hydroperoxo group in these complexes (Scheme 7).

2.3 EPR Spectroscopy

Fe-hydroperoxo systems have been studied by EPR spectroscopy. EPR data are summarized in Table 2.

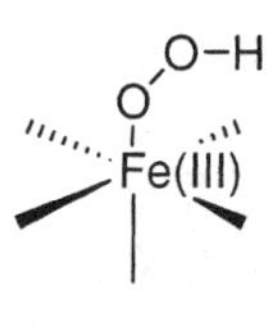

Scheme 7.

Table 2. EPR g_{max}, g_{inter}, g_{min} values for Fe(III)OOH $S = 1/2$ units. Parameters of the Griffith-Taylor model of the g tensor: a, b, c are the coefficients of the wave functions; $\Sigma = a^2 + b^2 + c^2$; V is the rhombic distortion energy; Δ is the axial distortion energy; both are in unit of λ, the spin-orbit coupling constant

Complex	Ref.	$-g_{min}$ (g_z)	g_{max} (g_y)	$-g_{inter}$ (g_x)	a	b	c	Σ	V/λ	Δ/λ
(BLM)Fe(OOH)	[19]	−1.94	2.26	−2.17	0.072	0.052	0.993	0.9930	3.04	−8.74
$[(trispicMeen)Fe(OOH)]^{2+}$	[9, 10]	−1.95	2.19	−2.12	0.055	0.039	0.992	0.9885	4.05	−11.31
$[(trispicen)Fe(OOH)]^{2+}$	[8]	−1.96	2.19	−2.14	0.053	0.041	0.994	0.9930	2.86	−11.19
$[(TPEN)Fe(OOH)]^{2+}$	[8]	−1.997	2.22	−2.15	0.056	0.040	0.999	1.0022	3.98	−11.29
$[(bztpen)Fe(OOH)]^{2+}$	[11]	−1.96	2.20	−2.16	0.055	0.046	0.995	0.9952	2.00	−10.45
$[(TPA)Fe(OOH)]^{2+}$	[18]	−1.96	2.19	−2.15	0.053	0.044	0.995	0.9939	2.18	−10.89
$[(N4Py)Fe(OOH)]^{2+}$	[12]	−1.98	2.17	−2.12	0.044	0.033	0.998	0.9989	4.34	−13.79
$[(Py5)Fe(OOH)]^{2+}$	[15]	−1.98	2.15	−2.13	0.040	0.035	0.998	0.9984	1.80	−13.89
$[Fe(bpy)_2(py)(OOH)]^{2+}$	[20]	−1.97	2.14	−2.11	0.040	0.033	0.995	0.9929	2.87	−14.31
$[Fe(phen)_2(py)(OOH)]^{2+}$	[20]	−1.97	2.13	−2.12	0.037	0.035	0.995	0.9929	0.95	−14.19
$[(PMA)Fe(OOH)]^{+}$	[16]	−1.94	2.22	−2.17	0.063	0.051	0.992	0.9900	1.90	−9.14
“[(Me,Me,m-xyl)Fe(OOH)]”	[17]	−1.93	2.23	−2.16	0.063	0.052	0.989	0.9870	2.50	−8.89

In every case these signals arise from a low-spin $S = 1/2$ Fe(III). These spectra are closer to axiality than their Fe(III) precursors. This was also found with $(BLM)Fe^{III}$ ($g = 2.45, 2.18, 1.89$) and $(BLM)Fe^{III}(OOH)$ ($g = 2.27, 2.17, 1.94$) [19]. In Fig. 2 the g values are represented as a function of g_{max}.

We observe that g_{min} slightly decreases and g_{inter} slightly increases when g_{max} increases. The differences $g_{max} - g_{min}$ and $g_{max} - g_{inter}$ increase when g_{max} increases.

It is possible to describe the g values for an LS Fe(III) by using the model of Griffith [21] as presented by Taylor [22]. This model takes into account spin-orbit coupling and distortion from octahedral geometry in such low spin Fe(III) complexes. In octahedral geometry, Fe(III) has a $^2T_{2g}$ ground state. The effects mentioned above lift the sixfold degeneracy of the $^2T_{2g}$ state and three Kramers doublets are obtained. In general the degeneracy is lifted enough to give well-separated Kramers doublets and only that lowest in energy is substantially populated at 100 K. Taylor has given convenient expressions of the two wave functions of the ground Kramers doublet and of the associated g values. The results are presented in Table 2. As Veselov et al. [23] we made the hypothesis that $g_x g_y g_z$ is positive. This hypothesis seems justified by the chemical meaning of the results. As described by Taylor [22] we selected the axes in such a way that $|V| < (2/3)|\Delta|$, where V is the rhombic distortion energy, Δ the axial distortion energy and λ the spin-orbit coupling parameter that has been assumed constant for all these LS Fe(III) systems. For $(BLM)Fe^{III}(OOH)$, one finds $g_z = -1.94$, $g_y = 2.27$, $g_x = -2.17$, $a = 0.072$,

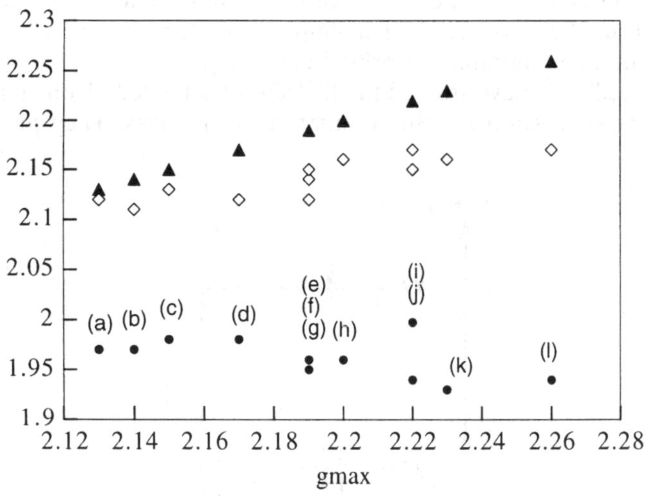

Fig. 2. g_{max} (▲), g_{inter} (◊), g_{min} (●) as a function of g_{max} for Fe(III)OOH $S = 1/2$ units: (a) $[Fe(phen)_2(py)(OOH)]^{2+}$; (b) $[Fe(bpy)_2(py)(OOH)]^{2+}$; (c) $[(Py5)Fe(OOH)]^{2+}$; (d) $[(N4Py)Fe(OOH)]^{2+}$; (e) $[(TPA)Fe(OOH)]^{2+}$; (f) $[(trispicen)Fe(OOH)]^{2+}$; (g) $[(trispicMeen)Fe(OOH)]^{2+}$; (h) $[(bztpen)Fe(OOH)]^{2+}$; (i) $[(TPEN)Fe(OOH)]^{2+}$; (j) $[(PMA)Fe(OOH)]^{+}$; (k) "[(Me,Me,*m*-xyl)Fe(OOH)]"; (l) (BLM)Fe(OOH)

$b = 0.051$, $c = 0.993$, $V/\lambda = 3.04$, $\Delta/\lambda = -8.74$ where the coefficients a, b, and c define the functions of the resonating Kramers doublet:

$$\begin{aligned} |+\rangle &= a|yz\alpha\rangle - ib|xz\alpha\rangle - c|xy\beta\rangle \\ |-\rangle &= -a|yz\beta\rangle - ib|xz\beta\rangle - c|xy\alpha\rangle \end{aligned} \quad (3)$$

The values obtained mean that the hole (and hence the unpaired electron) is in an orbital which has a dominant d_{xy} character, and that this orbital is strongly destabilized. The d_{xz} and d_{yz} orbitals would be lower in energy with the d_{xz} the lowest (Scheme 8).

As for model compounds, one can take the examples of $[(PMA)Fe^{III}(OOH)]^+$ with $g_z = -1.94$, $g_y = 2.22$, $g_x = -2.17$, $a = 0.063$, $b = 0.051$, $c = 0.992$, $V/\lambda = 1.90$, $\Delta/\lambda = -9.14$ and $[(N4py)Fe^{III}(OOH)]^{2+}$ with $g_z = -1.98$, $g_y = 2.17$, $g_x = -2.12$, $a = 0.044$, $b = 0.033$, $c = 0.998$, $V/\lambda = 4.34$, $\Delta/\lambda = -13.79$. These results show that $[(PMA)Fe^{III}(OOH)]^+$ and $(BLM)Fe^{III}(OOH)$ systems have similar tetragonal distortion energies (−9.14 and −8.74 in units of λ) but that $[(N4py)Fe^{III}(OOH)]^{2+}$ has a much larger one (−13.70) in absolute value. This is apparent in Fig. 3 which presents a diagram of V/λ as a function of $-\Delta/\lambda$.

This diagram suggests that these compounds can be divided into three types:

1. Those which possess at least 4 pyridine groups and exhibit a large $|\Delta|$
2. Those containing 3 pyridine groups which have an intermediate value of $|\Delta|$
3. The complexes with BLM and PMAH which have a smaller $|\Delta|$

A possible explanation is that the more accepting the Fe(III) is, the stronger the interaction is with the donor π^* antibonding orbital of the hydroperoxo group and thus the larger is $|\Delta|$. This implies that the hydroperoxo is along x or y since the most destabilized orbital is the d_{xy}.

Veselov et al. [23] have studied the ENDOR of activated bleomycin and give an important suggestion for the orientation of the axes. They proposed, by

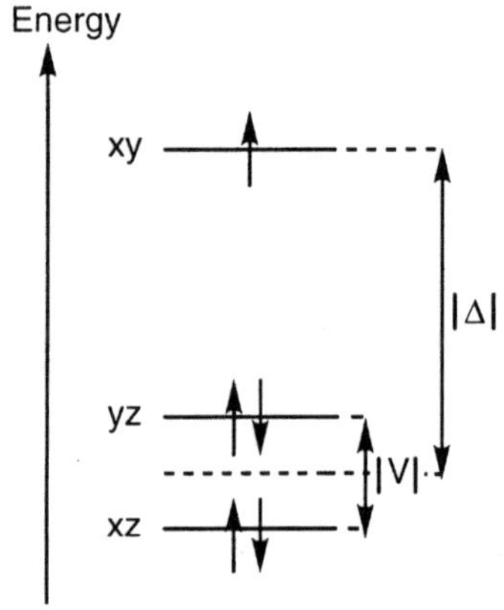

Scheme 8.

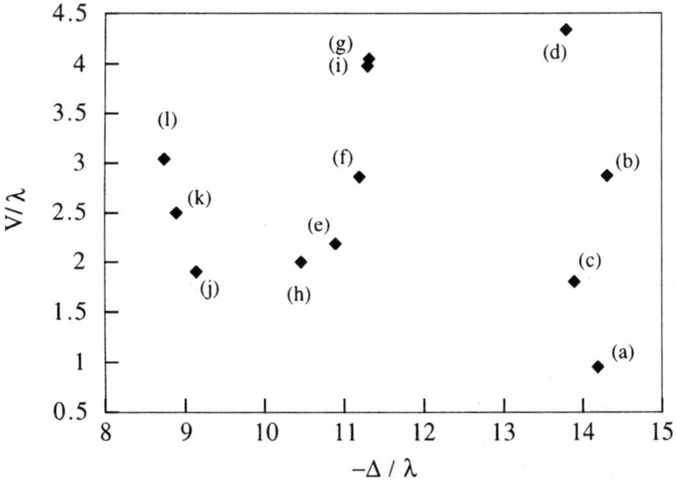

Fig. 3. Graphical representation of the Griffith-Taylor model of the g tensor of Fe(III)OOH $S = 1/2$ units. The rhombic distortion energy V is represented as a function of the opposite value of Δ, the axial distortion energy, both in units of λ, the spin-orbit coupling constant. The letter code for the complexes is the same as in Fig. 2

analogy with low spin heme systems [24], that the axis of g_{max} (y from the calculation above) is perpendicular to the conjugated pyrimidine plane and parallel to the Fe-peroxo axis. This is not in contradiction with the discussion above and leads to the following proposal for the orientation of the *g* tensor (Scheme 9).

In this scheme the *g* values reported are those of (BLM)Fe^{III}(OOH) but it is very likely that this diagram is valid for the model complexes as well.

EPR thus suggests that the transfer of electron density from the antibonding orbital of the hydroperoxo group must be more important for ligands of type 1

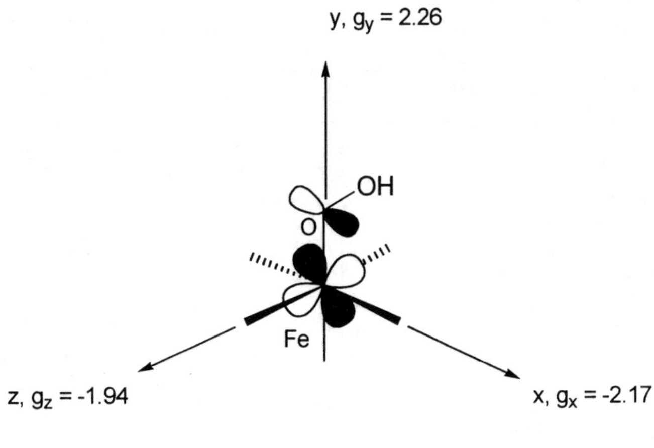

Scheme 9.

than for type 2 which in turn will present a transfer more important than type 3. Thus the Fe—O and O—O bonds must be stronger in the order 1 > 2 > 3, as far as the Π effects are concerned.

2.4 Resonance Raman Spectroscopy

The last conclusion can be compared to the results obtained by Resonance Raman using an excitation into the LMCT band. Data are given in Table 3.

The vibrations of a polyatomic species such as the FeOOH unit cannot be reduced to the stretchings of pairs of atoms. Nevertheless quite often a vibration includes a dominant contribution which is then used as a label. It is considered so in Table 3 but the approximate nature of such a description has to be remembered.

To estimate the adequacy of the description of a vibration by the diatomic model, it is useful to compare the shift observed upon isotopic substitution ($^{16}O/^{18}O$) with that predicted with the diatomic vibrator O—O or Fe—O formula

$$\begin{aligned} \nu_{F_2O} &= 1302.8\sqrt{f_{FeO}(\mu_{Fe}+\mu_O)} \\ \nu_{OO} &= 1302.8\sqrt{2f_{OO}\mu_O} \end{aligned} \qquad (4)$$

where f_i are the force constants in mdyn/Å and $\mu_i = 1/m_i$.

Several mononuclear cases have been studied and the results are pretty similar. With $[(\text{trispicMeen})Fe^{III}(OOH)]^{2+}$ [8] two frequencies are observed at

Table 3. Raman ν_{FeO} and ν_{OO} frequencies in cm^{-1} for Fe(III)OOH with $^{16}O^{16}O$ or $^{18}O^{18}O$. Calculated values are noted with *: a) diatomic vibrators: f_{OO} = 2.94 mdyn/Å; f_{FeO} = 2.93 mdyn/Å; b) diatomic vibrators: f_{OO} = 2.99 mdyn/Å; f_{FeO} = 2.79 mdyn/Å; c) diatomic vibrators: f_{OO} = 3.02 mdyn/Å; $f6_{FeO}$ = 2.86 mdyn/Å; d) diatomic vibrators: f_{OO} = 3.36 mdyn/Å; f_{FeO} = 1.86 mdyn/Å; e) polyatomic vibrators [25a]: f_{OO} = 3.29 mdyn/Å; f_{FeO} = 1.91 mdyn/Å and other constants non zero

Complex	$^{16}O^{16}O$		$^{18}O^{18}O$		Refs.
	ν_{OO}	ν_{FeO}	ν_{OO}	ν_{FeO}	
$[(N4Py)Fe(OOH)]^{2+}$	790	632	746 (−44)	616 (−16)	[13, 14]
	790*[a]	632*	745*(−45)	604*(−28)	
$[(TPA)Fe(OOH)]^{2+}$	789	626	–	–	[13]
$[(\text{trispicMeen})Fe(OOH)]^{2+}$	796	617	752 (−44)	600 (−17)	[8]
	796*[b]	617*	751*(−45)	590*(−27)	
$[(TPEN)Fe(OOH)]^{2+}$	796	617	–	–	[8]
$[(\text{trispicen})Fe(OOH)]^{2+}$	801	625	750 (−51)	602 (−23)	[8]
	801*[c]	625*	756*(−46)	597*(−27)	
Oxyhemet.	844	503	796 (−48)	479 (−24)	[25a, 26,
	844*[d]	503*	796*(−48)	481*(−22)	27]
	843*[e]	504*	796*(−47)	477*(−27)	
“$[Fe_2(\mu\text{-}Ph_4DBA)(OOH)]$”	843	–	797 (−46)	–	[7]

796 cm^{-1} and 617 cm^{-1} which shift upon $^{16}O/^{18}O$ substitution to 752 cm^{-1}(−44) and 600 cm^{-1}(−17) respectively. With $[(N4py)Fe^{III}(OOH)]^{2+}$ [13, 14] two frequencies are observed at 790 cm^{-1} and 632 cm^{-1} which shift upon substitution to 746 cm^{-1}(−44) and 616 cm^{-1}(−16) respectively. The highest frequency shift is of the order expected for a pure O—O frequency but the shift of the lowest frequency is lower than expected for a pure Fe—O vibration. The fact that the value of the Fe—O frequency is larger for $[(N4py)Fe^{III}(OOH)]^{2+}$ than for $[(trispicMeen)Fe^{III}(OOH)]^{2+}$ could confirm the analysis given in the EPR section above. However $[(N4py)Fe^{III}(OOH)]^{2+}$ presents a ν_{OO} frequency smaller than that of $[(trispicMeen)Fe^{III}(OOH)]^{2+}$ in apparent contradiction with the interpretation of the EPR. One has nevertheless to be cautious with such a qualitative model and more quantitative work is needed here.

As we will see later, analogous frequencies have also been detected for other Fe/peroxo type systems. Figure 4 summarizes the data published for Fe(III)OOH, Fe(III)O_2, and Fe(III)OOFe(III) units as $\nu_{FeO} = f(\nu_{OO})$. It is quite recognizable that the points in Fig. 4 form approximately one domain per type of unit.

What is surprising in the experimental values for the Fe(III)OOH low spin systems is that the ν_{FeO} is the largest and ν_{OO} the smallest of all frequencies known for Fe(III)-peroxo systems (upper left part of Fig. 4). For instance, as shown in Table 3, the diatomic model implies similar f_{FeO} and f_{OO} force constants (around 2.9 mdyn/Å) which are quite different from those that have been found for the Fe(III)OOH motif in hemerythrin by Brunold and Solomon (f_{FeO} = 1.91 mdyn/Å and f_{OO} = 3.29 mdyn/Å) [25a]. The reason for this is not clear to us. An NCA analysis for the mononuclear Fe(III)OOH is certainly needed. Such a calculation could lead to improved force constants. The influence of the FeOO angle has certainly to be studied as has the coupling between the stretching modes and the bending. We also found that H/D substitution has an effect on trispicen (801 cm^{-1} and 618 cm^{-1}) [8] which points toward the need for an NCA extended to the four atoms of the FeOOH unit. However it is likely that the force constants will remain of the order of those found by the simple diatomic model.

Interestingly, Roelfes et al. [14] have proposed that the unusually strong Fe—O bond may be mainly related to the low spin nature of the Fe(III) ion. This is very convincing since in other systems of Fig. 4, in which the Fe(III) is high spin, the Fe—O frequency is indeed much lower. In particular in the oxyhemerythrin motif Fe(III)OOH which contains a high spin Fe(III) the Fe—O frequency is 503 cm^{-1} [25–27] and the representative point in Fig. 4 is clearly separated from the low spin Fe(III)OOH ones.

Unfortunately Raman data have not been published for activated BLM. One can only propose from the model outlined above that in low spin activated BLM:

1. The Fe—O bond is strong due to the LS state of Fe(III)
2. The O—O bond is weak since, due to the poor electron accepting power of BLM, the electron transfer from the π^* antibonding peroxo orbital toward Fe(III) is weak

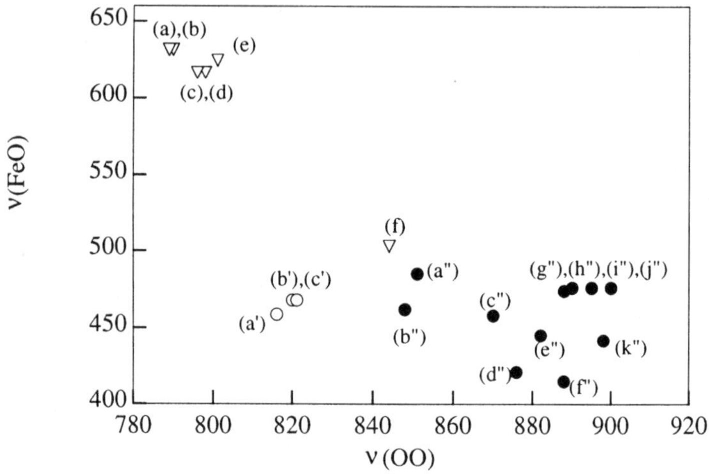

Fig. 4. Graphical representation of the Raman ν_{FeO} and ν_{OO} frequencies (in cm^{-1}) for Fe(III)OOH (∇), Fe(III)O_2 (○) and Fe(III)OOFe(III) (●) units. Note that each type of unit corresponds to one domain except the point corresponding to hemerythrin which is off the Fe(III)OOH domain (see text): (a) $[(TPA)Fe(OOH)]^{2+}$; (b) $[(N4Py)Fe(OOH)]^{2+}$; (c) $[(TPEN)Fe(OOH)]^{2+}$; (d) $[(trispicMeen)Fe(OOH)]^{2+}$; (e) $[(trispicen)Fe(OOH)]^{2+}$; (f) oxy-hemerythrin; (a′) $[(EDTA)Fe(O_2)]^{3-}$; (b′) $[(trispicMeen)Fe(O_2)]^{+}$; (c′) $[(TPEN)Fe(O_2)]^{+}$; (a″) ferritin; (b″) $[Fe_2(\mu\text{-}O_2)(\mu\text{-}O)(6\text{-}Me_3TPA)_2]^{2+}$; (c″) RNR mutant; (d″) $Fe_2(\mu\text{-}O_2)(\mu\text{-}OBz)\{HB(pz')_3\}_2$; (e″) $[Fe_2(\mu\text{-}O_2)(Ph_3PO)_8 \cdot 2H_2O]^{4+}$; (f″) $[Fe_2(\mu\text{-}O_2)(\mu\text{-}O_2CCH_2Ph)_2\{HB(pz')_3\}_2]$; (g″) $[Fe_2(\mu\text{-}O_2)(5,6\text{-}Me_2\text{-}HPTB)(NO_3)_2]^{+}$; (h″) $[Fe_2(\mu\text{-}O_2)(N\text{-}Et\text{-}HPTB)(NO_3)_2]^{+}$; (i″) $[Fe_2(\mu\text{-}O_2)(HPTB)(NO_3)_2]^{+}$; (j″) $[Fe_2(\mu\text{-}O_2)(N\text{-}Et\text{-}HPTB)(OBz)]^{2+}$; (k″) desaturase

3. The distal oxygen atom of the Fe(III)OOH group may be basic due to the negative ligands of Fe(III).

These features would support models of BLM reactivity that imply an O—O bond breakage upon protonation to generate an oxidizing Fe(V)O unit.

For the dimer system oxyhemet, the agreement between the observed values and those calculated by the simple diatomic model is striking. The frequency ν_{OO} is larger and the ν_{FeO} is smaller than in the mononuclear systems discussed above. In the function of hemerythrin, the Fe—O bond is broken and the O—O bond conserved, a difference with BLM which could ultimately be related to that in the spin state of the Fe(III) linked to the hydroperoxo group. For more details on hemerythrin, the reader is referred to a very complete work by Brunold and Solomon [25].

Some of these model compounds have shown a DNA cleavage activity similar to that of BLM [10, 28].

Recently Chen and Que studied the catalytic hydroxylation of alkanes by $[(bispicMe_2en)Fe^{III}(OOH)]^{2+}$ [29], an aminopyridine type system analogous to those discussed here. They propose that the reaction may involve the heterolytic breakage of the O—O bond of the Fe(III)(η^1-OOH) to form an Fe(V)O unit [29], in strong analogy to what is proposed for BLM [5].

The same authors demonstrate the possibility to do *cis*-dihydroxylation by a related Fe(III)OOH system using the ligand 6-Me_3TPA (see Fig. 5 for the ligand) [30]. To explain this very interesting reactivity the authors proposed that with this ligand a Fe(III)(η^2-OOH) would form, the O—O bond being broken later in the catalytic cycle (Scheme 10).

The authors did not give spectroscopic information concerning the Fe(III)(η^2-OOH) intermediate which, if it indeed exists, would be a new motif extremely interesting to explore.

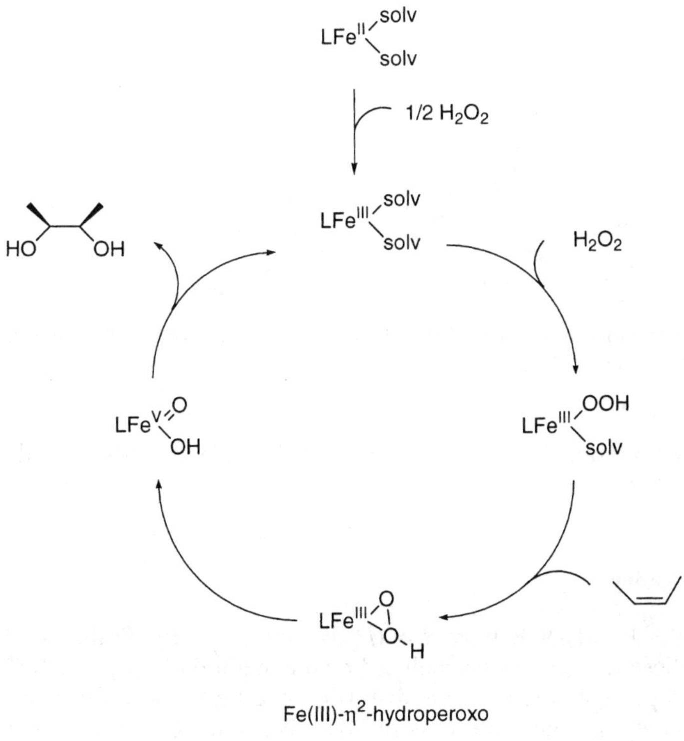

Scheme 10.

3 Fe-Alkylperoxo Units Fe(III)OOR

Several systems presenting the Fe(III)OOR group have been reported, most commonly with R = *tert*-butyl. The ligands of Fig. 5 were used as well as some ligands of Fig. 1.

3.1 UV-Visible Spectroscopy

As with the Fe(III)OOH-containing complexes, the Fe(III)OOR systems present a peroxo/Fe(III) LMCT band around 600 nm (Table 4). To our

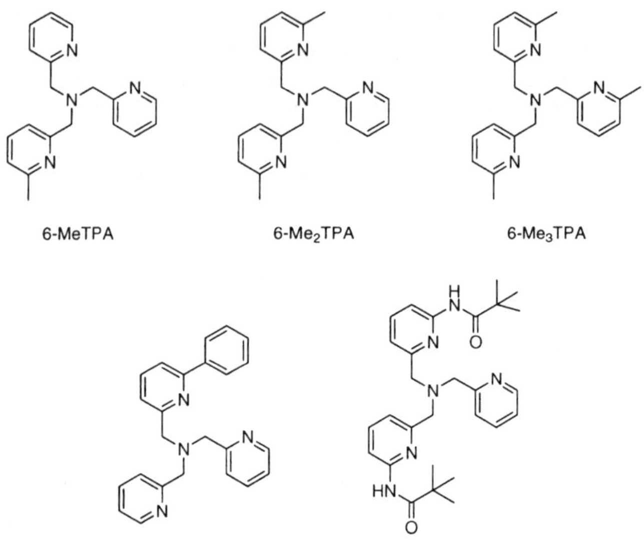

Fig. 5. Ligands which have allowed the characterization of Fe(III)OOR complexes, named according to literature

knowledge no rationalization of the characteristics of the LMCT band has been offered in this series.

3.2
EPR Spectroscopy

Some of the Fe(III)OOR have $S = 1/2$ as found by EPR (Table 5). The EPR signal of these species can be analyzed by the method of Taylor [22] (Table 5).

Figure 6 presents the rhombic distortion energy V/λ as a function of $-\Delta/\lambda$. Approximately the same trend as in Fig. 3 is found. The system with PMAH presents a small value of $|\Delta|$, those with phen or bpy a large value, and the TPA ones intermediate values. Some phen systems are found in the same area as the TPA ones. We propose that basically the same explanation is valid here with a strengthening of the Fe—O bond upon increase of the π-accepting power of the auxiliary ligand.

Very interestingly, Wada et al. [31] and Que et al. [32–34] have shown that Fe(III)OOR systems can be high spin ($S = 5/2$) with auxiliary ligands which exert a weak enough ligand field. The 6-MeTPA and 6-PhTPA ligands even give systems which present both spin states. In general the HS systems are found to be rhombic with $g = 4.3$ (Table 6).

There are notable exceptions in those systems, recently reported by Wada et al. [31] which exhibit an almost axial anisotropy (E/D close to 0.07). The Fe(III) precursor $[Fe(bppa)(t\text{-}BuCOO)]^{2+}$ of these last complexes has been crystallized. By X-ray diffraction it has been found that the Fe(III) ion is

Table 4. Wavelength (molar extinction coefficient) of the alkylperoxo → Fe(III) LMCT for Fe(III)OOR units

Complex	$\lambda_{max}(\varepsilon/M^{-1}\cdot cm^{-1})$	References
$[Fe(TPA)(H_2O)(OO\textit{t}\text{-}Bu)]^{2+}$	600 nm (2200) in MeCN	[33]
$[Fe(6\text{-}MeTPA)(H_2O)(OO\textit{t}\text{-}Bu)]^{2+}$	598 nm (2000) in MeCN	[33]
$[Fe(6\text{-}Me_2TPA)(H_2O)(OO\textit{t}\text{-}Bu)]^{2+}$	552 nm (2000) in MeCN	[33]
$[Fe(6\text{-}Me_3TPA)(H_2O)(OO\textit{t}\text{-}Bu)]^{2+}$	562 nm (2000) in MeCN	[33]
$[Fe(6\text{-}Me_3TPA)(OBz)(OO\textit{t}\text{-}Bu)]^{+}$	510 nm (2300) in CH_2Cl_2	[32]
$[Fe(6\text{-}Me_3TPA)(OBz)(OOcumyl)]^{+}$	506 nm (2300) in CH_2Cl_2	[32]
$[Fe(6\text{-}PhTPA)(OO\textit{t}\text{-}Bu)]^{2+}$	630 nm (?)	[34]
$[Fe(BPPA)(OO\textit{t}\text{-}Bu)]^{2+}$	613 nm (2000) in MeCN	[31]
$[Fe(BPPA)(OOcumyl)]^{2+}$	585 nm (2200) in MeCN	[31]
$[Fe(TPA)(HOCH_2Ph)(OO\textit{t}\text{-}Bu)]^{2+}$	598 nm (2000) in MeCN	[35]
$[Fe(bpy)_2(HOCH_2Ph)(OO\textit{t}\text{-}Bu)]^{2+}$	640 nm (1700) in MeCN	[36]
$[Fe(bpy)_2(HOCH_2Ph)(OOcumyl)]^{2+}$	627 nm (1700) in MeCN	[36]

heptacoordinated to the three pyridine nitrogen atoms, the tertiary amino nitrogen atom, two oxygen atoms from the pivalamido groups, and one oxygen atom from the $t\text{-}BuCOO^-$ anion. This is a rather unusual structure.

3.3 Resonance Raman Spectroscopy

Resonance Raman data have been published (Table 7).

Zang et al. [33] have shown that the low-spin and high-spin Fe(III)OOR have different spectra. The corresponding modes of vibrations are of mixed nature, implying also C-O and C-C stretchings.

3.4 Mass Spectrometry

The ions $[\{Fe(TPA)(OO\textit{t}\text{-}Bu)\}(ClO_4)]^+$ and $[\{Fe(TPA)(OO\textit{t}\text{-}Bu)\}(ClO_4)_3]^-$ have been detected by ESI-MS, demonstrating that the intermediate formed by reaction of *t*-BuOOH on $[Fe_2O(TPA)_2(H_2O)_2]^{4+}$ is a monomeric iron(III)-alkylperoxo complex [35]. Several other alkylperoxo complexes have been identified using this technique [31, 33].

3.5 Reactivity

Lange et al. [34] have studied the reactivity of the complex of 6-PhTPA with *t*-BuOOH.

In a previous publication the authors showed that an Fe(III)OOR group undergoes a homolytic splitting into Fe(IV)O and •OR [38]. The ligand

Table 5. EPR g_{max}, g_{inter}, g_{min} values for Fe(III)OOR $S = 1/2$ units, $^{(*)}$ indicates that the compound presents both spin states (see also Table 6). Parameters of the Griffith-Taylor model of the g tensor: a, b, c are the coefficients of the wave functions; $\Sigma = a^2 + b^2 + c^2$; V is the rhombic distortion energy; Δ is the axial distortion energy; both are in unit of λ the spin-orbit coupling constant

Complex	References	$-g_{min}$ (g_z)	g_{max} (g_y)	$-g_{inter}$ (g_x)	a	b	c	Σ	V/λ	Δ/λ
$[Fe(TPA)(H_2O)(OO\textit{t}\text{-}Bu)]^{2+}$	[33]	−1.98	2.19	−2.14	0.048	0.037	0.998	1.001	3.49	−12.39
$[Fe(6\text{-}MeTPA)(H_2O)(OO\textit{t}\text{-}Bu)]^{2+(*)}$	[33]	−1.97	2.20	−2.12	0.053	0.034	0.996	0.9966	5.44	−12.40
$[Fe(TPA)(HOCH_2Ph)(OO\textit{t}\text{-}Bu)]^{2+}$	[35]	−1.97	2.19	−2.12	0.050	0.034	0.996	0.9961	4.96	−12.57
$[Fe(PMA)(OO\textit{t}\text{-}Bu)]^{+}$	[37]	−1.93	2.28	−2.18	0.077	0.055	0.991	0.9919	2.89	−8.10
$[Fe(bpy)_2(HOCH_2Ph)(OO\textit{t}\text{-}Bu)]^{2+}$	[36]	−1.98	2.18	−2.12	0.046	0.032	0.998	0.9994	4.97	−13.55
$[Fe(bpy)_2(py)(OO\textit{t}\text{-}Bu)]^{2+}$	[20]	−1.97	2.19	−2.12	0.050	0.034	0.996	0.9961	4.96	−12.57
$[Fe(phen)_2(py)(OO\textit{t}\text{-}Bu)]^{2+}$	[20]	−1.96	2.165	−2.12	0.047	0.037	0.993	0.9908	3.19	−12.39
$[Fe(phen)_2(MeOH)(OO\textit{t}\text{-}Bu)]^{2+}$	[20]	−1.97	2.16	−2.19	0.044	0.030	0.995	0.9934	5.56	−14.29
$[Fe(6\text{-}PhTPA)(OO\textit{t}\text{-}Bu)]^{2+(*)}$	[34]	−1.97	2.19	−2.12	0.050	0.034	0.996	0.9961	4.96	−12.57

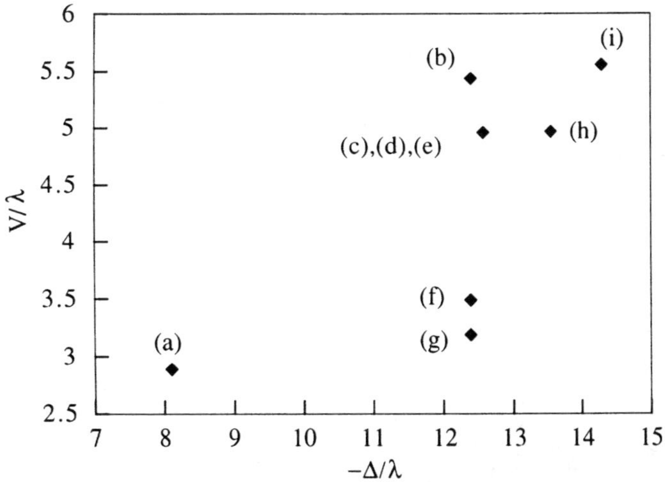

Fig. 6. Graphical representation of the Griffith-Taylor model of the g tensor of Fe(III)OOR $S = 1/2$ units. The rhombic distortion energy V is represented as a function of the opposite value of Δ, the axial distortion energy, both in unit of λ, the spin-orbit coupling constant: (a) $[(PMA)Fe(OO\mathit{t}\text{-}Bu)]^{+}$; (b) $[Fe(6\text{-}MeTPA)(H_2O)(OO\mathit{t}\text{-}Bu)]^{2+}$; (c) $[Fe(6\text{-}PhTPA)(H_2O)(OO\mathit{t}\text{-}Bu)]^{2+}$; (d) $[Fe(bpy)_2(py)(OO\mathit{t}\text{-}Bu)]^{2+}$; (e) $[Fe(TPA)(HOCH_2Ph)(OO\mathit{t}\text{-}Bu)]^{2+}$; (f) $[Fe(TPA)(H_2O)(OO\mathit{t}\text{-}Bu)]^{2+}$; (g) $[Fe(phen)_2(py)(OO\mathit{t}\text{-}Bu)]^{2+}$; (h) $[Fe(bpy)_2(py)(OO\mathit{t}\text{-}Bu)]^{2+}$; (i) $[Fe(phen)_2(MeOH)(OO\mathit{t}\text{-}Bu)]^{2+}$

6-PhTPA was built to test the insertion of the oxygen atom into the conveniently oriented aromatic C—H bond. Indeed the formation of the phenol group was observed (Scheme 11). Isotopic labeling experiments using $RO^{18}OH$ showed the oxygen atom of the phenol group was the ^{18}O [34]. This is indeed a remarkable result.

Table 6. EPR data for high-spin $S = 5/2$ Fe(III)OOR complexes, $^{(*)}$ indicates that the compound presents both spin states (see Table 5)

Complex	EPR data	Refs.
$[Fe(6\text{-}MeTPA)(H_2O)(OO\mathit{t}\text{-}Bu)]^{2+(*)}$	4.3	[33]
$[Fe(6\text{-}PhTPA)(H_2O)(OO\mathit{t}\text{-}Bu)]^{2+(*)}$	4.3	[34]
$[Fe(6\text{-}Me_2TPA)(H_2O)(OO\mathit{t}\text{-}Bu)]^{2+}$	4.3	[33]
$[Fe(6\text{-}Me_3TPA)(H_2O)(OO\mathit{t}\text{-}Bu)]^{2+}$	4.3	[33]
$[Fe(6\text{-}Me_3TPA)(OBz)(OO\mathit{t}\text{-}Bu)]^{+}$	4.3	[32]
$[Fe(6\text{-}Me_3TPA)(OBz)(OOcumyl)]^{+}$	4.3	[32]
$[Fe(BPPA)(OO\mathit{t}\text{-}Bu)]^{2+}$	7.58, 5.81, 4.25, 1.82 ($E/D = 0.067$)	[31]
$[Fe(BPPA)(OOcumyl)]^{2+}$	7.76, 5.65, 4.20, 1.78 ($E/D = 0.070$)	[31]

Table 7. Raman frequencies in cm^{-1} for $Fe(III)^{16}O^{16}OR$ or $Fe(III)^{18}O^{16}OR$ units

Complex	$Fe—^{16}O^{16}O—R$	$Fe—^{18}O^{16}O—R$	Refs.
$[Fe(TPA)(HOCH_2Ph)(OO\mathit{t}\text{-}Bu)]^{2+}$	490, 696, 796 (LS)	–	[35]
$[Fe(TPA)(H_2O)(OO\mathit{t}\text{-}Bu)]^{2+}$	490, 696, 796 (LS)	488, 638, 657, 672, 778	[33]
$[Fe(6\text{-}MeTPA)(H_2O)(OO\mathit{t}\text{-}Bu)]^{2+}$	488, 682, 790 (LS) 469, 648, 842, 878 (HS)	–	[33]
$[Fe(6\text{-}Me_2TPA)(H_2O)(OO\mathit{t}\text{-}Bu)]^{2+}$	470, 647, 843, 881 (HS)	–	[33]
$[Fe(6\text{-}Me_3TPA)(H_2O)(OO\mathit{t}\text{-}Bu)]^{2+}$	468, 637, 842, 877 (HS)	463, 612, 829, 867	[33]
$[Fe(6\text{-}Me_3TPA)(OBz)(OO\mathit{t}\text{-}Bu)]^{+}$	464, 648, 844, 880 (HS)	–	[32]
$[Fe(6\text{-}Me_3TPA)(OBz)(OOcumyl)]^{+}$	432, 554, 650, 880 (HS)	432, 626, 804, 842	[32]
$[Fe(BPPA)(OO\mathit{t}\text{-}Bu)]^{2+}$	469, 629, 838, 873 (HS)	–	[31]
$[Fe(BPPA)(OOcumyl)]^{2+}$	493, 639, 838, 878 (HS)	–	[31]
$[Fe(bpy)_2(HOCH_2Ph)(OO\mathit{t}\text{-}Bu)]^{2+}$	678, 808 (LS)	–	[36]
$[Fe(bpy)_2(HOCH_2Ph)(OOcumyl)]^{2+}$	696, 805 (LS)	–	[36]
$[Fe(6\text{-}PhTPA)(H_2O)(OO\mathit{t}\text{-}Bu)]^{2+}$	502, 683 (LS) 469, 644, 841, 873 (HS)	–	[34]

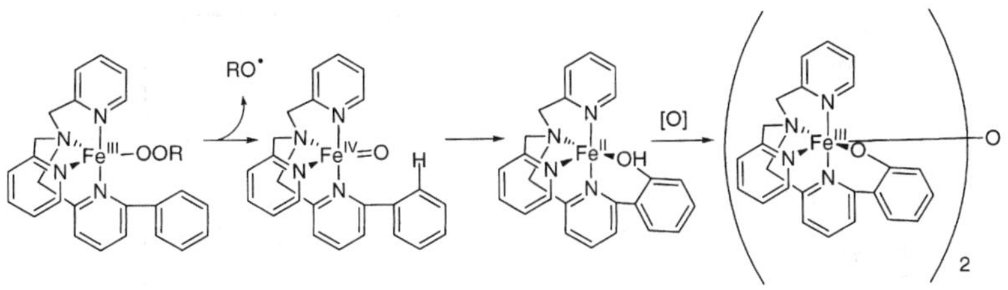

Scheme 11.

4
Fe-Peroxo Units Fe(III)O_2

With the $EDTA^{4-}$ ligand, an interesting FeO_2 species has been obtained [39–41]. Analogous species were observed by deprotonation of the Fe(III)OOH species formed with the ligands trispicMeen [42], TPEN [8], and Rtpen [11] (see above).

4.1
UV-Visible Spectroscopy

UV-vis data are reported in Table 8. The LMCT band in the visible region for $[(EDTA)Fe(O_2)]^{3-}$ occurs at 520 nm [39–41].

This complex has been studied in detail by Neese and Solomon [41]. With the aminopyridine ligands, a LMCT band is observed around 750 nm, at a lower energy than that of $[(EDTA)Fe(O_2)]^{3-}$. This is likely to be related to the stronger π acidity of the aminopyridine type ligands. For these ligands the deprotonation of the Fe(III)OOH upon addition of a base was detected by the spectacular change in color from purple to blue as demonstrated by the UV-visible spectra of $[(trispicMeen)Fe(OOH)]^{2+}$ and $[(trispicMeen)Fe(O_2)]^{+}$ (Fig. 7) [42].

4.2
EPR Spectroscopy

The $[(EDTA)Fe(O_2)]^{3-}$ complex has been found to be high spin rhombic (Table 9).

Table 8. Wavelength (molar extinction coefficient) of the peroxo → Fe(III) LMCT for Fe(III)O_2 units

Complex	$\lambda_{max}(\varepsilon/M^{-1}\cdot cm^{-1})$	Refs.
$[(EDTA)Fe(\eta^2\text{-}O_2)]^{3-}$	520 nm (520) in water	[39, 40, 41]
$[(trispicMeen)Fe(\eta^2\text{-}O_2)]^{+}$	740 nm (500) in MeOH	[42]
$[(TPEN)Fe(\eta^2\text{-}O_2)]^{+}$	755 nm (450) in MeOH	[8]
$[(ettpen)Fe(\eta^2\text{-}O_2)]^{+}$	747 nm (500) in MeOH	[11]

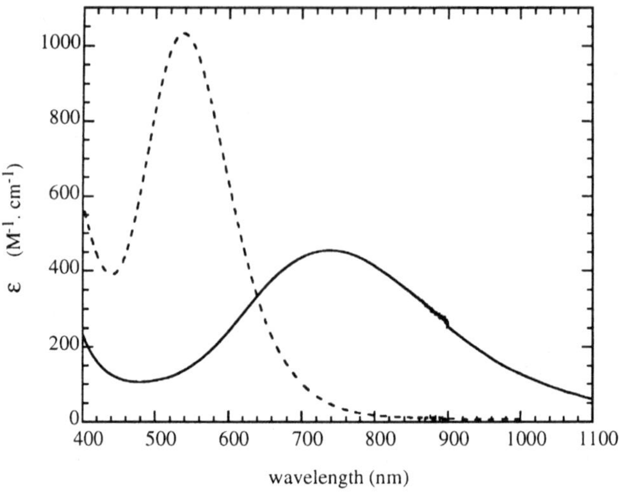

Fig. 7. UV-vis spectra of $[(\text{trispicMeen})\text{Fe(OOH)}]^{2+}$ (*dashed line*) and $[(\text{trispicMeen})\text{Fe}(O_2)]^+$ (*bold line*) in MeOH (from Simaan et al. [42])

The ZFS parameters are found to be $E/D = 0.21$ and $D = -1 \pm 0.25\ \text{cm}^{-1}$ [41]. The aminopyridine type systems have a different EPR signature: for instance, $[(\text{trispicMeen})\text{Fe}(O_2)]^+$ [42] (Table 9) has g values which indicate $E/D = 0.08$. The D value has not been determined.

Table 9. EPR data for high-spin $S = 5/2$ $\text{Fe(III)}O_2$ complexes

Complex	EPR data	Refs.
$[(\text{EDTA})\text{Fe}(\eta^2\text{-}O_2)]^{3-}$	$E/D = 0.21$ and $D = -1 \pm 0.25\ \text{cm}^{-1}$	[40, 41]
$[(\text{trispicMeen})\text{Fe}(\eta^2\text{-}O_2)]^+$	$g = 7.5, 5.9$ ($E/D = 0.08$)	[42]
$[(\text{TPEN})\text{Fe}(\eta^2\text{-}O_2)]^+$	$g = 10, 8.1, 5.6, 3.2, 2.4$ ($E/D = 0.1$)	[8]
$[(\text{bztpen})\text{Fe}(\eta^2\text{-}O_2)]^+$	$g = 7.60, 5.74$	[11]

4.3 Resonance Raman Spectroscopy

Raman spectra have been measured and proved the η^2-coordination mode of the peroxo group (Table 10).

The NCA treatment of FeO_2 taking into account f_{OO} for the O—O bond and f_{FeO} for the Fe—O bonds has been published [41]. The expressions given below use the angle OFeO = 2α as a parameter (Scheme 12).

The number of frequencies is $3n-6 = 3$. They are indicated below:

$$\begin{aligned} \nu_{\text{FeOas}} &= 1302.8\sqrt{[f_{\text{FeO}}(\mu_O + \mu_{\text{Fe}}(1-\cos 2\alpha))]} \\ \nu_{\pm} &= 1302.8\sqrt{[(A \pm \sqrt{B})/2]} \end{aligned} \tag{5}$$

Table 10. Raman ν_{OO}, ν_{FeO}, and ν_{FeOas} frequencies in cm^{-1} for $Fe(III)O_2$ with $^{16}O^{16}O$ or $^{18}O^{18}O$. Calculated values are noted with *: a) diatomic vibrators: $f_{OO} = 3.14$ mdyn/Å; $f_{FeO} = 1.55$ mdyn/Å; b) polyatomic vibrators calculation [41]: $f_{OO} = 3.02$ mdyn/Å; $f_{FeO} = 1.55$ mdyn/Å; c) diatomic vibrators: $f_{OO} = 3.17$ mdyn/Å; $f_{FeO} = 1.61$ mdyn/Å; d) polyatomic vibrators calculation: $f_{OO} = 3.015$ mdyn/Å; $f_{FeO} = 1.60$ mdyn/Å

Complex	$^{16}O^{16}O$			$^{18}O^{18}O$			Refs.
	ν_{OO}	ν_{FeO}	ν_{FeOas}	ν_{OO}	ν_{FeO}	ν_{FeOas}	
$[(EDTA)Fe(\eta^2\text{-}O_2)]^{3-}$	816	459	–	776 (−40)	446 (−13)	–	[40, 41]
	816*[a]	459*	–	770* (−46)	439* (−20)	–	
	820*[b]	462*	421*	774* (−46)	445* (−17)	−398* (−23)	
$[(trispicMeen)Fe(\eta^2\text{-}O_2)]^{+}$	820	468	–	776 (−44)	454 (−14)	–	[8]
	820*[c]	469*	–	773* (−47)	448* (−19)	–	
	821*[d]	468*	427*	775* (−46)	451* (−17)	405* (−22)	
$[(TPEN)Fe(\eta^2\text{-}O_2)]^{+}$	821	468	–	–	–	–	[8]

with

$$A = f_{FeO}(\mu_O + \mu_{Fe}(1 + \cos 2\alpha)) + 2f_{OO}\mu_O$$
$$B = [f_{FeO}(\mu_O + \mu_{Fe}(1 + \cos 2\alpha)) - 2f_{OO}\mu_o]^2 + 8f_{FeO}f_{OO}\mu_O^2 \sin^2 \alpha$$

One may notice that for $\alpha = 0$ the vibrations can be considered as ν_{FeOas}, ν_{FeOs}, ν_{OO}. For a non-zero angle, some mixing occurs but this remains small if the angle is small, in such a way that ν_+ can be considered as ν_{OO} and ν_- as ν_{FeOs}.

The resonance Raman data for $[(EDTA)Fe^{III}(O_2)]^{3-}$ have been reported as $\nu_{OO} = 816$ cm^{-1} and $\nu_{FeOs} = 459$ cm^{-1}. Upon $^{16}O/^{18}O$ substitution, those frequencies shift to 776 cm^{-1} and 446 cm^{-1}. Taking $f_{OO} = 3.02$ mdyn/Å and $f_{FeO} = 1.55$ mdyn/Å one obtains $\nu_{OO} = 820$ cm^{-1} and $\nu_{FeOs} = 462$ cm^{-1} for $^{16}O^{16}O$ and $\nu_{OO} = 774$ cm^{-1} and $\nu_{FeOs} = 445$ cm^{-1} for $^{18}O^{18}O$ [41]. The agreement is satisfying. The vibrations are rather "pure" since using the diatomic model one calculates the following shifted frequencies $\nu_{OO} = 770$ cm^{-1} and $\nu_{FeO} = 439$ cm^{-1} with $^{18}O^{18}O$ (see Table 10).

Another example is $[(trispicMeen)Fe(O_2)]^{+}$. It was found that $\nu_{OO} = 820$ cm^{-1} and $\nu_{FeOs} = 468$ cm^{-1} and that upon $^{16}O/^{18}O$ substitution, they shift to 776 cm^{-1} and 456 cm^{-1} [8]. Taking $f_{OO} = 3.015$ mdyn/Å and

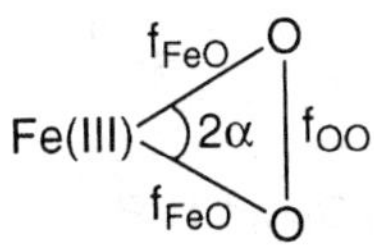

Scheme 12.

$f_{FeO} = 1.6$ mdyn/Å one obtains $\nu_{OO} = 821$ cm^{-1} and $\nu_{FeOs} = 468$ cm^{-1} for $^{16}O^{16}O$ and $\nu_{OO} = 775$ cm^{-1} and $\nu_{FeOs} = 451$ cm^{-1} for $^{18}O^{18}O$. This suggests that the Fe—O bonds with the neutral pyridine ligand are stronger than with $EDTA^{4-}$ which makes sense if one considers the electron acceptor role of pyridine. One remarks that in Fig. 4, the FeO_2 systems occupy a well defined zone which is different from the FeOOH and FeOOFe systems. In comparison with FeOOH, the weaker Fe—O bond in FeO_2 systems may be due to their HS state. Moreover the stronger O—O bond could imply a lower oxidative activity. Indeed it has been reported that $[(EDTA)Fe(O_2)]^{3-}$ is unreactive toward organic substrates at pH > 10.5 while it becomes very reactive when the pH decreases [39]. These observations strongly suggest that the oxidant species is or arises from a protonated peroxo species.

The geometry proposed for $[(EDTA)Fe(O_2)]^{3-}$ is depicted in Scheme 13 [41] and for $[(trispicMeen)Fe(O_2)]^{+}$ in Scheme 14 [42].

That with the ligand trispicMeen is more tentative and one cannot dismiss the heptadentate structure.

4.4 Mass Spectrometry

Formation of the Fe(III)-peroxo complexes has been confirmed by ESI-MS for the following aminopyridine systems: $[(trispicMeen)Fe(O_2)]^{+}$ (*m/z* 435) [42],

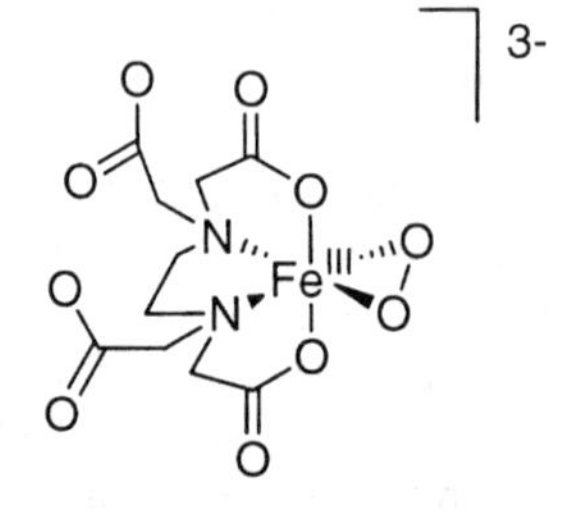

Scheme 13.

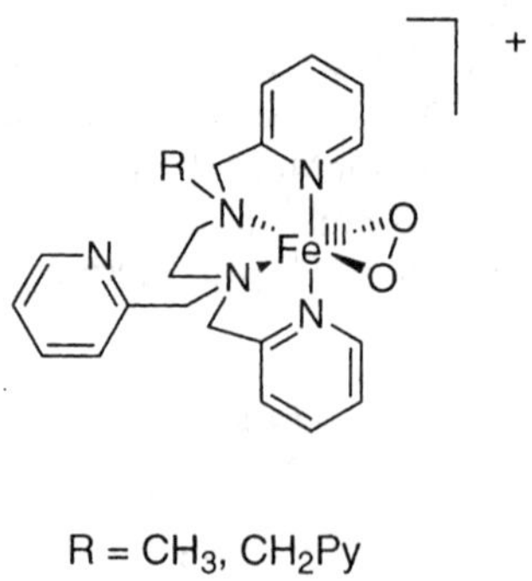

Scheme 14.

$[(TPEN)Fe(O_2)]^+$ (*m/z* 512.3) [8], and $[(bztpen)Fe(O_2)]^+$ (*m/z* 511.2) [11]. It has also been observed that this peak disappears as soon as the blue coloration disappears [8, 42].

5 Peroxo Bridged Diiron Units Fe(III)OOFe(III)

The ligands allowing the preparation of Fe(III)—O—O—Fe(III) motifs are represented in Fig. 8. They are of two types:

1. Non-assembling ligands such as $HB(pz')_3^-$ or $Mes_2ArCO_2^-$
2. Assembling ligands which use either a phenolato or an alkoxo bridge

5.1 X-Ray Diffraction

In three cases a peroxo dimer has been crystallized and an X-ray diffraction study has revealed the structure [43–45]. In the complex $Fe_2(O_2)(O_2CCH_2Ph)_2\{HB(pz')_3\}_2$ prepared by Kim and Lippard [43], the fragment FeOOFe has a *cis-gauche* conformation (dihedral angle = 53°, Fig. 9).

The Fe—O distances are 1.881(6) Å, 1.877(6) Å and the O—O distance is 1.406(8) Å. The two Fe(III) ions are bridged by two carboxylates and one peroxo dianion.

Figure 10 reproduces the structure of $[Fe_2(O_2)(Ph\text{-}bimp)(OBz)]^{2+}$ found by Ookubo et al. [44]. This complex uses the assembling heptadentate, monoanionic, ligand Ph-bimp$^-$. The two Fe(III) ions are bridged by the phenolato oxygen atom, one carboxylato group, and one peroxo. The fragment FeOOFe has a *cis* almost planar conformation (dihedral angle = 10°, Fig. 10).

The Fe—O distances are 1.944(4) Å, 1.864(4) Å and the O—O distance is 1.427(6) Å.

Figure 11 gives the structure of $[Fe_2(O_2)(N\text{-}Et\text{-}HPTB)(Ph_3PO)_2]^{3+}$ [45]. The assembling ligand is an heptadentate monoanion. The two Fe(III) ions are only doubly bridged here, by a peroxo and an alkoxo group. This imposes on the FeOOFe fragment a *cis* planar conformation (dihedral angle = 0°). The Fe—O distances are 1.880(4) Å and the O—O distance is 1.416(7) Å.

5.2 UV-Visible Spectroscopy

UV-vis data are collected in Table 11 for some of the compounds (see Table 11 for references).

The most thorough optical study has been published by Brunold et al. for $Fe_2(O_2)(OBz)_2\{HB(pz')_3\}_2$ [46]. The main features related to peroxo/Fe(III) LMCT occur at 14,600 cm^{-1} and 27,200 cm^{-1} (ε = 3800 M^{-1} cm^{-1}). When the peroxo group bridges the two Fe(III) ions, the π^* HOMOs on the peroxo group are split into in-plane π^*_σ and out-of-plane π^*_π orbitals. The two main LMCT

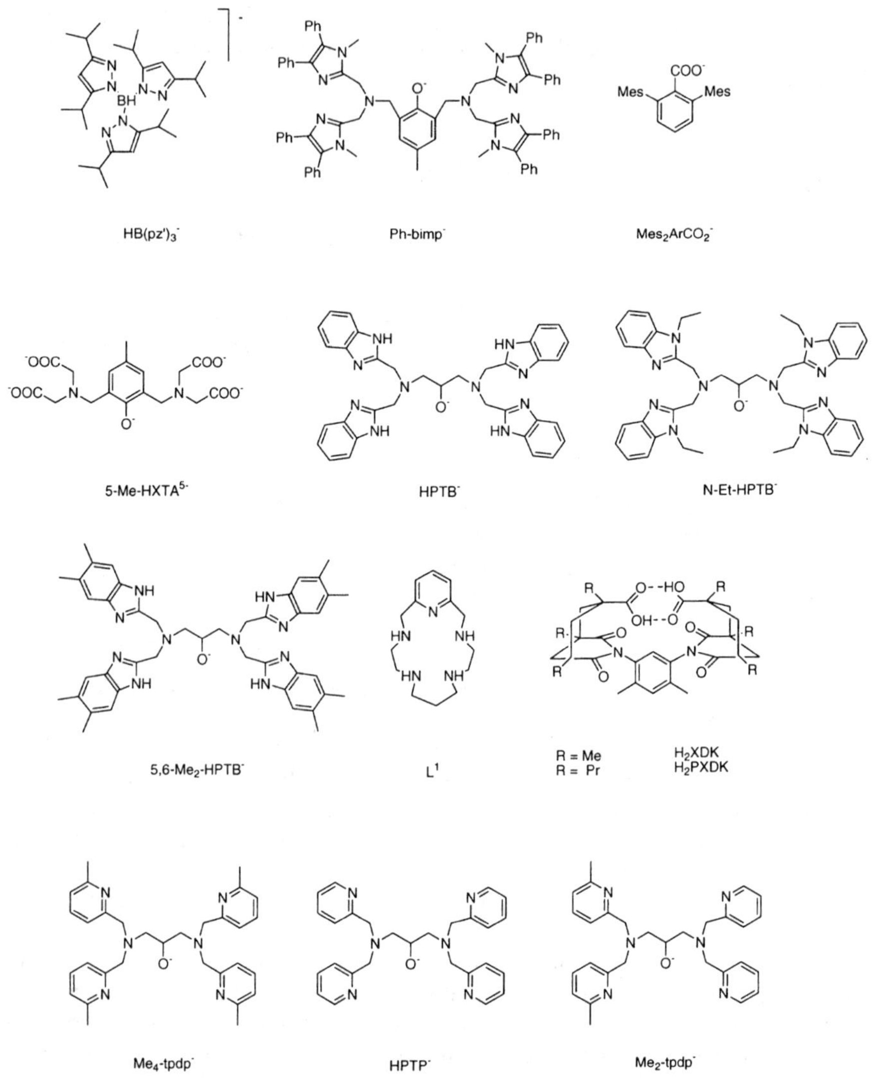

Fig. 8. Ligands which have allowed the characterization of Fe(III)OOFe(III) complexes, named according to literature

bands arise from transition between these two orbitals and the Fe *d* orbitals [46].

5.3 Resonance Raman Spectroscopy

The Raman data are summarized in Table 12. The ν_{OO} frequency occurs between 848 cm^{-1} and 900 cm^{-1}.

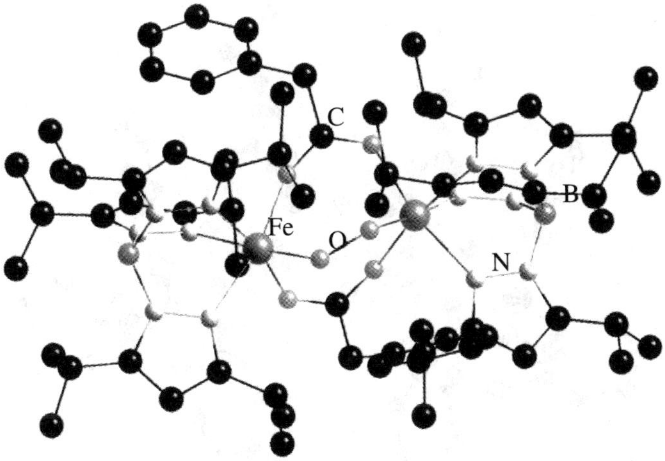

Fig. 9. X-ray diffraction structure of $Fe_2(\mu\text{-}O_2)(\mu\text{-}O_2CCH_2Ph)_2\{HB(pz')_3\}_2$ (from Kim and Lippard [43])

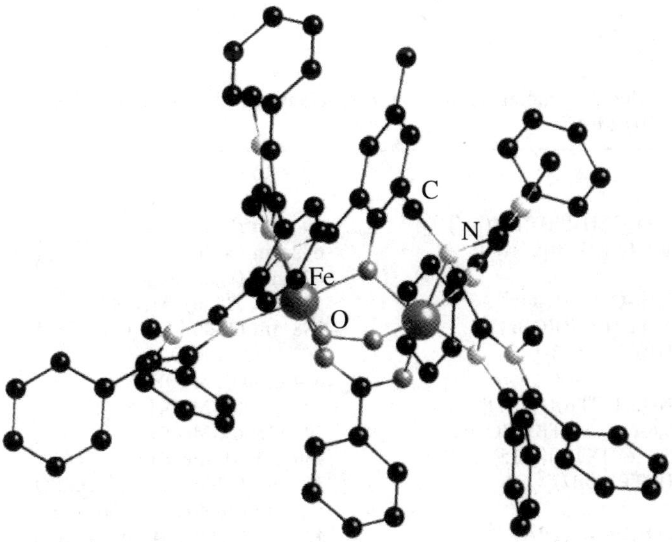

Fig. 10. X-ray diffraction structure of $[Fe_2(\mu\text{-}O_2)(\text{Ph-bimp})(\text{OBz})]^{2+}$ (found by Ookubo et al. [44])

This frequency is higher than in the Fe(III)OOH and Fe(III)O_2 systems (Fig. 4). In Fig. 4 one may also notice that the zone occupied by the FeOOFe systems is ill-defined. The Fe—O frequency is between 462 cm^{-1} and 481 cm^{-1} for the systems which are likely to possess a *cis* planar conformation [47–49] and is smaller (around 420 cm^{-1}) for the two systems which have a *cis-gauche*

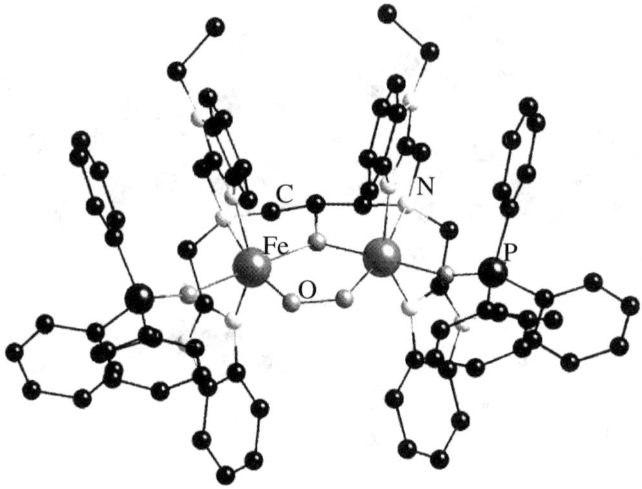

Fig. 11. X-ray diffraction structure of $[Fe_2(\mu\text{-}O_2)(N\text{-}Et\text{-}HPTB)(Ph_3PO)_2]^{3+}$ (found by Dong et al. [45])

Table 11. Wavelength (molar extinction coefficient) of the peroxo → Fe(III) LMCT for Fe(III)OOFe(III) units

Complex	λ_{max} (ε/$M^{-1}\cdot cm^{-1}$)	Refs.
$[Fe_2(\mu\text{-}O_2)(\mu\text{-}O_2CCH_2Ph)_2\{HB(pz')_3\}_2]$	694 nm (2650)	[43]
$[Fe_2(\mu\text{-}O_2)(\mu\text{-}OBz)_2\{HB(pz')_3\}_2]$	682 nm (3450)	[50]
	679 nm (?), toluene	
$[Fe_2(Ph_3PO)_8(\mu\text{-}O_2)\cdot 2H_2O]^{4+}$	576 nm (3540), MeCN	[51]
$[Fe_2(\mu\text{-}O_2)(N\text{-}Et\text{-}HPTB)(OBz)]^{2+}$	588 nm (1500), CH_2Cl_2	[52]
$[Fe_2(\mu\text{-}O_2)(HPTB)(NO_3)_2]^{+}$	560 (2200), H_2O	[47]
	604 (1600), MeOH	
$[Fe_2(\mu\text{-}O_2)(N\text{-}Et\text{-}HPTB)(NO_3)_2]^{+}$	606 (1500), MeOH	[47]
$[Fe_2(\mu\text{-}O_2)(5,6\text{-}Me_2\text{-}HPTB)(NO_3)_2]^{+}$	600 (1500), MeOH	[47]
$[Fe_2(\mu\text{-}O_2)(5\text{-}Me\text{-}HXTA)(OBz)]^{2-}$	480 (2370), MeOH	[53]
$[Fe_2(\mu\text{-}O_2)(HPTP)(OBz)]^{2+}$	562 nm (3200), CH_2Cl_2/DMSO	[48]
	572 nm (2060), CH_3CN/DMSO	
$[Fe_2(\mu\text{-}O)(\mu\text{-}O_2)(6\text{-}Me_3TPA)_2]^{2+}$	494 nm (1100), 648 nm (1200)	[49]
	846 nm (230), CH_3CN	
$[Fe_2(\mu\text{-}O_2)(Ph\text{-}bimp)(OBz)]^{2+}$	500–800 nm (1700), CH_3CN	[44]
"$[Fe_2(Mes_2ArCO_2)_4(\mu\text{-}O_2)]$"	540 nm (2300), CH_2Cl_2	[54]
"$[Fe_2(\mu\text{-}O_2)(Me_4\text{-}tpdp)(OBz)]$"	616 nm (2000), CH_2Cl_2	[55]
"$[Fe_2(\mu\text{-}O_2)(Me_4\text{-}tpdp)(CF_3COO)]$"	618 nm (1000), CH_2Cl_2	[55]
"$[Fe_2(\mu\text{-}O_2)(Me_2\text{-}tpdp)(CF_3COO)]$"	603 nm (1400), CH_2Cl_2	[55]
"$[Fe_2(\mu\text{-}O_2)(\mu\text{-}OBz)(\mu\text{-}XDK)(ImH)(OBz)]$"	660 nm (≈170), THF	[56]
"$[Fe_2(\mu\text{-}O_2)(\mu\text{-}O_2CC(CH_3)_3)(\mu\text{-}PXDK)$ $(N\text{-}MeIm)_2(O_2CC(CH_3)_3)]$"	670 nm (≈170), THF	[56]
"$[Fe_2(L^1)_2(\mu\text{-}O_2)]$"	540 nm (187), water	[57]

Table 12. Raman ν_{OO}, ν_{FeO}, and ν_{FeOas} frequencies in cm^{-1} for Fe(III)O_2Fe(III) with $^{16}O^{16}O$ or $^{18}O^{18}O$

Complex	$^{16}O^{16}O$			$^{18}O^{18}O$ (Raman shifts in cm^{-1})			Refs.
	ν_{OO}	ν_{FeO}	ν_{FeOas}	ν_{OO}	ν_{FeO}	ν_{FeOas}	
$[Fe_2(\mu\text{-}O_2)(\mu\text{-}O_2CCH_2Ph)_2\{HB(pz')_3\}_2]$	888	415	–	842 (−46)	404 (−11)	–	[43]
$[Fe_2(\mu\text{-}O_2)(\mu\text{-}OBz)_2\{HB(pz')_3\}_2]$	876	418	–	827 (−49)	409 (−9)	–	[50a]
$[Fe_2(Ph_3PO)_8(\mu\text{-}O_2)\cdot 2H_2O]^{4+}$	882	445	565	848 (−34)	–	–	[51]
$[Fe_2(\mu\text{-}O_2)(HPTB)(NO_3)_2]^{+}$	895	476	–	854 (−41)	457 (−19)	–	[47]
$[Fe_2(\mu\text{-}O_2)(N\text{-}Et\text{-}HPTB)(NO_3)_2]^{+}$	890	476	–	838 (−52)	460 (−16)	–	[47]
$[Fe_2(\mu\text{-}O_2)(5{,}6\text{-}Me_2\text{-}HPTB)(NO_3)_2]^{+}$	888	474	–	–	–	–	[47]
$[Fe_2(\mu\text{-}O_2)(N\text{-}Et\text{-}HPTB)(OBz)]^{2+}$	900	476	–	850 (−50)	460 (−16)	–	[48]
$[Fe_2(\mu\text{-}O_2)(HPTP)(OBz)]^{2+}$	877, 893	453, 481	–	834	444		[48]
$[Fe_2(\mu\text{-}O)(\mu\text{-}O_2)(6\text{-}Me_3TPA)_2]^{2+}$	848	462	–	802 (−46)	441 (−21)	–	[49]
"$[Fe_2(Mes_2ArCO_2)_4(\mu\text{-}O_2)]$"	885	–	–	871 (−14)	–	–	[54]
"$[Fe_2(\mu\text{-}O_2)(Me_4\text{-}tpdp)(OBz)]$"	891, 918	450	486	857, 889	442 (−8)	479 (−7)	[55]
$[Fe_2(\mu\text{-}O_2)(5\text{-}Me\text{-}HXTA)(OBz)]^{2-}$	884	–	–	–	–	–	[53]
Desaturase	898	442	490	845 (−53)	425 (−17)	471 (−19)	[60]
RNR mutant	870	458	499	824 (−46)	442 (−16)	477 (−22)	[59]
Ferritin	851	485	499	800 (−51)	468 (−17)	487 (−12)	[58]

conformation [43, 50a]. In these two last cases, the Fe—O frequencies are the lowest known to date for all the systems in this review.

In their study, Brunold et al. [46] specifically addressed the question of the large value of the O—O stretching and proposed that it is due to mechanical coupling with Fe—O stretchings and that intrinsically the strength of the O—O bond (f_{OO} = 3.1 mdyn/Å) is similar to that in oxyhemet (f_{OO} = 3.29 mdyn/Å) or FeO_2 (f_{OO} = 3.02 mdyn/Å) systems.

Raman data have been recently published for enzymes which present a peroxo intermediate which is very likely of the μ-1,2-type. In ferritin [58], the following Raman frequencies have been observed (Table 12): ν_{OO} = 851 cm^{-1}, ν_{FeOs} = 485 cm^{-1}, ν_{FeOas} = 499 cm^{-1}. The ν_{OO} and ν_{FeOs} values are included in Fig. 4. They are close to the *cis* planar model system prepared with 6-Me_3TPA [49]. Data reported for a RNR mutant [59] (ν_{OO} = 870 cm^{-1}, ν_{FeOs} = 458 cm^{-1}, ν_{FeOas} = 499 cm^{-1}) and a Desaturase [60] (ν_{OO} = 898 cm^{-1}, ν_{FeOs} = 442 cm^{-1}, ν_{FeOas} = 490 cm^{-1}) are different from each other. In Fig. 4 they are intermediate between values measured for model complexes, although one may argue that the signals for desaturase are closer to the *cis-gauche* model systems and that for the RNR mutant is closer to *cis* planar. It seems that the *cis-gauche* $Fe_2(O_2)(O_2CCH_2Ph)_2\{HB(pz')_3\}_2$ chemical model is very close to the peroxo intermediate in Desaturase.

6 Conclusion

Iron-peroxo or hydroperoxo systems are important intermediates between the initial Fe(II)-dioxygen adduct and the "activated" form of the catalytic site in many enzymes [2, 5].

In Scheme 15, mononuclear iron systems in which peroxo forms have been identified to date are summarized.

The systems with a question mark have been proposed to imply the intermediate of the type indicated but to our knowledge this intermediate has not been spectroscopically characterized.

The η^2-peroxo form has been identified with heme systems [61, 62], $EDTA^{4-}$ [39–41], and polydentate aminopyridine ligands [8, 11, 42]. The η^1-peroxo may exist in some heme systems but it is not yet proven spectroscopically. These peroxo forms can be protonated. In case of aminopyridine or BLM, the monoprotonated form Fe(III)(η^1-OOH) can be isolated and spectroscopically studied (see above). In heme systems or $EDTA^{4-}$, the Fe(III)(η^1-OOH) seems basic and undergoes a second protonation to lead to Fe(V)O, which prohibits its observation. With aminopyridine and BLM it is possible that a second protonation occurs but information is lacking. The low spin nature of Fe(III)(η^1-OOH) seems decisive in explaining the strength of the Fe—O bond and the weakness of the O—O bond, preludes to the evolution of Fe(III)(η^1-OOH) into a Fe(V)O type structure by breakage of the O—O bond.

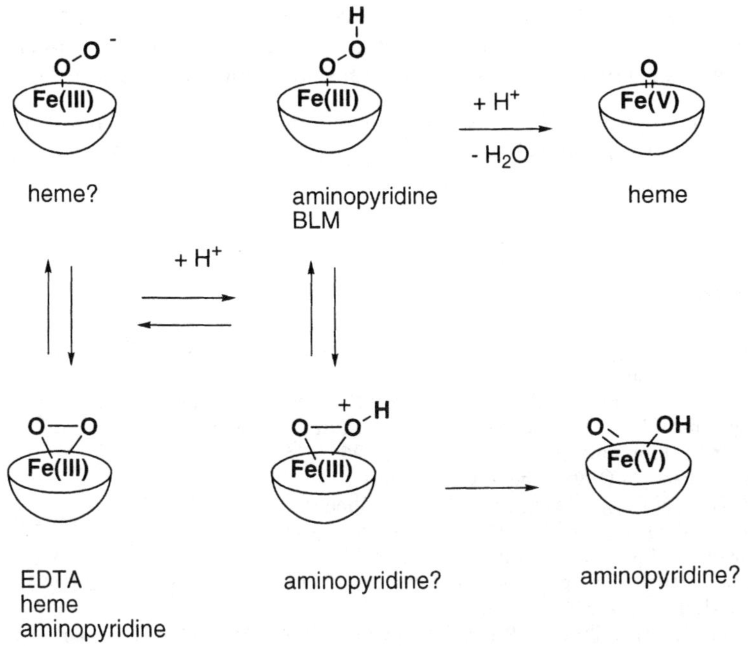

Scheme 15.

Although to date spectroscopic evidence has been given only for Fe(III)(η^2-OO) and Fe(III)(η^1-OOH) species, the existence of the "complementary" Fe(III)(η^1-OO) and Fe(III)(η^2-OOH) species is quite possible and their spectroscopic detection would be very valuable.

Finally, the peroxo bridged Fe(III)OOFe(III) systems are also fascinating. How these change to the Fe(IV)O_2Fe(IV) form which has been shown to be the active species in MMO is a very interesting question [63]. One possibility is an internal rearrangement which would be analogous to that which is observed in copper chemistry (see the chapter by Tolman in this book).

Another possibility has been suggested recently by Brunold et al. (Scheme 16) [46].

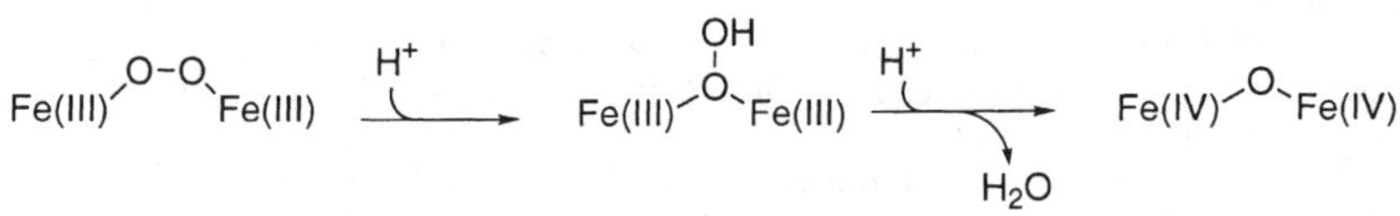

Scheme 16.

It implies the protonation of a *cis* μ-1,2 peroxo group to give a μ-η^1 hydroperoxo system which upon a second protonation leads to the Fe(IV)Fe(IV) dinuclear site. An alternative leading to incorporation into the

$Fe(IV)O_2Fe(IV)$ form, of only one of the two oxygen atoms of O_2, has recently been proposed by Lee and Lipscomb [64]. Certainly spectroscopy on chemical models and proteins will allow one to track the details of this transformation.

Note added in Proof. During the preparation of this manuscript one more example of Fe-Peroxo Unit $Fe(III)O_2$ with the N4Py ligand has been submitted [65]. It has appeared after the completion of the present review.

7 References

1. Lippard SJ, Berg JM (1994) Principles of bioinorganic chemistry. University Science Books, Mill Valley
2. Sono M, Roach MP, Coulter ED, Dawson JH (1996) Chem Rev 96: 2841
3. Ferguson-Miller S, Babcock GT (1996) Chem Rev 96: 2889
4. Wallar BJ, Lipscomb JD (1996) Chem Rev 96: 2625
5. Que L, Ho RYN (1996) Chem Rev 96: 2607
6. Sam JW, Tang XJ, Peisach J (1994) J Am Chem Soc 116: 5250
7. Mizoguchi TJ, Lippard SJ (1998) J Am Chem Soc 120: 11022
8. Simaan AJ et al., Eur J Inorg Chem, in press
9. Bernal I, Jensen IM, Jensen KB, McKenzie CJ, Toftlund H, Tuchagues JP (1995) J Chem Soc Dalton Trans. 3667
10. Mialane P, Nivorojkine A, Pratviel G, Azéma L, Slany M, Godde F, Simaan A, Banse F, Kargar-Grisel T, Bouchoux G, Sainton J, Horner O, Guilhem J, Tchertanova L, Meunier B, Girerd JJ (1999) Inorg Chem 38: 1085
11. Jensen KB, McKenzie CJ, Nielsen LP, Pedersen JZ, Svendsen HM (1999) Chem Commun 1313
12. Lubben M, Meetsma A, Wilkinson EC, Feringa B, Que L (1995) Angew Chem Int Ed Engl 34: 1512
13. Ho RYN, Roelfes G, Feringa BL, Que L (1999) J Am Chem Soc 121: 264
14. Roelfes G, Lubben M, Chen K, Ho RYN, Meestma A, Genseberger S, Hermant RM, Hage R, Mandal SK, Young VG, Zang Y, Kooijman H, Spek AL, Que L, Feringa BL (1999) Inorg Chem 38: 1929
15. deVries ME, LaCrois RM, Roelfes G, Kooijman H, Spek AL, Hage R, Feringa BL (1997) Chem Commun 1549
16. Lippai I, Magliozzo RS, Peisach J (1999) J Am Chem Soc 121: 780
17. Sauer-Masarwa A, Herron N, Fendrick CM, Busch DH (1993) Inorg Chem 32: 1086
18. Kim C, Chen K, Kim J, Que L (1997) J Am Chem Soc 119: 5964
19. Burger RM, Horwitz SB, Peisach J, Wittenberg JB (1979) J of Biol Chem 254: 12,299
20. Sobolev AP, Babushkin DE, Talsi EP (1996) Mendeleev Commun 236
21. Griffith JS (1961) The theory of transition metal ions. Cambridge University Press, London
22. Taylor CPS (1977) Biochimica et Biophysica Acta 491: 137
23. Veselov A, Sun H, Sienkiewicz A, Taylor H, Burger RM, Scholes CP (1995) J Am Chem Soc 117: 7508
24. Hori H (1971) Biochimica et Biophysica Acta 251: 227
25. (a) Brunold TC, Solomon EI (1999) J Am Chem Soc 121: 8277; (b) Brunold TC, Solomon EI (1999) J Am Chem Soc 121: 8288
26. Shiemke AK, Loehr TM, Sanders-Loehr J (1984) J Am Chem Soc 106: 4951
27. Kurtz DM, Shriver DF, Klotz IM (1976) J Am Chem Soc 98: 5033
28. Guajardo RJ, Hudson SE, Brown SJ, Mascharak PK (1993) J Am Chem Soc 115: 7971
29. Chen K, Que L (1999) Chem Commun 1375
30. Chen K, Que L (1999) Angew Chem Int Ed Engl 38: 2227

31. Wada H, Ogo S, Watanabe Y, Mukai M, Kitagawa T, Jitsukawa K, Masuda H, Einaga H (1999) Inorg Chem 38: 3592
32. Zang Y, Elgren TE, Dong Y, Que L (1993) J Am Chem Soc 115: 811
33. Zang Y, Kim J, Dong Y, Wilkinson EC, Appleman EH, Que L (1997) J Am Chem Soc 119: 4197
34. Lange SJ, Miyake H, Que L (1999) J Am Chem Soc 121: 6330
35. Kim J, Larka E, Wilkinson EC, Que L (1995) Ang. Chem Int Ed Engl 34: 2048
36. Ménage S, Wilkinson EC, Que L, Fontecave M (1995) Angew Chem Int Ed Engl 34: 203
37. Nguyen C, Guajardo RJ, Mascharak PK (1996) Inorg Chem 35: 6273
38. MacFaul PA, Ingold KU, Wayner DDM, Que L (1997) J Am Chem Soc 119: 10,594
39. Walling C, Kurz M, Schugar HJ (1970) Inorg Chem 9: 931
40. Ahmad S, McCallum JD, Shiemke AK, Appelman EH, Loehr TM, Sanders-Loehr J (1988) Inorg Chem 27: 2230
41. Neese F, Solomon EI (1998) J Am Chem Soc 120: 12,829
42. Simaan AJ, Banse F, Mialane P, Boussac A, Un S, Kargar-Grisel T, Bouchoux G, Girerd JJ (1999) Eur J Inorg Chem 993
43. Kim K, Lippard SJ (1996) J Am Chem Soc 118: 4914
44. Ookubo T, Sugimoto H, Nagayama T, Masuda H, Sato T, Tanaka K, Maeda Y, Okawa H, Hayashi Y, Uehara A, Suzuki M (1996) J Am Chem Soc 118: 701
45. Dong Y, Yan S, Young VG, Que L (1996) Angew Chem Int Ed Engl 35: 618
46. Brunold TC, Tamura N, Kitajima N, Moro-oka Y, Solomon EI (1998) J Am Chem Soc 120: 5674
47. Brennan BA, Chen Q, Juarez-Garcia C, True AE, O'Connor CJ, Que L (1991) Inorg Chem 30: 1937
48. Dong Y, Ménage S, Brennan BA, Elgren TE, Jang HG, Pearce LL, Que L (1993) J Am Chem Soc 115: 1851
49. Dong Y, Zang Y, Shu L, Wilkinson EC, Que L, Kauffmann K, Münck E (1997) J Am Chem Soc 119: 12,683
50. (a) Kitajima N, Fukui H, Moro-oka Y (1990) J Am Chem Soc 112: 6402; (b) Kitajima N, Tamura N, Amagai H, Fukui H, Moro-Oka Y, Mizutani Y, Kitagawa T, Mathur R, Heerwegh K, Reed CA, Randall CR, Que L, Tatsumi K (1994) J Am Chem Soc 116: 9071
51. Sawyer DT, McDowell MS, Spencer L, Tsang PKS (1989) Inorg Chem 28: 1166
52. Ménage S, Brennan BA, Juarez-Garcia C, Münck E, Que L (1990) J Am Chem Soc 112: 6423
53. Murch BP, Bradley FC, Que L (1986) J Am Chem Soc 108: 5027
54. Hagadorn JR, Que L, Tolman WB (1998) J Am Chem Soc 120: 13,531
55. Hayashi Y, Kayatani T, Sugimoto H, Suzuki M, Inomata K, Uehara A, Mizutani Y, Kitagawa T, Maeda Y (1995) J Am Chem Soc 117: 11,220
56. Herold S, Lippard SJ (1997) J Am Chem Soc 119: 145
57. Kimura E, Kodama M, Machida R, Ishizu K (1982) Inorg Chem 21: 595
58. Moënne-Loccoz P, Krebs C, Herlihy K, Edmonson, DE, Theil EC, Huynh BH, Loehr TM (1999) Biochemistry 38: 5290
59. Moënne-Loccoz P, Baldwin J, Ley BA, Loehr TM, Bollinger JM (1998) Biochemistry 37: 14,659
60. Broadwater JA, Ai J, Loehr TM, Sanders-Loehr J, Fox BG (1998) Biochemistry 37: 14,664
61. Friant P, Goulon J, Fischer J, Ricard L, Schappacher M, Weiss R, Momenteau M (1985) Nouveau J Chimie 9: 33
62. Burstyn JN, Roe JA, Miksztal AR, Shaevitz BA, Lang G, Valentine JS (1988) J Am Chem Soc 110: 1382
63. Shu L, Nesheim JC, Kauffmann K, Münck E, Lipscomb JD, Que L (1997) Science 275: 515
64. Lee SY, Lipscomb JD (1999) Biochemistry 38: 4423
65. Ho RYN, Roelfes G, Hermant R, Hage R, Feringa BL, Que L (1999) Chem Commun 2161

Copper-Dioxygen and Copper-Oxo Species Relevant to Copper Oxygenases and Oxidases

Allan G. Blackman[1], William B. Tolman[2]

[1] Department of Chemistry, University of Otago, P.O. Box 56, Dunedin, New Zealand
[2] Department of Chemistry and Center for Metals in Biocatalysis, University of Minnesota, 207 Pleasant St., SE, Minneapolis, MN 55455, USA
[1] *E-mail: blackman@alkali.otago.ac.nz*
[2] *E-mail: tolman@chem.umn.edu*

Copper proteins mediate both the transport and activation of dioxygen in a number of biological systems. The active sites of these proteins comprise mononuclear, dinuclear, and trinuclear copper centers, with the copper ions displaying a variety of coordination numbers and stereochemistries. Information regarding the mechanism by which these proteins activate dioxygen has been obtained by studies of the reactions of small molecule model copper complexes with dioxygen and its derivatives. Superoxo, peroxo, and bis(μ-oxo) intermediates in these reactions have recently been characterized by X-ray crystallography and this article concentrates on the structures of these intermediates, along with several Cu/O_2 complexes that have been well characterized by spectroscopic methods. The oxygenase-type reactivities of a number of copper complexes on reaction with dioxygen are also discussed.

Keywords: Copper complexes, Dioxygen, X-ray structures, C-H activation, Bioinorganic

List of Abbreviations

BArF	tetrakis[3,5-bis(trifluoromethyl)phenyl]borate
DMF	dimethylformamide
EPR	electron paramagnetic resonance
ESI	electrospray ionization
EXAFS	extended X-ray absorption fine structure
LMCT	ligand-to-metal charge transfer
MCD	magnetic circular dichroism
Me_2im	1,2-dimethylimidazole
pzH	pyrazole
SQUID	superconducting quantum interference device
UV/vis	ultra-violet/visible

1 Introduction

The biological importance of copper has been realized since the early years of the nineteenth century, when copper was first isolated from a number of plant species, although its exact function was not understood at that time [1]. Since then, copper proteins and enzymes have been identified in a variety of bacteria, fungi, plant and animal species, mediating such processes as electron transfer, dioxygen transport, dioxygen activation, reduction of nitrogen oxides to elemental nitrogen, and the disproportionation of superoxide ion. Of particular importance are hemocyanin, a dioxygen-carrier protein, and the copper oxidases and oxygenases, which either couple substrate oxidation to the reduction of dioxygen to hydrogen peroxide or water (Eq. 1) or incorporate O_2-derived oxygen atoms into substrate (Eq. 2), respectively:

$$O_2 \begin{cases} \xrightarrow{+2H^+ + 2e^-} H_2O_2 \\ \xrightarrow{+4H^+ + 4e^-} 2H_2O \end{cases} \tag{1}$$

$$\text{R–H} \xrightarrow[\text{Cu oxygenase}]{O_2} \text{R–OH} \tag{2}$$

Detailed structural information on these proteins has been obtained through X-ray crystallographic and spectroscopic studies, revealing similarities in the types of ligands bound to the active site copper ions (e.g., preponderance of histidine imidazolyls) but great diversity in the metal site nuclearities and

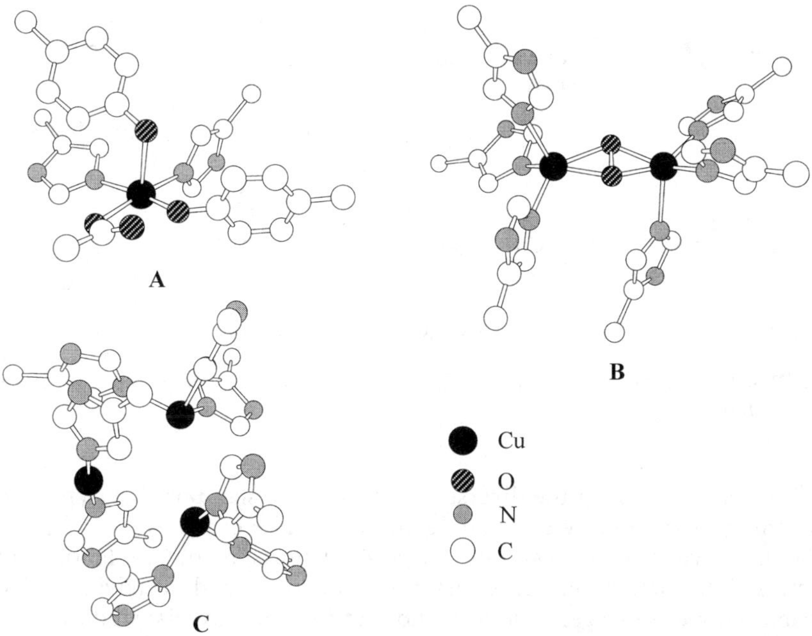

Fig. 1A–C. X-ray structures of the active sites of: **A** galactose oxidase at pH 4.5 (PDB Identification Code 1GOF) [2]; **B** oxyhemocyanin (PDB Identification Code 1NOL) [3]; **C** ascorbate oxidase (PDB Identification Code 1AOZ) [4]

coordination geometries (cf. selected examples in Fig. 1) [2–4]. These structural differences are paralleled by differences in mechanism and function. Thus, while the presence of one or more reduced, histidine-coordinated Cu(I) sites that react with atmospheric dioxygen is a common feature, the subsequent course of dioxygen reduction and substrate oxidation varies considerably among hemocyanin and the copper oxidases and oxygenases [5].

With the goal of unraveling important structure/function relationships pertinent to the larger problem of dioxygen binding and activation by metallobiomolecules [6–9], synergistic biochemical and synthetic modeling approaches have been taken in order to understand how dioxygen interacts with the reduced sites of these proteins and what factors control the fate of the resultant intermediates. Dioxygen may be envisaged to bind to a single metal ion following 1- or 2-electron reduction as either superoxide (O_2^-) or peroxide (O_2^{2-}), and can coordinate in an "end-on" η^1 or "side-on" η^2 fashion (Fig. 2). Protonation may occur to yield a hydroperoxide ligand, for example, or subsequent coordination of the bound ligand to a second copper ion can result in a number of bridging modes, such as μ-η^1, μ-η^1:η^1 (*cis* or *trans*), or μ-η^2:η^2. Still further bonding modes are potentially available in tri- and tetranuclear complexes. The O—O bond may also be cleaved on coordination to the copper ion, resulting in complexes containing μ-oxo bridges derived from dioxygen.

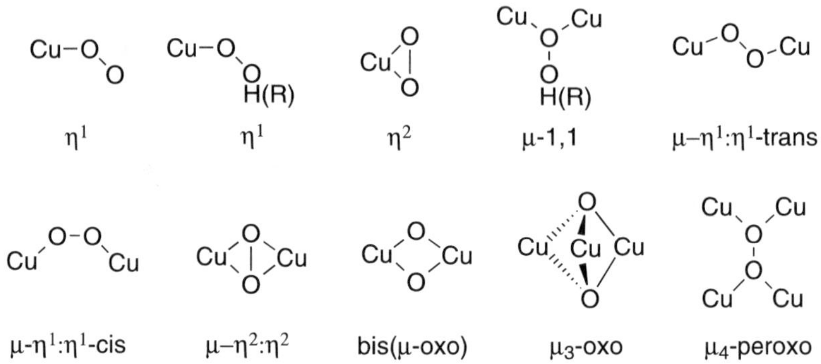

Fig. 2. Possible coordination modes of dioxygen derivatives in mono- and polynuclear copper complexes

Interconversions among the diverse structures represent possible steps in the mechanisms traversed by the oxidase and oxygenase enzymes.

In this review we concentrate on recent efforts to obtain fundamental chemical information on the structures, properties, and reactivities of the possible copper-dioxygen adducts and derived species via studies of the reactivity of copper complexes with O_2, H_2O_2, or related species. While such studies are often complicated by the fact that copper complexes of superoxide and peroxide are generally relatively unstable, the combined use of low-temperature handling techniques, ligand design principles, and improved analytical methods have recently led to great advances. Most importantly, X-ray crystal structures of reactive model complexes have facilitated detailed correlation of structural and spectroscopic features, ultimately enabling entry into biorelevant mechanistic studies. Here we will emphasize complexes that have been structurally well defined by X-ray diffraction methods. These are listed in Table 1 along with important metric and spectroscopic data, while Fig. 3 gives the structures of the ligands discussed in this article. Recent reviews on systems characterized by other methods or on related topics are also available [10–19].

2 Structures of Copper-Dioxygen and Copper-Oxo Model Complexes

2.1 Mononuclear Complexes

The initial interaction of a dioxygen molecule with a copper ion in a metalloenzyme undoubtedly results in monodentate η^1 or "side-on" η^2 coordination of a reduced dioxygen species. Thus, the synthesis and characterization of mononuclear copper complexes containing superoxide or peroxide and their derivatives is important in order to gain insight into the initial O_2-binding process. Such syntheses have proved to be non-trivial due to a propensity for either attack of a second Cu(I) ion to yield dimeric species or for rapid decomposition to occur, so few mononuclear Cu/O_2 complexes have been unequivocally structurally characterized. In a strategy with ample precedent in the porphyrin literature [20], dimerization reactions may be inhibited by using sterically bulky multidentate ligands. In some cases, these ligands may also provide a protected cleft in which the bound O_2-species may be stabilized by interactions with ligand substituents. This approach has been increasingly exploited over the past decade, resulting in several reports of the successful synthesis and characterization of important mononuclear Cu/O_2 complexes.

2.1.1 *Superoxide Complexes*

Monodentate copper-superoxide complexes $[LCuO_2]^+$ (L = TMPA or BQPA) have been characterized spectroscopically and shown by kinetic studies to be precursors to dimeric peroxo species, but isolation and structural characterization has not been possible [21, 22]. The spectroscopic characterization of the deep-red diamagnetic superoxo complex $[Cu(L^1H_2)(O_2)(NEt_3)]$ obtained on treatment of a Cu(I) precursor with O_2 at −50 °C to −70 °C has also recently been reported [23]. In an exciting paper, reaction of a TMPA derivative, tppa, with $[Cu(CH_3CN)_4]ClO_4$ and O_2 in MeOH at −80 °C was reported to yield the monodentate superoxide complex $[Cu(tppa)(O_2)]ClO_4$ [24]. The electronic spectrum of the reaction mixture consisted of two bands in the d-d region at 657 nm and 803 nm and a shoulder at 315 nm which was assigned to a $O_2^- \rightarrow Cu^{2+}$ charge transfer transition; these features are similar to those reported previously for the TMPA and BQPA systems [21, 22]. The reaction mixture was EPR silent at −80 °C and this was proposed to be due to strong antiferromagnetic coupling between the unpaired electrons on the Cu atom and the superoxide ligand. No vibrational spectroscopic data to support the superoxide formulation were reported. Storage of the reaction mixture at low temperature for several days gave a crystalline product that was subjected to room-temperature crystal structural analysis. The data were interpreted to show the presence of a monodentate coordinated superoxide ligand stabilized by hydrogen bonding to the amide protons of the tripodal ligand (Fig. 4).

However, the structure of the supposed superoxo complex was found by other workers to be incorrect; redetermination using the original X-ray data showed it to be $[Cu(tppa)OH]ClO_4$ [25] (Fig. 5). The correct direct-methods solution showed the position of all non-hydrogen atoms with the exception of the unbound oxygen atom (originally reported to have unreasonably large thermal parameters) and no peak corresponding to this atom was observed in the difference Fourier map after several cycles of least-squares refinements. Instead, a peak corresponding to the hydrogen atom of a hydroxide ligand was found and the complex was therefore formulated as a unique example of a Cu^{2+}-OH species in which the hydroxide ligand engages in intramolecular hydrogen bonding entirely different from that postulated for the putative superoxide complex.

Table 1. Metric and spectroscopic parameters for Cu-dioxygen, oxo and alkyl/acyl/hydroperoxo complexes that have been characterized by X-ray crystallography

Complex[a]	d(Cu—O)/Å	d(O—O)/Å	d(Cu···Cu)/Å	λ_{max}/nm (ε/M^{-1} cm^{-1})	ν(O—O)/cm^{-1}[b]	Refs.
[Cu(HB(3-*t*-Bu-5-*i*-Prpz)$_3$)(O_2)]	1.84 (1)	1.22 (3)		352 (2330) 510 (230)	1111 (1062)	[28]
[Cu(HB(3,5-*i*-Pr_2pz)$_3$)(OOCMe_2Ph)]	1.816 (4) 1.814 (4)[c]	1.460 (6) 1.454 (6)		348 (4000) 569 (3600) 811 (400)	Not reported	[32]
[Cu(bppa)(OOH)]ClO_4	1.888 (4)	1.460 (6)		380 (890) 660 (150) 830 (250)	856 (810)	[36]
[Cu_2(XYL-O-)(*m*-ClC_6H_4C(O)O_2)] (ClO_4)$_2$ · CH_3CN	1.948 (8) 1.971 (9)	1.463 (12)	3.197	395 (5500–5900) 650 (420)	Not reported	[38]
[{Cu(TPA)}$_2$(O_2)](PF_6)$_2$ · 5Et_2O	1.852 (5)	1.432 (6)	4.359 (1)	435 (1700) 524 (11300) 615 (5800) 1035 (180)	832 (788)	[44, 46]
[{Cu(HB(3,5-*i*-Pr_2pz)$_3$)}$_2$(O_2)]	1.903 (11) 1.927 (9)	1.412 (12)	3.560 (3)	349 (21000) 551 (790)	741 (698)	[31, 52]
[{Cu(d_{21}-Bn_3TACN)}$_2$(O)$_2$] (SbF_6)$_2$ · 7(CH_3)$_2$CO · 2CH_3CN	1.808 (5) 1.803 (5)	2.287 (2)	2.794 (2)	318 (12000) 430 (14000)	602,608 (583) (Cu—O)	[64]
[Cu_2(d_{28}-*i*-Pr_4DTNE)(μ-O)$_2$] (SbF_6)$_2$ · 3CH_2Cl_2	1.823 (4) 1.836 (4) 1.824 (4) 1.827 (4)	2.351 (2)	2.783 (1)	316 (13000) 414 (14000)	600 (582,574) (Cu—O)	[65]

$[\{Cu(L_{ME})\}_2(O)_2](CF_3SO_3)_2 \cdot 4CH_2Cl_2$	1.814 (6) 1.809 (6) 1.796 (6) 1.804 (6)	2.344 (1)	2.743 (1)	306 (21000) 401 (28000)	610 (587) (Cu—O)	[67]
$[(L_{TM})_3Cu_3O_2](CF_3SO_3)_3$ [d]	1.83 (1) 1.83 (1) 1.98 (1) 2.01 (1) 2.01 (1) 2.04 (2)	2.37	2.641 (3) 2.704 (3)	290 (12500) 355 (15000) 480 (1400) 620 (800)	Not reported	[85]
$[Cu_4(L^2)_2(O_2)(OMe)_2(ClO_4)]$ $ClO_4 \cdot MeOH$	1.961 (2) 1.939 (2)	1.453 (4)	2.994 (2) 3.030 (2) 4.184 (3) 4.317 (2)	284 (15900) 384 (9700) 587 (610)	878 (841)	[87, 90]

[a] Ligand abbreviations are given in Fig. 3.
[b] Numbers in parentheses refer to the ^{18}O-labeled derivative.
[c] Two independent molecules in the asymmetric unit.
[d] Two independent molecules in the asymmetric unit; data only reported for one.

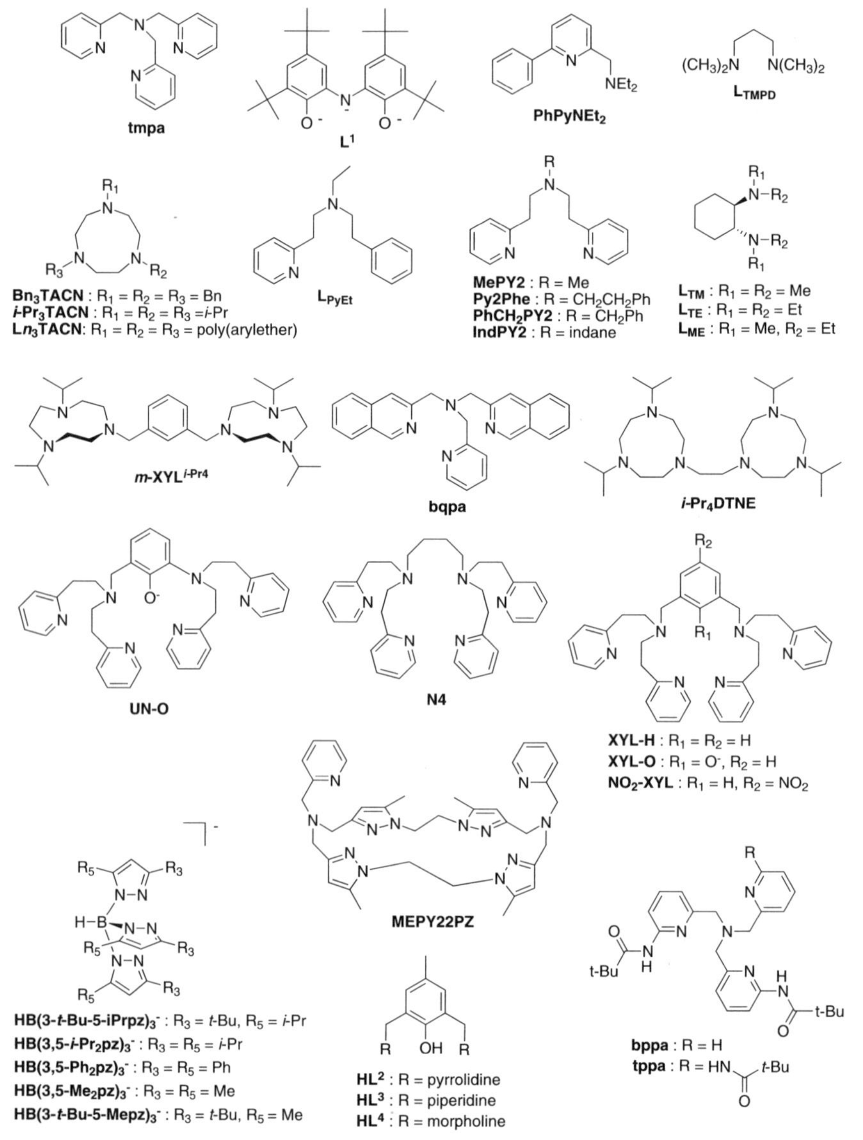

Fig. 3. Ligand structures and abbreviations

As part of this same study, the hydroxide complex was synthesized independently and the X-ray structure of this material was found to be identical to that obtained from the previously published data, confirming unequivocally that the original formulation of the *crystalline* material as a superoxide complex was in error. Debate about the nature of the solution species in the reaction mixture still continues [26, 27], but in the absence of corroborating vibrational spectral data or a bona fide X-ray crystal structure,

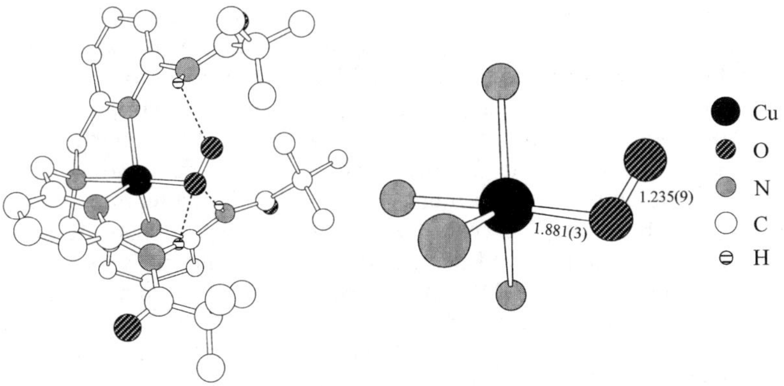

Fig. 4. X-ray structure of the purported monodentate superoxide complex cation $[Cu(tppa)(O_2)]^+$ [24]. Selected bond lengths are given in Å

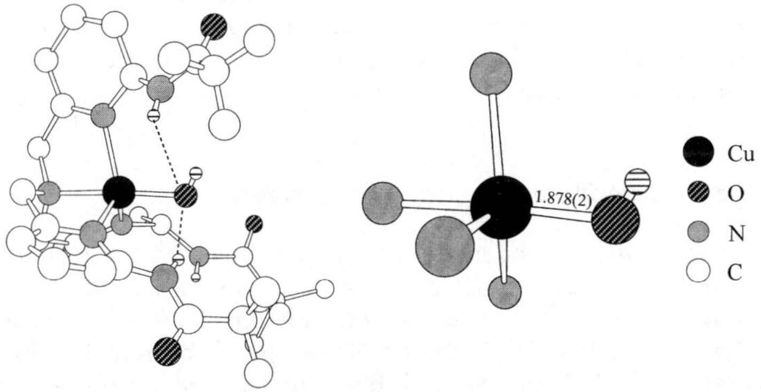

Fig. 5. X-ray structure of the $[Cu(tppa)OH]^+$ cation [25]. Selected bond lengths are given in Å

the original report of the generation of $[Cu(tppa)(O_2)]ClO_4$ must remain under suspicion.

A crystallographically characterized example of superoxide coordinated to Cu in an η^2 fashion was reported in 1994 [28]. The complex was prepared by exposure of a solution of [Cu(HB(3-*t*-Bu-5-*i*-Prpz)$_3$)(DMF)] to O_2 in CH_2Cl_2 at −50 °C. Manometry suggested a 1:1 Cu:O_2 ratio, and reversible O_2 binding was demonstrated in toluene by vacuum cycling. A red-brown solid was obtained from the reaction mixture at −20 °C, elemental analysis of which suggested the formula [Cu(HB(3-*t*-Bu-5-*i*-Prpz)$_3$)(O_2)]. The IR spectrum of the solid exhibited a band at 1112 cm^{-1}, which was assigned to ν(O—O) of a superoxide ion. Also, the resonance Raman spectrum of the complex prepared using $^{16}O_2$ showed a band at 1111 cm^{-1} which shifted the expected amount for an O—O oscillator to 1062 cm^{-1} on isotopic substitution with $^{18}O_2$. The solid was diamagnetic, as evidenced by solid-state SQUID magnetic susceptibility measurements and by its sharp ^1H NMR spectrum at −40 °C, suggesting a strong antiferromagnetic interaction between the superoxide ligand and the copper atom. A poor quality data set was obtained from weakly diffracting crystals which nonetheless confirmed the identity of the complex (Fig. 6).

A distorted square-pyramidal geometry applies about the copper atom, with the "side-on" bound superoxide ligand occupying two basal plane positions. While η^1-binding appears to be preferred when the supporting ligand is tetradentate (see above), η^2 binding is favored by the tridentate supporting ligand in [Cu(HB(3-*t*-Bu-5-*i*-Prpz)$_3$)(O$_2$)]. The tendency for Cu^{II} to adopt 5-coordinate geometries thus is a key structural determinant in Cu/O$_2$ chemistry. The O—O bond distance of 1.22(3) Å confirms its formulation as a superoxide complex, being very similar to that found in the closely related superoxo complex [Co(HB(3-*t*-Bu-5-Mepz)$_3$)(O$_2$)] (1.262(8) Å) [29] and much shorter than the O—O bond distance in the side-on bound peroxo complex [Mn(3,5-*i*-Pr$_2$pzH)(HB(3,5-*i*-Pr$_2$pz)$_3$)(O$_2$)] (1.43(1) Å) [30]. The successful isolation of monomeric [Cu(HB(3-*t*-Bu-5-*i*-Prpz)$_3$)(O$_2$)] was presumably dependent on the bulky nature of the supporting N-donor ligand which obviated formation of a superoxo- or peroxo-bridged dimer. The presence of the *tert*-butyl group in the 3-position appears essential to stop dimerization, as the same reaction using either HB(3,5-*i*-Pr$_2$pz)$_3^-$ or HB(3,5-Ph$_2$pz)$_3^-$ as the supporting ligand gives μ-η^2:η^2-peroxo species (see below) [31].

2.1.2
Alkylperoxide, Acylperoxide, and Hydroperoxide Complexes

The syntheses of the monodentate alkylperoxide complexes [Cu(HB(3,5-*i*-Pr$_2$pz)$_3$)(OOR)] (R = *t*-Bu or CMe$_2$Ph) were reported in 1993 [32]. The complexes were obtained from reaction of the bis(μm-hydroxo)-bridged dimer [{Cu(HB(3,5-*i*-Pr$_2$pz)$_3$)}$_2$(OH)$_2$] with excess *tert*-butyl- or cumyl-hydroperoxide in pentane at −20 °C to give dark blue microcrystalline solids that were stable below −20 °C for a week, but which decomposed at room temperature. The X-ray structure of the compound with R = CMe$_2$Ph (Fig. 7) confirmed the

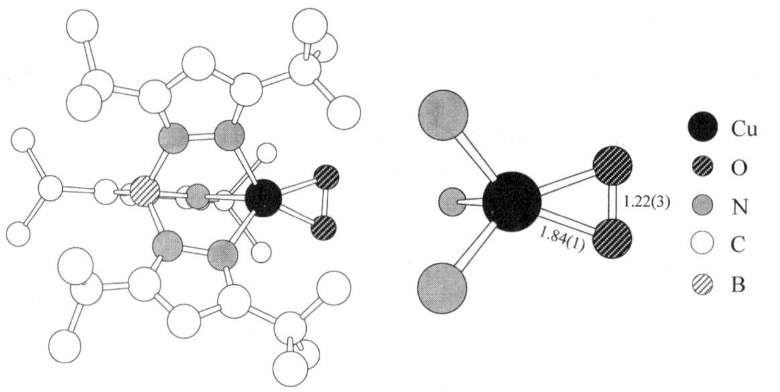

Fig. 6. X-ray structure of the η^2-superoxo complex [Cu(HB(3-*t*-Bu-5-*i*-Prpz)$_3$)(O$_2$)] [28]. Selected bond lengths are given in Å

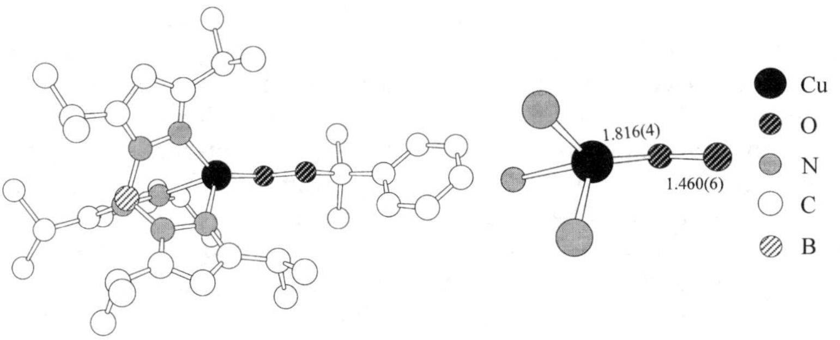

Fig. 7. X-ray structure of the alkylperoxide complex [Cu(HB(3,5-*i*-Pr_2pz)$_3$)(OOCMe_2Ph)] [32]. Selected bond lengths are given in Å

monomeric formulation of the complex and showed a distorted tetrahedral geometry about the copper ion, with a short (1.816(4) Å) Cu—O bond and an O—O bond distance in the cumylperoxide ligand of 1.460(6) Å. The monodentate cumylperoxide ligand is coordinated in a bent fashion, with a Cu—O—O bond angle of 112.1(4)°. These structural parameters are very similar to those observed in a series of crystallographically characterized Co(III) alkylperoxo complexes [33–35].

The only structurally characterized copper-hydroperoxo complex was reported in 1998 using the sterically hindered tripodal ligand bppa (Fig. 8) [36]. Addition of a large (20-fold) excess of aqueous H_2O_2 to acetonitrile solutions of [Cu(bppa)]ClO_4 or [Cu(bppa)(CH_3CO_2)]ClO_4 gave [Cu(bppa)(OOH)]ClO_4 which was characterized by X-ray crystallography, EPR, electronic and resonance Raman spectroscopy, and ESI mass spectrometry. The copper atom adopts a trigonal bipyramidal geometry, and the hydroperoxide ligand is bound in a bent fashion with a Cu—O—O bond angle of 114.5°. The O—O bond length (1.460(6) Å) is only slightly shorter than that found in H_2O_2

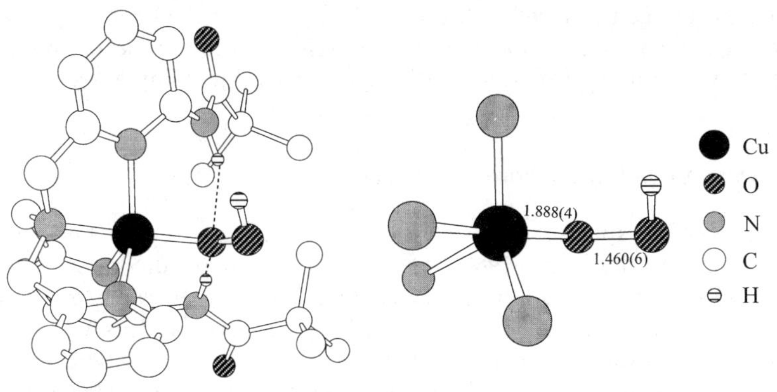

Fig. 8. X-ray structure of the hydroperoxide complex cation [Cu(bppa)(OOH)]$^+$ [36]. Selected bond lengths are given in Å

(1.490 Å), supporting the assignment of this as a hydroperoxide rather than a superoxide ligand. The remarkable stability of $[Cu(bppa)(OOH)]ClO_4$ (no change in the electronic absorption spectrum over one month at room temperature in MeCN) appears to be due to a combination of the bulkiness of the hydrophobic bppa ligand and the presence of two internal hydrogen bonds from the amide N-H protons to the coordinated oxygen atom of the hydroperoxide ligand. The electronic spectrum of $[Cu(bppa)(OOH)]ClO_4$ consists of two low-intensity d-d bands at 830 nm and 660 nm and a more intense band around 380 nm which was assigned to a ligand-to-metal charge-transfer transition. Of critical importance for confirming the structure of the complex was the resonance Raman spectrum (excitation wavelength 441.6 nm), which showed a peak at 856 cm^{-1} that shifted to 810 cm^{-1} when $H_2{}^{18}O_2$ was used in the synthesis. The EPR spectrum of $[Cu(bppa)(OOH)]ClO_4$ is typical of trigonal bipyramidal Cu^{2+} and the ESI mass spectrum shows peaks at m/z 584 ($[Cu(bppa)(OOH)]^+$) and 784 ($\{[Cu(bppa)(OOH)](ClO_4)_2\}^-$) which moved to m/z 588 and 788 respectively when $H_2{}^{18}O_2$ was used.

Synthesis of the monodentate mononuclear acylperoxo complexes $[Cu(TMPA)(m\text{-}ClC_6H_4C(O)O_2)](PF_6) \cdot 0.5CH_2Cl_2$ and $[Cu(Me_2im)_3(m\text{-}ClC_6H_4C(O)O_2)](PF_6)$ was achieved by treatment of the corresponding peroxo dimers with *m*-CPBA at −90 °C [37]. Magnetic and EPR data obtained on the room-temperature stable complexes were consistent with their formulation as monomers, while IR data suggested monodentate, rather than bidentate coordination of the peracid ligand.

2.2 Binuclear Complexes

By far the most common copper-peroxo and -oxo complexes to be structurally characterized are binuclear, the reduction of dioxygen by two copper ions to yield (peroxo)dicopper(II) or bis(oxo)dicopper(III) species being particularly thermodynamically favorable. As in the case of the mononuclear compounds described above (Sect. 2.1), the number, disposition, and type of N-donors in the supporting ligand as well as the steric influences of the ligand substituents have been shown to be important for controlling the course of the Cu^I/O_2 reactions and allowing isolation of discrete, crystalline complexes.

2.2.1 *μ-1,1-Acylperoxo and Hydroperoxo Complexes*

Although not a species derived directly from dioxygen, the synthesis and characterization in 1987 of the acylperoxo dicopper complex $[Cu_2(XYL\text{-}O\text{-})(m\text{-}ClC_6H_4C(O)O_2)](ClO_4)_2 \cdot CH_3CN$ (Fig. 9) represented a major advance in the field of copper-dioxygen chemistry as this was the first reported X-ray structure of a copper-peroxo species of any type [38]. The complex was prepared at −80 °C either by treatment of the hydroxo-bridged dimer $[Cu_2(XYL\text{-}O)(OH)]^{2+}$ with *m*-CPBA or acylation of the peroxo-bridged dimer $[Cu_2(XYL\text{-}O)(O_2)]^+$ (see below) with *m*-chlorobenzoyl chloride. The

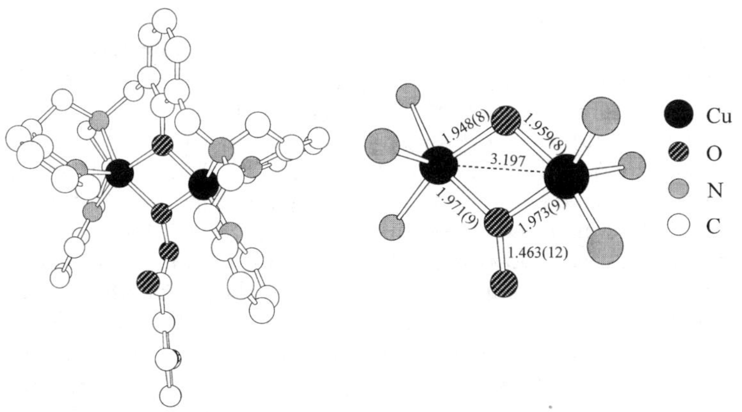

Fig. 9. X-ray structure of the acylperoxide complex cation $[Cu_2(XYL\text{-}O\text{-})(m\text{-}ClC_6H_4C(O)O_2)]^{2+}$ [38]. Selected bond lengths are given in Å

crystal structure of the complex showed the acylperoxide ligand to be bound such that the terminal oxygen atom bridged the two copper ions, resulting in a Cu···Cu separation of 3.197 Å.

Although no X-ray structures are available, a strong case for the existence of μ-1,1 hydroperoxo dimers has been made on the basis of spectroscopic data. Treatment of the hydroxo-bridged dimer $[Cu_2(XYL\text{-}O\text{-})(OH)]^{2-}$ with excess H_2O_2 at −80 °C, reaction of the peroxo-bridged dimer $[Cu_2(XYL\text{-}O\text{-})(O_2)]^+$ with $HBF_4 \cdot Et_2O$ at −80 °C and oxygenation of $[Cu_2(XYL\text{-}OH)]^{2+}$ at −80 °C all yielded the same product, identified as the μ-1,1-hydroperoxo species $[Cu_2(XYL\text{-}O\text{-})(OOH)]^{2+}$ on the basis of the similarity of its spectral features with those of the structurally characterized acylperoxo complex $[Cu_2(XYL\text{-}O\text{-})(m\text{-}ClC_6H_4C(O)O_2)](ClO_4)_2 \cdot CH_3CN$ [39]. Both the acylperoxo and hydroperoxo complexes exhibit intense absorptions at 395 nm in the UV/vis spectrum, as well as weaker d-d bands at longer wavelengths. Further support for the μ-1,1-hydroperoxo formulation of $[Cu_2(XYL\text{-}O\text{-})(OOH)]^{2+}$ was obtained from comparison of EXAFS data for this complex with those for the structurally characterized complex $[Cu_2(XYL\text{-}O\text{-})(OH)]^{2+}$ [40]. While this hydroperoxo complex was only stable at low temperatures, slight modification of the XYL-O ligand [41] allowed isolation of a room-temperature stable hydroperoxo complex $[Cu_2(UN\text{-}O\text{-})(OOH)](PF_6)_2 \cdot 1/2Et_2O$ [42]. A band at 892 cm^{-1} in the resonance Raman spectrum of this complex was assigned to ν(O—O) and was observed to shift to 840 cm^{-1} on substitution with $^{18}O_2$ [43].

2.2.2
μ-η^1:η^1-Peroxo Complexes

The first copper-dioxygen complex to be structurally characterized was reported in 1988 [44]. Reaction of $[Cu(TMPA)(RCN)]PF_6$ (R = Me, Et) with O_2 at −80 °C in EtCN or CH_2Cl_2 gave a color change from orange to deep purple.

Manometry showed the reaction stoichiometry to be $Cu:O_2 = 2:1$ and the binding of O_2 was shown to be reversible by vacuum/warming cycling experiments. Crystals of $[\{Cu(TMPA)\}_2(O_2)](PF_6)_2 \cdot 5Et_2O$ were obtained on storage of the reaction mixture at −85 °C (Fig. 10).

The structure consists of two $Cu(TMPA)^{2+}$ moieties bridged by peroxide in a *trans*-1,2 fashion with a Cu—Cu distance of 4.359 Å. This binding mode apparently is preferred because of the tetradentate supporting N-donor ligand; due to the aforementioned propensity for Cu(II) to adopt 5-coordinate geometries (here, distorted trigonal bipyramidal), monodentate coordination of peroxide is favored. Importantly, the O—O bond length of 1.432(6) Å confirms the assignment of a bridging peroxide ligand. The EPR-silence of the product is indicative of a strong antiferromagnetic coupling between the copper atoms mediated by the bridging peroxo ligand. Full spectroscopic characterization for the complex was reported in subsequent papers [45, 46]. Three major bands were observed in the UV/vis spectrum at 615 ($\varepsilon = 5800\ M^{-1}\ cm^{-1}$), 524 ($\varepsilon = 11300\ M^{-1}\ cm^{-1}$) and 435 nm ($\varepsilon = 1700\ M^{-1}\ cm^{-1}$), and these were assigned to $O_2^{2-} \rightarrow Cu^{2+}$ charge transfer transitions. A less intense band at 1035 nm ($\varepsilon = 180\ M^{-1}\ cm^{-1}$) was assigned to a d-d transition. Bands in the resonance Raman spectrum at 832 cm^{-1} and 561 cm^{-1} were observed to shift to 788 cm^{-1} and 535 cm^{-1} on substitution of $^{16}O_2$ by $^{18}O_2$. These bands were assigned to ν(O—O) and a symmetric ν(Cu—O), respectively, on the basis of their isotope shifts and characteristic frequencies. The synthesis of $[\{Cu(TMPA)\}_2(O_2)](PF_6)_2$ represented an important advance in copper-dioxygen chemistry as it showed for the first time that the correct choice of ligand combined with low-temperature conditions could allow the isolation of copper-dioxygen species. Moreover, the definitive correlation of spectroscopic and X-ray crystal structural data for this compound has allowed subsequent identification of the *trans*-1,2-O_2^{2-} unit in other copper complexes, all of which have in common tetradentate ligation of an N-donor supporting ligand [47–49]. Of great interest amongst these is the remarkably stable purple species formed on oxygenation of the dicopper(I) complex of the macrocyclic ligand MEPY22PZ [49]. This complex shows intense bands at 525 nm and 625 nm and a weak shoulder at 438 nm in the

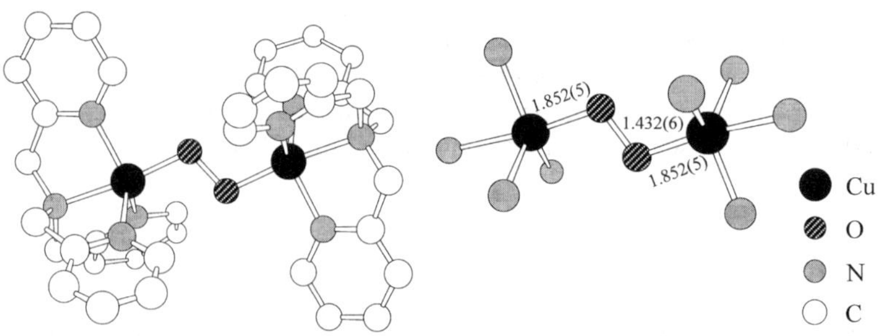

Fig. 10. X-ray structure of the peroxo complex cation $[\{Cu(TMPA)\}_2(O_2)]^{2+}$ [44]. Selected bond lengths are given in Å

UV/vis spectrum, while the resonance Raman spectrum shows a band assigned to ν(O—O) at 844 cm^{-1} which shifts to 798 cm^{-1} on substitution with $^{18}O_2$. The authors proposed this to be an intramolecular *trans*-1,2-peroxo species, although molecular mechanics calculations showed that formation of an intermolecular species was also feasible. The stability of the peroxo species ($t_{1/2}$ = 250 s in MeOH at 21 °C) was ascribed to the relative inaccessibility of the macrocyclic cavity in which the metal ions are coordinated.

2.2.3
μ-η^2:η^2-Peroxo Complexes

In 1988, the synthesis of a peroxide-bridged Cu^{2+} complex from the reaction of [{Cu(HB(3,5-Me_2pz)$_3$)}$_2$O] with an equimolar amount of H_2O_2 at −35 °C was reported [50]. The product, proposed to be [{Cu(HB(3,5-Me_2pz)$_3$)}$_2$(O_2)] on the basis of elemental analysis and mass spectral data, was diamagnetic, giving no EPR signal and exhibiting a sharp ^{1}H NMR spectrum. The resonance Raman spectrum showed a band at 725 cm^{-1} that shifted to 686 cm^{-1} when $H_2^{18}O_2$ was used to prepare the complex. While the isotope shift supported the presence of an O—O bond in the complex, the band energies were rather low for a peroxo unit (typically ν(O—O) > 800 cm^{-1} in transition metal peroxide compounds) [51]. The UV/vis spectrum of the dark purple solid was remarkably similar to that of oxyhemocyanin, with bands at 530 nm (ε = 840 M^{-1} cm^{-1}) and 338 nm (ε = 20,800 M^{-1} cm^{-1}). In view of this similarity and analogous resonance Raman and magnetic properties, the complex was advanced as a structural model of this protein, albeit in the absence of crystal structural data needed to confirm the mode of binding of peroxide to the metal centers. These critical data were available the following year, when the same group reported the synthesis of [{Cu(HB(3,5-*i*-Pr_2pz)$_3$)}$_2$ (O_2)] by both O_2 addition to the Cu(I) precursor [Cu(HB(3,5-*i*-Pr_2pz)$_3$)] and reaction of H_2O_2 with the dimeric Cu(II) complex [{Cu(HB(3,5-*i*-Pr_2pz)$_3$)}$_2$ $(OH)_2$] [52]. The spectral features of the peroxo product were similar to those reported for the dimethyl analogue above, but enhanced stability and crystallinity conferred by isopropyl substitution on the ligands allowed further characterization by X-ray crystallography (Fig. 11).

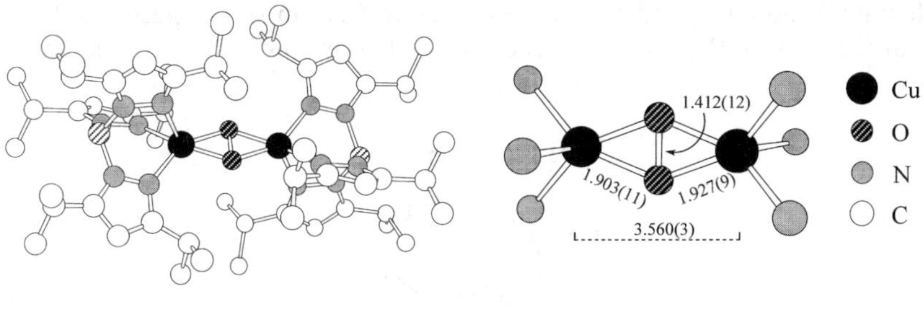

Fig. 11. X-ray structure of the peroxo complex [{Cu(HB(3,5-*i*-Pr_2pz)$_3$)}$_2$(O_2)] [52]. Selected bond lengths are given in Å

The structure revealed peroxide bridging the two copper ions (3.56 Å apart) in a symmetrical, planar η^2:η^2 fashion with all Cu—O distances the same within experimental error and an O—O distance of 1.412(12) Å. Although seen previously in complexes of lanthanum [53] and uranium [54], this planar (μ-η^2:η^2-peroxo)dimetal unit was a first in transition metal chemistry. Full spectroscopic characterization of the $[\{Cu(HB(3,5\text{-}R_2pz)_3)\}_2(O_2)]$ (R = Me, *i*-Pr, Ph) complexes has been described, revealing, among other things, close agreement with the properties of oxyhemocyanin [31, 55]. Final confirmation of the structural congruity between the complexes and the protein active site came with the report of the X-ray structure of oxyhemocyanin [3]. The correct prediction through exploratory synthetic chemistry of the presence of the (μ-η^2:η^2-peroxo)dicopper core in oxyhemocyanin ranks as one of the greatest successes of bioinorganic model chemistry.

Since the initial discoveries using the $HB(3,5\text{-}R_2pz)_3^-$ ligand system, additional (μ-η^2:η^2-peroxo)dicopper complexes have been identified by spectroscopic means [56–58]. There are also a number of peroxo complexes which, while displaying resonance Raman spectral properties consistent with a μ-η^2:η^2-peroxo formulation (e.g. ν(O—O) = 747 cm^{-1}, shifted to 707 cm^{-1} on substitution by $^{18}O_2$ [59]) also show slightly anomalous UV/vis spectra. In addition to bands around 360 nm and 530 nm typical of μ-η^2:η^2-peroxo complexes, they also display a moderately intense band in the range 420–490 nm [60, 61]. This has been proposed to be due to a bent "butterfly" μ-η^2:η^2-peroxo core (Fig. 12), with the extra band being due to an LMCT transition which is formally forbidden in the planar μ-η^2:η^2-peroxo core [62]. The possibility that the extra band was due to the isomeric bis(μ-oxo) dicopper(III) core (see below) was discounted on the basis of a resonance Raman excitation profile [59].

Interconversion between two different types of peroxo cores has been observed in complexes of the binucleating ligand N4. Stopped-flow spectrophotometric studies of the oxygenation of the Cu(I) complex $[(N4)Cu_2]^{2+}$ at −94 °C gave evidence for the initial formation of a complex containing the μ-η^1:η^1-peroxo core, which then rapidly rearranged to the bent "butterfly" μ-η^2:η^2-peroxo core [63]. Kinetic data suggested that this rearrangement could not occur directly, and a mechanism involving the intermediacy of either the Cu(I) starting material or a transient superoxo Cu(II) species was proposed.

All of the complexes containing the μ-η^2:η^2-peroxo core incorporate a tridentate *N*-donor ligand that allows η^2-coordination of the peroxide to give 5-coordinate Cu(II) ions, and each has UV/vis spectral features and an

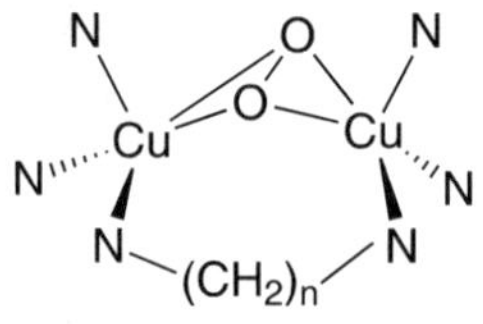

Fig. 12. The bent "butterfly" μ-η^2:η^2-peroxo core

anomolously low O—O stretching frequency now recognized to be signatures of the (μ-η^2:η^2-peroxo)dicopper unit. In addition to being present in oxyhemocyanin, the weakened O—O bond implied by the generally low ν(O—O) value suggests that this bonding motif may lie on the reaction pathway of dioxygen O—O bond cleavage in enzymatic and other catalytic oxidations (see below).

2.2.4
Bis(μ-Oxo) Complexes

While the presence of Cu(I) and Cu(II) in metalloproteins is well established, it has long been thought that complexes of Cu(III) are unlikely to exist in biological systems due to the high redox potential required to access this oxidation state. However, this view must now be called into question with the recent synthesis and structural characterization of several bis(μ-oxo)dicopper(III) complexes [14, 18, 64–67]. In all cases, the complexes were prepared via oxidation of a Cu(I) precursor with molecular oxygen, thus demonstrating the viability of generating high-valent copper species in biological systems. The X-ray structures of three compounds in this class are shown in Figs. 13–15. All three are thermally unstable molecules that decompose via ligand-degradation pathways involving rate-controlling C—H bond scission characterized by large H/D kinetic isotope effects (see below). As a result, in order to slow decay sufficiently so that crystals suitable for analysis by X-ray diffraction could be grown, complexes with perdeuterated ligand substituents were often used.

The bis(μ-oxo)dicopper cores in these structures are quite congruent despite differences in the nature of the supporting N-donor ligands. Thus, the Cu—O distances (average = 1.81 Å) are short compared to typical Cu(II)—O distances (1.9–2.0 Å); bond-valence sum analyses suggested that the short Cu—O and —N distances in these compounds are indicative of a Cu(III) oxidation level [68]. Long O···O distances (average = 2.327 Å) indicate the

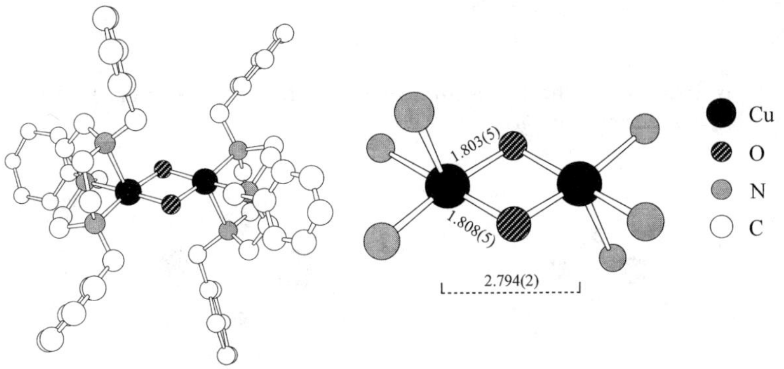

Fig. 13. X-ray structure of the bis(μ-oxo) complex cation $[\{Cu(d_{21}\text{-}Bn_3TACN)\}_2(\mu\text{-}O)_2]^{2+}$ [64]. Selected bond lengths are given in Å

absence of a bond between these atoms and the Cu···Cu separation (average = 2.773 Å) is much shorter than those found in other structurally characterized binuclear copper-peroxo and superoxo species (Table 1). Overall, the $[Cu_2(\mu\text{-}O)_2]^{2+}$ unit is structurally similar to other such cores with other metals (e.g., Fe [69], Mn [70], Ni, and Co [71]). The core is supported by tridentate *N*-donor macrocycles in $[\{Cu(d_{21}\text{-}Bn_3TACN)\}_2(\mu\text{-}O)_2](SbF_6)_2 \cdot 7(CH_3)_2CO \cdot 2CH_3CN$ (Fig. 13) [64] and $[Cu_2(d_{28}\text{-}i\text{-}Pr_4DTNE)(\mu m\text{-}O)_2](SbF_6)_2 \cdot 3CH_2Cl_2$ (Fig. 14) [65], but different relative orientations are adopted by the square pyramidal metal ions; in the Bn_3TACN case, the axial N-donors are disposed in an anti conformation, whereas in the *i*-Pr_4DTNE instance they are forced by the ethylene tether to be syn. In both structures, close contacts between ligand substituent C—D bonds implicate Cu—O···D bonding interactions that may lie along the decomposition pathways traversed by these bis(μ-oxo) complexes (see below). In the structurally characterized complex $[\{Cu(L_{ME})\}_2(\mu\text{-}O)_2](CF_3SO_3)_2 \cdot 4CH_2Cl_2$ (Fig. 15) [67] and analogous complexes containing the peralkylated cyclohexanediamine ligands L_{TM} and L_{TE}, bidentate N-donors lead to square planar metal ion geometries. Interestingly, use of the more flexible ligand L_{TMPD} gave a bis(μ-oxo) dimer that exhibited very different reactivity towards oxidizable substrates. EXAFS data suggested that this complex contained 5-coordinate copper centers [72].

Spectroscopic studies have revealed important signatures of the bis(μ-oxo)dicopper core in these and related compounds [14, 66, 68], which in conjunction with theoretical calculations [73–77] have enabled the delineation

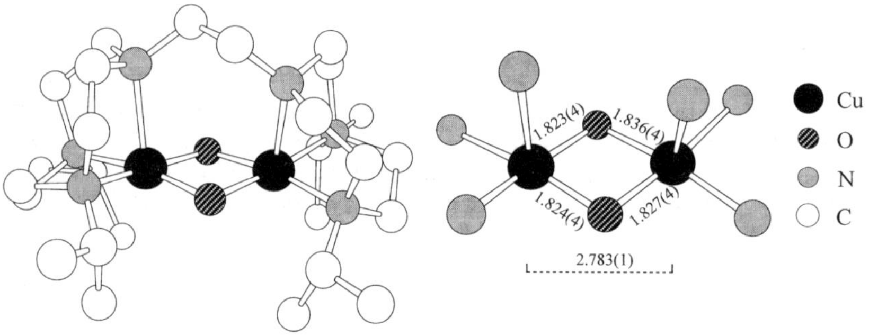

Fig. 14. X-ray structure of the bis(μ-oxo) complex cation $[Cu_2(d_{28}\text{-}i\text{-}Pr_4DTNE)(\mu\text{-}O)_2]^{2+}$ [65]. Selected bond lengths are given in Å

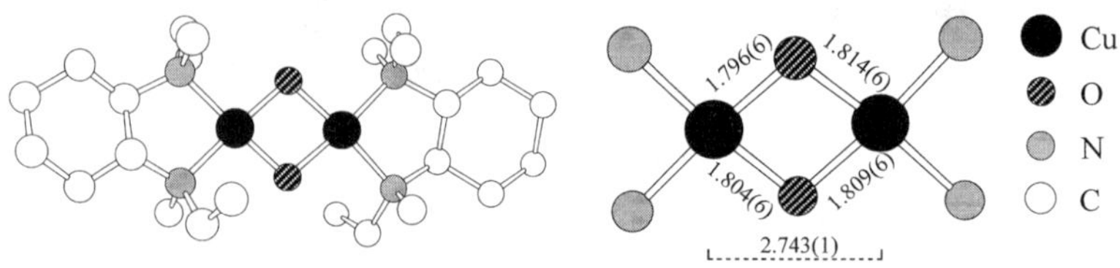

Fig. 15. X-ray structure of the bis(μ-oxo) complex cation $[\{Cu(L_{ME})\}_2(\mu\text{-}O)_2]^{2+}$ [67]. Selected bond lengths are given in Å

of key aspects of its electronic structure (see Note Added in Proof). A pair of intense optical absorbance features at ~320 and ~430 nm with $\varepsilon = 12{,}000$–$16{,}000\ M^{-1}\ cm^{-1}$ involve charge-transfer (presumably oxo → Cu(III)) within the core, quite different from the pattern observed for other peroxo- or superoxo-copper compounds. Consistent with the UV/vis assignment, resonance Raman spectra obtained using excitation wavelengths within the ~430 nm absorption band exhibited a strongly enhanced feature at ~600 cm^{-1} that shifted by 23–29 cm^{-1} upon ^{18}O-substitution. Detailed studies of the Raman spectra of a series of complexes with differing supporting ligands and isotopic compositions, as well as ab initio calculations on model systems, support assignment of this feature to a symmetric vibration of the Cu_2O_2 rhomb [78]. The bis(μ-oxo)dicopper compounds are uniformly EPR silent and exhibit ^1H NMR signals in the 0–10 ppm region, clearly reflecting their diamagnetism. On the basis of the geometric parameters and ab initio calculations, a bis(μ-oxo)dicopper(III) valence bond description was suggested, with the caveat that the system is highly covalent due to strong bonding interactions between O p and Cu d orbitals in the HOMO (Fig. 16). A Cu X-ray absorption K-edge study provided experimental support for this notion [66].

The bis(μ-oxo)dicopper(III) and previously discussed (μ-η^2:η^2-peroxo) dicopper(II) cores are thus isomers potentially related by O—O bond making and breaking coupled with electron transfer to and from the copper ions, respectively. Experimental confirmation of the interconversion between the two isomeric cores was accomplished for the system in which *i*-Pr_3TACN was the supporting ligand [64]. It was subsequently shown that oxygenation of $[Cu(i\text{-}Pr_3TACN)(CH_3CN)]^+$ yielded either or both isomers depending on the solvent, counterion, or temperature. For example, resonance Raman and UV/vis spectroscopy showed that in CH_2Cl_2 the peroxo form predominated, but in concentrated THF solutions (>1 mmol l^{-1}) the bis(μ-oxo) core was generated instead. Interestingly, the cores could be interconverted in solvent mixing experiments; addition of THF (50-fold excess v/v) to a concentrated (1.2 mmol l^{-1}) CH_2Cl_2 solution of the peroxo complex yielded the bis(μ-oxo)

Fig. 16. The HOMO of anti $[\{(NH_3)_3Cu\}_2(\mu\text{-}O)_2]^{2+}$ calculated at the RHF/STO-3G* level [68]

form, and the opposite protocol reversed the process. Rapid equilibration was initially implicated by stopped-flow kinetics investigations of the oxygenation reaction in acetone solution, where spectroscopy indicated the presence of a mixture of both cores. More recently, studies of dilute solutions of the system in CH_2Cl_2, THF, or mixtures of the two supported equilibration of the cores in these solvents as well, and revealed how the equilibrium behaved as a function of the solvent and temperature [79]. The combined data are consistent with closely similar thermodynamic stabilities for the isomeric $[Cu_2(\mu\text{-}O)_2]^{2+}$ and $[Cu_2(\mu\text{-}\eta^2{:}\eta^2\text{-}O_2)]^{2+}$ cores supported by *i*-Pr_3TACN, as well as a low barrier for their interconversion. These conclusions have been rationalized by calculations at various levels of theory [73–77]. The reversible isomerization of the $[Cu_2(\mu\text{-}O)_2]^{2+}$ and $[Cu_2(\mu\text{-}\eta^2{:}\eta^2\text{-}O_2)]^{2+}$units has obvious consequences in the biological production and processing of dioxygen (see below).

Using the information gleaned from the combined structural/spectroscopic studies of the crystallographically characterized bis(μ-oxo)dicopper complexes in Figs. 13–15 as a basis, additional examples of such species have been studied. Oxygenation of the Cu(I) complex of the binucleating xylyl-based ligand *m*-$XYL^{i\text{-}Pr4}$ resulted in the formation of both intra- and intermolecular products, with the product distribution depending on the solution concentration (Scheme 1). In dilute solution (<0.1 mmol l^{-1}), the μ-$\eta^2{:}\eta^2$-peroxo product was formed predominantly and reacted further to hydroxylate the arene ring, but as the concentration of the Cu(I) starting material was increased, intermolecular reaction with dioxygen resulted in the formation of the "dimer of dimers" bis(μ-oxo) product, decomposition of which ultimately led to *N*-dealkylation of the TACN ring [80].

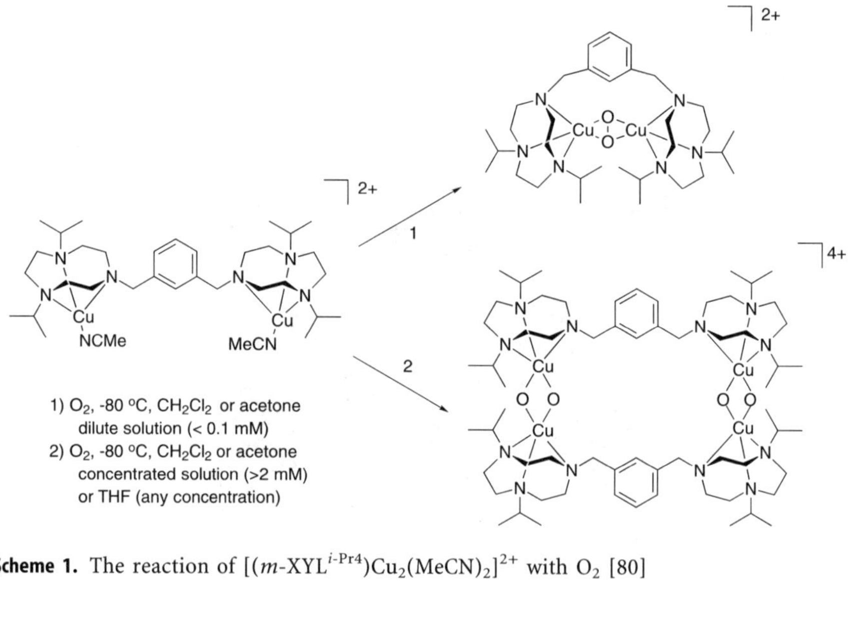

Scheme 1. The reaction of $[(m\text{-}XYL^{i\text{-}Pr4})Cu_2(MeCN)_2]^{2+}$ with O_2 [80]

Bis(μ-oxo) complexes have been prepared using the bidentate N-donor ligands $PhPyNEt_2$ [81] and L_{PyEt} [82] and substituted derivatives thereof, by oxygenation of the corresponding Cu(I) precursors. The oxygenated solutions showed a UV/vis spectral feature typical of the bis(μ-oxo) core at ~402 nm while the resonance Raman spectra contained a band at 607 cm^{-1} which shifted to ~580 cm^{-1} on substitution of $^{18}O_2$. The decomposition of both sets of complexes resulted in ligand hydroxylation (see below).

Successful attempts have been made to increase the stability of the bis(μ-oxo)dicopper(III) unit by utilizing aryl-ether dendrimer ligands based on TACN [83]. The bis(μ-oxo) copper complexes were prepared by oxygenation of the Cu(I) precursors $[(Ln_3TACN)Cu(MeCN)]PF_6$ at −78 °C. The oxygenation rates were found to decrease as the size of the dendritic ligand increased; the Cu(I) complex of the largest dendrimer failed to react with O_2 over 16 h under these conditions. This behavior was ascribed to the steric bulk of the dendrimer ligand, which inhibits interaction of the O_2 molecule with the encapsulated Cu(I) ions. The stability of the bis(μ-oxo)copper complexes was also found to depend on the size of the dendrimer ligand, with the largest complex studied having a half life of 3075 s at −10 °C, compared with the half life of 7 s exhibited by $[\{(Bn_3TACN)Cu\}_2(\mu\text{-}O)_2]^{2+}$ at this temperature. Kinetic data measured for the decomposition reactions of the bis(μ-oxo) complexes suggested that the increased stability of the larger complexes was essentially entropic in nature and could be attributed to the inaccessibility of the benzylic hydrogens, the abstraction of which has been postulated to be an important step in the decomposition reactions of such complexes (see below) [84].

Another system in which the bis(μ-oxo) and μ-η^2:η^2-peroxo cores have been found to coexist both in the solid state and in solution has recently been described [85, 86]. UV/vis and resonance Raman spectral features attributable to both cores were observed in solution from the product of the low temperature oxygenation of [(MePY2)Cu(MeCN)]BArF in CH_2Cl_2, with the μ-η^2:η^2-peroxo complex predominating (>80%). The crystal structure of the oxygenation product "$[\{Cu(MePY2)\}_2O_2](BArF)_2$" showed metric parameters for the dicopper core (O—O = 1.67 Å, Cu···Cu = 3.45 Å) which were intermediate between those values typical of the bis(μ-oxo) and μ-η^2:η^2-peroxo cores, suggesting that the crystalline material was in fact a solid solution of complexes containing both cores. Such behavior has also been observed in crystals isolated from oxygenation mixtures of $[Cu(i\text{-}Pr_3TACN)(MeCN)]^+$ in CH_2Cl_2 [14].

2.3 Trinuclear Complexes

The copper proteins ascorbate oxidase, ceruloplasmin, and laccase, which carry out the 4-electron reduction of O_2 to H_2O, have all been found to contain a trinuclear copper center at the active site [16] and attempts have been made to prepare tricopper-dioxygen species as models for these proteins. The synthesis of the trinuclear model complex $[(L_{TM})_3Cu_3O_2](CF_3SO_3)_3$ was reported in 1996 from the reaction of the Cu(I) precursor $[(L_{TM})Cu(MeCN)]^+$

with O_2 at −80°C in CH_2Cl_2 [87]. Manometry showed a 3:1 Cu/O_2 ratio for this reaction and the dark brown product displayed intense bands in the UV/vis spectrum at 290 nm and 355 nm and less intense bands at 480 nm and 620 nm. No resonance Raman data for the complex were reported. The structure of the product was confirmed by X-ray crystallography and a diagram of the cation is shown in Fig. 17.

The structure consists of three copper ions, each in an approximately square-planar environment, coordinated to two nitrogens from the chelating ligand and two oxygen atoms. The O···O distance (2.37 Å) shows that the O—O bond of dioxygen has been cleaved and that the structure is best formulated as containing two μ_3-oxo bridges. A Cu(II, II, III) formulation of this complex was suggested by the Cu—O bond lengths within the trimer, with two sets of Cu—O distances (2.01(1) Å and 1.98(1) Å) being longer than the third (1.83(1) Å). This was also supported by subsequent magnetism and MCD data, and molecular orbital calculation results, which also showed that the Cu(II) ions were ferromagnetically coupled [88]. The synthesis of this complex again shows that the Cu(III) oxidation state can be accessed simply with the use of molecular oxygen as oxidant.

2.4 Tetranuclear Complexes

A remarkable room-temperature stable tetranuclear Cu(II) complex containing a μ_4-bridging peroxide ligand was isolated as dark green crystals from the reaction of $Cu(ClO_4)_2 \cdot 6H_2O$ with 4-methyl-2,6-bis(pyrrolidinomethyl) phenol (HL^2) in MeOH in the presence of triethylamine and 3,5-di-*tert*-butylcatechol [89]. The crystal structure of $[Cu_4(L^2)_2(O_2)(OMe)_2(ClO_4)]ClO_4 \cdot MeOH$ (Fig. 18) shows the four copper ions to be in square-pyramidal environments, each coordinated to a phenoxide, perchlorate, and methoxide oxygen atom, and a pyrrolidine nitrogen atom. Coordination about each

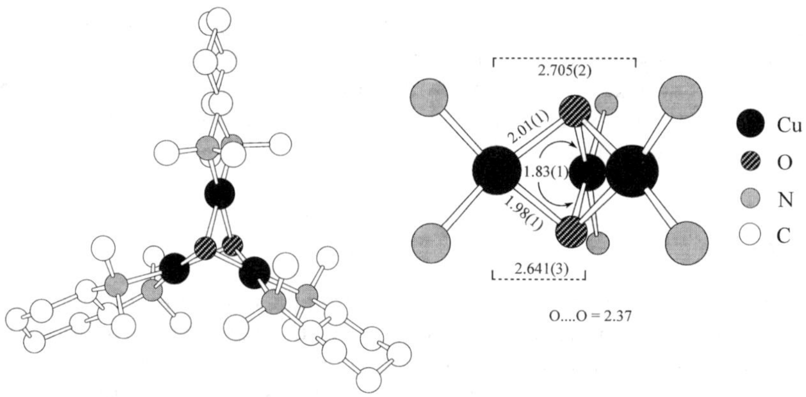

Fig. 17. X-ray structure of the trinuclear complex cation $[(L_{TM})_3Cu_3O_2]^{3+}$ [87]. Selected bond lengths are given in Å

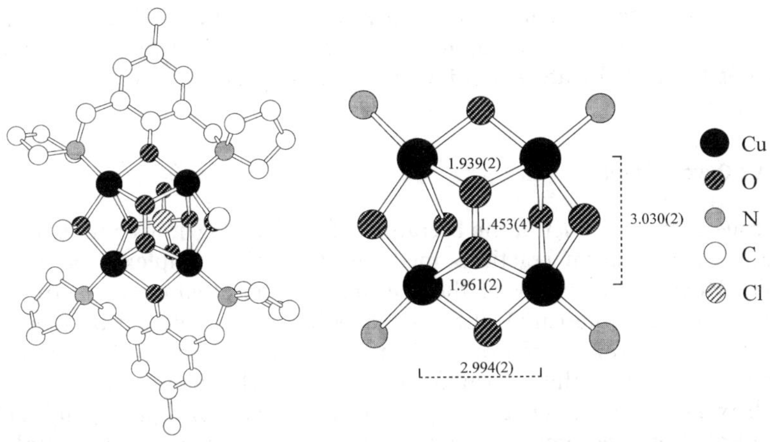

Fig. 18. X-ray structure of the tetranuclear complex cation $[Cu_4(L^2)_2(O_2)(OMe)_2(ClO_4)]^+$ [89]. Selected bond lengths are given in Å

copper ion is completed by a μ_4-bridging peroxo ligand (O—O = 1.453(4) Å) which lies above the plane of the copper ions. This μ_4-bridging peroxo structural motif has also been observed in multinuclear complexes of Mo [90] and Fe [91]. The Cu—O bond lengths involving the peroxo ligand are 1.961(2) Å and 1.939(2) Å.

A subsequent paper described the synthesis of two further tetranuclear peroxo complexes using the piperidine- and morpholine-appended phenol ligands HL^3 and HL^4, and reported spectroscopic data on all three complexes [92]. The complexes exhibited similar UV/vis spectra, with bands around 285 nm, 390 nm, and 580 nm. These were assigned to ligand transitions, a combination of phenolate → Cu(II) and peroxo → Cu(II) charge transfers, and peroxo → Cu(II) charge transfer transitions, respectively. The resonance Raman spectra of the three complexes showed bands assigned to ν(O—O) at 878 cm^{-1}, 898 cm^{-1}, and 888 cm^{-1}. Magnetic data collected on the structurally characterized complex suggested very strong antiferromagnetic coupling between the copper ions.

3 Reactivity of Oxygenase Model Complexes

The oxygenation reactions mediated by such copper proteins as tyrosinase and dopamine β-hydroxylase involve incorporation of an O_2-derived oxygen atom into an aromatic or aliphatic substrate [15, 16]. Such reactions necessarily involve substrate C—H activation and are generally proposed to occur via attack by an activated oxygen species. While oxidations of organic compounds by copper salts are well known [93], detailed mechanistic information such as that which would be obtained by studies of well-characterized copper-oxygen species is needed in order to understand and control selectivities [11].

Progress toward those goals is summarized below, with specific emphasis on recent results involving oxidations of ligand hydrocarbyl units during decompositions of the aforementioned copper-oxygen complexes.

3.1 Aromatic Oxygenation

Numerous examples exist in the literature of both endogenous and exogenous aromatic hydroxylation reactions mediated by Cu(I) complexes on reaction with O_2. However, in many of these examples, the attempted spectroscopic characterization of the Cu/O_2 intermediates has been thwarted by their high reactivity and/or low concentration in the reaction mixtures [94–96], and in only a few cases have the intermediates been well characterized.

Hydroxylation of the benzene ring of the xylyl-based bridging ligand XYL-H was observed on low-temperature oxygenation of the dicopper(I) complex of this ligand and an intermediate having an absorbance maximum at 435 nm was observed spectroscopically [97]. Subsequent studies showed that the incorporated oxygen atom derived from molecular oxygen (Scheme 2) [98].

Detailed kinetic and thermodynamic studies suggested that the mechanism of the hydroxylation reaction was akin to that of electrophilic aromatic substitution, in that electron-donating substituents on the benzene ring gave increased rates of reaction, and that no deuterium isotope effect was observed. Operation of a radical mechanism was precluded by the observation that addition of radical traps to the reaction mixture did not affect the rate of reaction and it was assumed that a bent "butterfly" μ-η^2:η^2-peroxo species effected the hydroxylation reaction. However, the discovery that the μ-η^2:η^2-peroxo and bis(μ-oxo) cores can interconvert [64] prompted a reinvestigation

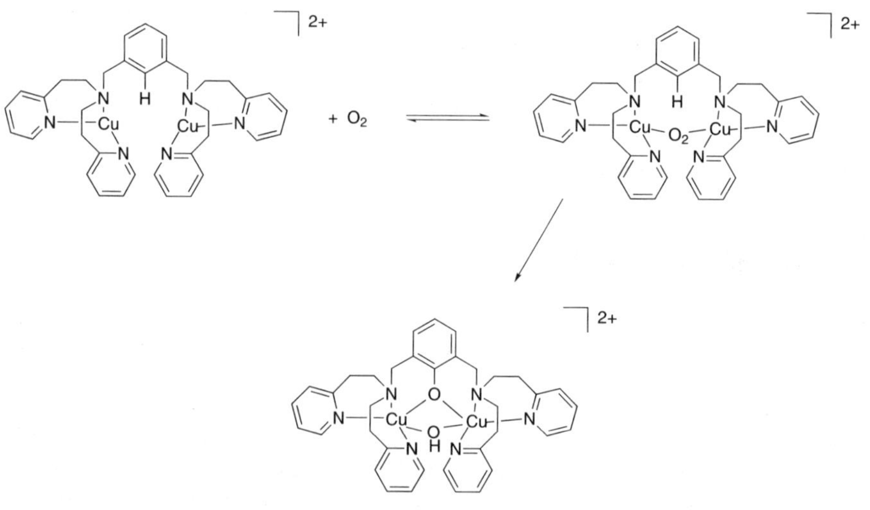

Scheme 2. The reaction of $[Cu_2(XYL\text{-}H)]^{2+}$ with O_2 [98]

of this system with a view to attempting to distinguish whether a μ-η^2:η^2-peroxo or bis(μ-oxo) species was responsible for the hydroxylation reaction [59]. To this end, the kinetics of reaction of the Cu(I) complex of the nitro-substituted xylyl ligand NO_2-XYL were investigated. In this system, resonance Raman data suggested a maximum bis(μ-oxo):μ-η^2:η^2-peroxo ratio of 0.0013 which would require the bis(μ-oxo) species to be $\sim$1000 times more reactive than the μ-η^2:η^2-peroxo species to explain the observed kinetics. The authors contended that this reactivity difference was not supported by MO calculations and hence they proposed that the μ-η^2:η^2-peroxo species reacted directly to give the hydroxylated product.

As mentioned above (Sect. 2.2.4), oxygenation of the 3-coordinate Cu(I) complex [Cu(PhPyNEt$_2$)(MeCN)](SbF$_6$) in acetone or THF solution at –70 °C gave an EPR-silent yellow species that exhibited spectral features typical of a bis(μ-oxo) core (UV/vis:λ_{max} 404 nm, resonance Raman:607 cm^{-1}, shifted to 580 cm^{-1} on substitution with $^{18}O_2$) [81]. Hydroxylation at the ortho position of the phenyl substituent was observed when the oxygenated solution was allowed to warm to room temperature, with the yield for this process being 70% of theoretical. Identical behavior was observed when the oxygenation was carried out at room temperature, and ^{18}O labeling confirmed the incorporated oxygen atom derived from molecular O_2 (Scheme 3).

No isotope effect was observed on deuteration of the phenyl ring, arguing against C—H bond breaking being rate-determining. The rate of decay of the yellow species was found to be dependent on the nature of substituents on the phenyl group, with electron-withdrawing substituents giving slower rates and electron-donating substituents increasing the rate of reaction to the point where the intermediate yellow species could not be observed. Although the amount of μ-η^2:η^2-peroxo species present in the reaction mixture was suggested to be very small (<10% on the basis of resonance Raman data), the bis(μ-oxo) dimer could not be unequivocally proposed as the only active species in the hydroxylation reaction, due to the possibility of a rapid equilibrium between the bis(μ-oxo) and μ-η^2:η^2-peroxo cores.

3.2
Aliphatic Oxygenation

Model complexes that carry out aliphatic ligand hydroxylation are much less numerous than those that hydroxylate aromatic substrates. Such transforma-

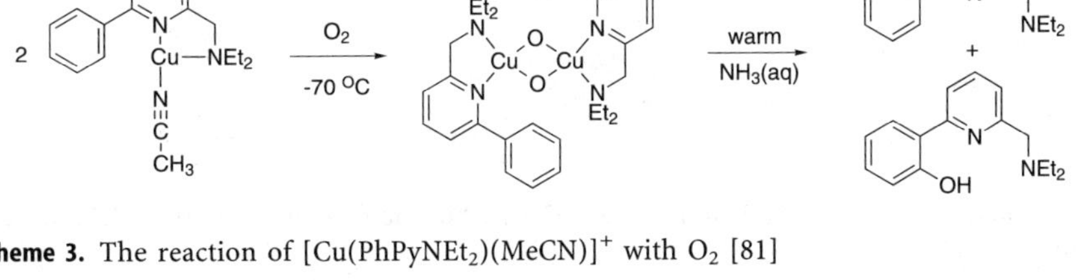

Scheme 3. The reaction of [Cu(PhPyNEt$_2$)(MeCN)]$^+$ with O_2 [81]

tions have been manifested in oxidative *N*-dealkylation reactions of coordinated ligands (Schemes 4 and 5) [84, 99], studies of which have shown that the incorporated oxygen atom derives from molecular O_2 by way of a Cu/O_2 species.

Detailed mechanistic information has been obtained from a study of the *N*-dealkylation reactions of a series of $[Cu_2(\mu\text{-}O)_2]^{2+}$ complexes supported by *N*-substituted TACN ligands [84]. Decomposition of $[\{(Bn_3TACN)Cu\}_2(\mu\text{-}^{18}O)_2]^{2+}$ resulted in the formation of ^{18}O-labeled benzaldehyde, and a double-labeling experiment showed that the O-atom transfer from the bis(μ-oxo) bridge was an intramolecular process. Large kinetic isotope effects (26–40 at −40 °C) were observed for the N-dealkylation reactions, suggesting C—H bond breaking was rate-determining. A Hammett study indicated the copper core to be electrophilic ($\rho = -0.8$) and this was supported by the observation that the bis(μ-oxo) complexes did not react with strong acids. Alternative mechanisms involving stepwise H-atom abstraction by the intact bis(μ-oxo) core followed by a rapid "rebound" collapse of the carbon radical (Path a, Scheme 6), or a concerted C—H scission and O-atom transfer (Path b, Scheme 6) were proposed.

The first example of quantitative ligand hydroxylation at the benzylic position of a Cu complex was reported in 1995 (Scheme 7) [100]. Reaction of a CH_2Cl_2 solution of the crystallographically characterized Cu(II) complex $[Cu(Py2Phe)(OClO_3)_2]$ under Ar with equimolar amounts of benzoin and triethylamine, and subsequent treatment with O_2 gave the dinuclear product $[Cu_2(Py2Phe\text{-}O)_2](ClO_4)_2 \cdot 2H_2O$.

X-ray analysis of this complex showed that the ligand had been hydroxylated regiospecifically at the benzylic position and ^{18}O labeling studies

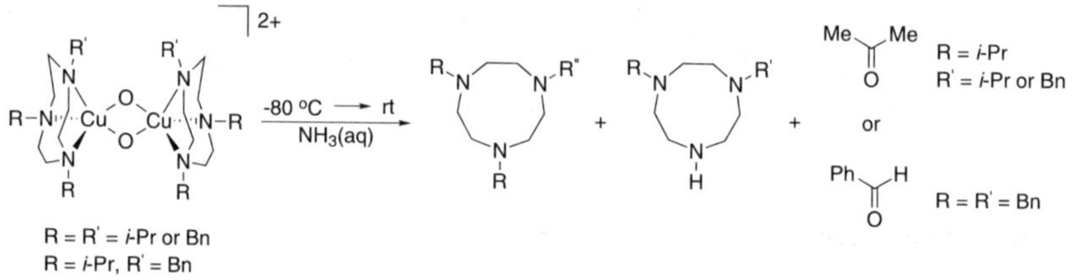

Scheme 4. Oxidative *N*-dealkylation of the TACN ligand in $[\{Cu(R_3TACN)\}_2(\mu\text{-}O)_2]^{2+}$ complexes [84]

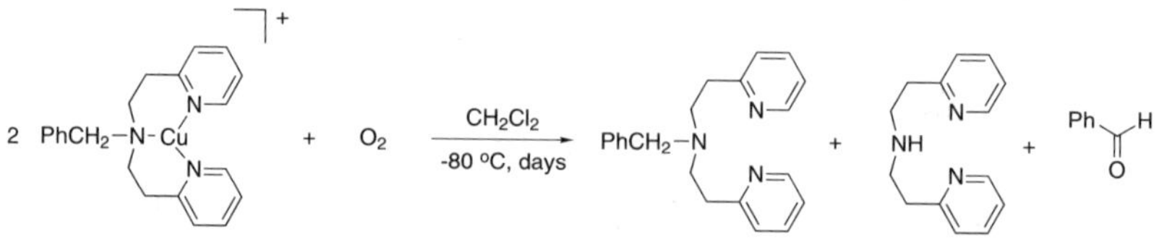

Scheme 5. Oxidative *N*-dealkylation of the $PhCH_2PY2$ ligand on reaction of $[Cu(PhCH_2PY2]^+$ with O_2 [99]

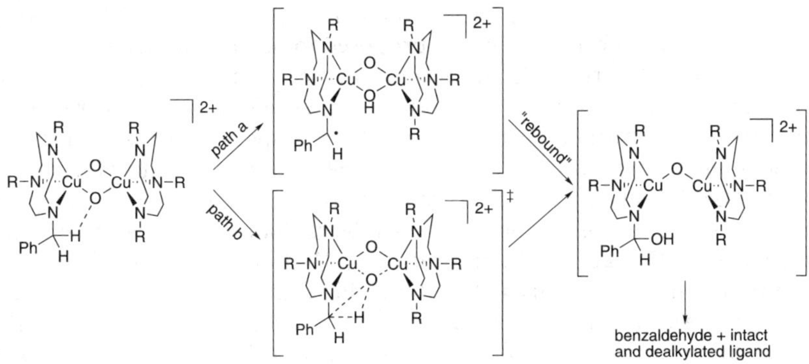

Scheme 6. Proposed mechanism of *N*-dealkylation of the TACN ligand in $[\{Cu(Bn_3TACN)\}_2(\mu\text{-}O)_2]^{2+}$[84]

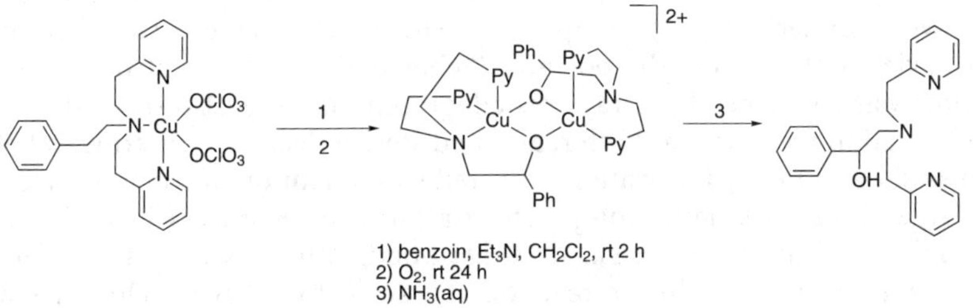

Scheme 7. Benzylic hydroxylation of the Py2Phe ligand [100]

showed the incorporated oxygen atom derived from molecular oxygen. The yield of hydroxylated ligand was quantitative. Manometry showed a 1:1 Cu/O_2 reaction stoichiometry and it was shown that the presence of benzoin, base and O_2 was necessary for ligand hydroxylation to occur. Similar behavior was observed when an equimolar mixture of the Cu(I) complex $[Cu(MeCN)_4]PF_6$ and Py2Phe in CH_2Cl_2 was treated with O_2, although in this case the yield of hydroxylated ligand was 50%. The authors proposed that the hydroxylation reaction occurred via the intermediacy of a μ-η^2:η^2-peroxo species. Further support for this came in a subsequent paper that reported detailed spectroscopic and kinetic studies of the hydroxylation reaction [101]. Bubbling O_2 through a CH_2Cl_2 solution of $[Cu(Py2Phe)]PF_6$ at −80 °C gave a solution that was EPR-silent, and which exhibited UV/vis maxima at 362 nm and 526 nm. The resonance Raman spectrum (excitation wavelength 514.5 nm) of a frozen (77 K) acetone solution of the oxygenated product showed a band at 746 cm^{-1} which moved to 704 cm^{-1} on substitution of $^{18}O_2$. These spectroscopic data are all consistent with formation of a dimeric μ-η^2:η^2-peroxo species and no evidence for the formation of any dimeric bis(μ-oxo) species was observed. Decay of the μ-η^2:η^2-peroxo species was found to be a first order process, and was ascribed to correspond to the rate of ligand hydroxylation. A relatively small kinetic isotope effect was measured for this process

(k_H/k_D = 5.9 at −80 °C, 1.8 at −40 °C) and cleavage of the O—O bond within the μ-η^2:η^2-peroxo species was hence proposed to be the rate-determining step in the hydroxylation reaction. A mechanism was proposed whereby a bis(μ-oxo) species formed from O—O bond cleavage effected the hydroxylation either directly, or through formation of a Cu^{II}—$O^{\cdot}$ species. In a later study, use of a bidentate ligand closely related to Py2Phe resulted in exclusive formation of a bis(μ-oxo) intermediate on oxygenation of the Cu(I) precursor. Rate data obtained for the decomposition of this intermediate showed a large kinetic isotope effect (35.4 at −80°C) indicative of H-atom abstraction being rate-determining, and it was argued that the bis(μ-oxo) species, rather than a peroxo species, was the active agent in the hydroxylation reaction [82]. Hydroxylation at the benzylic position of a calix[4]arene-based ligand analogous to Py2Phe through treatment of its Cu(I) complex with O_2 has also been reported [102].

Reaction of the Cu(II) complex [Cu(IndPY2)(OSO_2CF_3)$_2$] in CH_2Cl_2 under Ar with 2 equivalents of benzoin/triethylamine and subsequent treatment with O_2 gave a dimeric Cu(II) complex in which the indane ligand had been hydroxylated at the benzylic position (Scheme 8) [103]. The structure of the product was confirmed by X-ray crystallography, and this also showed that the hydroxylation reaction was both regio- and stereoselective. Deuterium labeling studies showed the hydroxylation occurred with retention of configuration and the large observed kinetic isotope effect argued for a process involving rate-determining cleavage of a benzylic C—H bond. Attempts to characterize the Cu/O_2 species involved in the reaction at −80 °C by UV/vis techniques were unsuccessful, owing to the high reactivity of this species.

4 Conclusions

While there remains much work to be done in order to understand the exact mechanisms by which Cu-containing proteins utilize dioxygen, the results outlined herein have allowed us to gain a much greater understanding of the fundamentals of the chemistry of copper-dioxygen species. The use of low temperature techniques has been central to the isolation and characterization

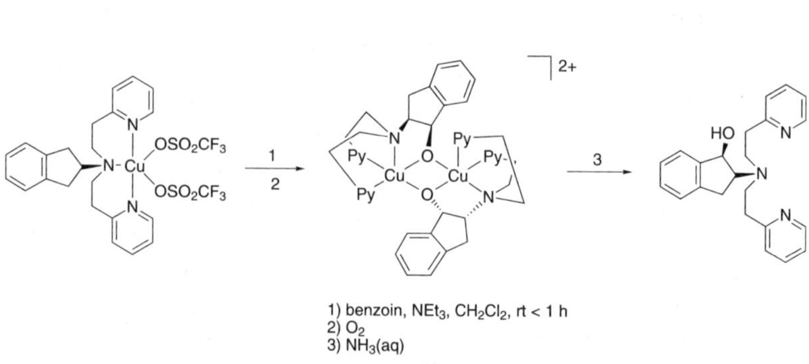

Scheme 8. Regio- and stereoselective hydroxylation of the IndPY2 ligand [103]

of Cu/O_2 species. In this respect, the X-ray structures that have been obtained have been of paramount importance, allowing (in most cases) unequivocal characterization of numerous different binding modes of the reduced forms of O_2. These different binding modes appear to be dependent on the nature of the ancillary ligand on the copper ion; tetradentate ligands favor monodentate coordination of the O_2 derivative, while tridentate ligands enforce bidentate coordination, due to the propensity of Cu(II) to be 5-coordinate.

The isolation and characterization of Cu(III)-oxo species and the observation of the interconversion between the $[Cu_2(\mu\text{-}O)_2]^{2+}$ and $[Cu_2(\mu\text{-}\eta^2\text{:}\eta^2\text{-}O_2)]^{2+}$ cores has provided clear experimental precedence for a way that the O—O bond in O_2 may be broken and formed. It seems likely that future studies of the $[Cu_2(\mu\text{-}O)_2]^{2+}/[Cu_2(\mu\text{-}\eta^2\text{:}\eta^2\text{-}O_2)]^{2+}$ interconversion will bear relevance not only to other biological oxygen-activating systems [104], but also to the water oxidation center of Photosystem II, in which O—O coupling is performed [105].

The question of whether the $[Cu_2(\mu\text{-}O)_2]^{2+}$ or $[Cu_2(\mu\text{-}\eta^2\text{:}\eta^2\text{-}O_2)]^{2+}$ core is responsible for oxygenation reactions has still not been answered unequivocally, and more work in this area is required. The quest for greater stability of the Cu/O_2 intermediates has begun to be addressed, and this too represents an important area of future study (see Note Added in Proof). Such enhanced stability would allow greater ease of study of such compounds, and will undoubtedly be achieved by using ligands having special electronic or steric features designed to minimize decomposition reactions and/or provide thermodynamic stabilization.

While much information regarding the structures and reactivities of dinuclear Cu/O_2 complexes as models for such proteins as hemocyanin and tyrosinase has recently been obtained, our understanding of mononuclear and trinuclear systems is not similarly well developed. For example, the crystal structure of peptidylglycine α-hydroxylating monooxygenase (PHM), the oxygenase domain of peptidyl α-amidating monooxygenase (PAM), shows the presence of two mononuclear copper sites separated by 11 Å, one at which O_2-binding occurs while the other mediates electron transfer [106]. The mechanism by which C—H bond activation occurs at a mononuclear copper site in this protein is unclear. A novel mechanism involving monocopper-dioxygen species has been proposed for the structurally similar protein dopamine β-monooxygenase (DβM) that warrants evaluation via studies of mononuclear model complexes [15, 107]. Similarly, there is still only one crystallographically characterized example of a trinuclear Cu/O_2 complex, and, given the importance of trinuclear sites in copper proteins such as the multicopper oxidases, the synthesis of trinuclear complexes as models for these is obviously of high priority. We await the results of these future studies with interest.

Acknowledgements. AGB gratefully acknowledges a period of sabbatical leave at the University of Minnesota, granted by the University of Otago. WBT thanks the NSF, NIH, and the Alfred P Sloan and Camille and Henry Dreyfus foundations for support of work from his laboratory described herein. We are grateful to Professors Ken Karlin and Peter Comba for supplying us with preprints of submitted papers.

Note Added in Proof. Since submission of this manuscript, several reports describing the spectroscopic characterization of copper-dioxygen complexes have appeared [108–110]. Discussion of copper-dioxygen chemistry has been included in a recent review article [111]. A dendrimer that incorporates PY2 ligands coordinated to Cu(I) ions has been reported to bind dioxygen [112] and a detailed mechanistic study of the reaction of a dicopper(I) complex with dioxygen has been published [113]. A detailed electronic structural study of bis(μ-oxo)dicopper(III) compounds in which UV-vis absorption and resonance Raman spectral analysis are emphasized has appeared [114]. Finally, a room-temperature stable (μ-η^2:η^2-peroxo)dicopper complex ($t_{1/2}$ = 25 h in CH_2Cl_2 at 25 °C) supported by a binucleating hexapyridyl ligand has been structurally characterized by X-ray crystallography [115]. The $[Cu_2(O_2)]^{2+}$ unit is slightly bent (by ~17°) and has O—O and Cu···Cu distances of 1.485(8) Å and 3.477(7) Å, respectively.

5 References

1. Mellor JW (1923) A comprehensive treatise on inorganic and theoretical chemistry, vol III. Longmans, London
2. Ito N, Phillips SEV, Stevens C, Ogel ZB, McPherson MJ, Keen JN, Yadav KDS, Knowles PF (1991) Nature 350: 87
3. Magnus KA, Hazes B, Ton-That H, Bonaventura C, Bonaventura J, Hol WGJ (1994) Proteins: Structure, Function and Genetics 19: 302
4. Messerschmidt A, Ladenstein R, Huber R, Bolognesi M, Avigliano L, Petruzzelli R, Rossi A, Finazzi-Agró A (1992) J Mol Biol 224: 179
5. Messerschmidt A (1998) In: Sinnott M (ed) Comprehensive biological catalysis: a mechanistic reference, vol 3. Academic Press, San Diego, p 401
6. Que L Jr, Ho RYN (1996) Chem Rev 96: 2607
7. Wallar BJ, Lipscomb JD (1996) Chem Rev 96: 2625
8. Feig AL, Lippard SJ (1994) Chem Rev 94: 759
9. Ortiz de Montellano PR (ed) (1995) Cytochrome P450: structure, mechanism and biochemistry. Plenum, New York
10. Kopf M-A, Karlin, KD (1999) Models of copper enzymes and heme-copper oxidases. In: Meunier B (ed) Biomimetic oxidations, chap 7. Imperial College Press, London
11. Karlin KD, Zuberbühler AD (1999) Formation, structure and reactivity of copper dioxygen complexes. In: Reedijk J, Bouwman E (eds) Bioinorganic catalysis, chap14. Marcel Dekker, New York
12. Fontecave M, Pierre J-L (1998) Coord Chem Rev 170: 125
13. Karlin KD, Kaderli S, Zuberbühler AD (1997) Acc Chem Res 30: 139
14. Tolman WB (1997) Acc Chem Res 30: 227
15. Klinman J (1996) Chem Rev 96: 2541
16. Solomon EI, Sundaram UM, Machonkin TE (1996) Chem Rev 96: 2563
17. Kitajima N, Moro-oka Y (1994) Chem Rev 94: 737
18. Holland PL, Tolman WB (1999) Coord Chem Rev 190–192: 855
19. Karlin KD, Tyeklár Z (eds) (1993) Bioinorganic chemistry of copper. Chapman and Hall, New York
20. Momenteau MM, Reed CA (1994) Chem Rev 94: 659
21. Karlin KD, Wei N, Jung B, Kaderli S, Zuberbühler A (1991) J Am Chem Soc 113: 5868
22. Karlin KD, Wei N, Jung B, Kaderli S, Niklaus P, Zuberbühler AD (1993) J Am Chem Soc 115: 9506
23. Chaudhuri P, Hess M, Weyhermüller, Wieghardt K (1999) Angew Chem Int Ed Engl 38: 1095
24. Harata M, Jitsukawa K, Masuda H, Einaga H (1994) J Am Chem Soc 116: 10817

25. Berreau LM, Mahapatra S, Halfen JA, Young VG Jr, Tolman WB (1996) Inorg Chem 35: 6339
26. Harata M, Jitsukawa K, Masuda H, Einaga H (1996) Chem Lett 813
27. Harata M, Jitsukawa K, Masuda H, Einaga H (1998) Bull Chem Soc Jpn 71: 637
28. Fujisawa K, Tanaka M, Moro-oka Y, Kitajima N (1994) J Am Chem Soc 116: 12079
29. Egan JW Jr, Haggerty BS, Rheingold AL, Sendlinger SC, Theopold KH (1990) J Am Chem Soc 112: 2445
30. Kitajima N, Komatsuzaki H, Hikichi S, Osawa M, Moro-oka Y (1994) J Am Chem Soc 116: 11,596
31. Kitajima N, Fujisawa K, Fujimoto C, Moro-oka Y, Hashimoto S, Kitagawa T, Toriumi K, Tatsumi K, Makamura A (1992) J Am Chem Soc 114: 1277
32. Kitajima N, Katayama T, Fujisawa K, Iwata Y, Moro-oka Y (1993) J Am Chem Soc 115: 7872
33. Chavez FA, Briones JA, Olmsted MM, Mascharak PK (1999) Inorg Chem 38: 1603
34. Chavez FA, Rowland JM, Olmsted MM, Mascharak PK (1998) J Am Chem Soc 120: 9015
35. Chavez FA, Nguyen CV, Olmsted MM, Mascharak PK (1996) Inorg Chem 35: 6282
36. Wada A, Harata M, Hasegawa K, Jitsukawa K, Masuda H, Mukai M, Kitagawa T, Einaga H (1998) Angew Chem Int Ed Engl 37: 798
37. Sanyal I, Ghosh P, Karlin KD (1995) Inorg Chem 34: 3050
38. Ghosh P, Tyeklár Z, Karlin KD, Jacobson RR, Zubieta J (1987) J Am Chem Soc 109: 6889
39. Karlin KD, Ghosh P, Cruse RW, Farooq A, Gultneh Y, Jacobson RR, Blackburn NJ, Strange RW, Zubieta J (1988) J Am Chem Soc 110: 6769
40. Karlin KD, Hayes JC, Gultneh Y, Cruse RW, McKown JW, Hutchinson JP, Zubieta J (1984) J Am Chem Soc 106: 2121
41. Nasir MS, Karlin KD, McGowty D, Zubieta J (1991) J Am Chem Soc 113: 698
42. Mahroof-Tahir M, Murthy NN, Karlin KD, Blackburn NJ, Shaikh SN, Zubieta J (1992) Inorg Chem 31: 3001
43. Root DE, Mahroof-Tahir M, Karlin KD, Solomon EI (1998) Inorg Chem 37: 4838
44. Jacobson RR, Tyeklár Z, Farooq A, Karlin KD, Liu S, Zubieta J (1988) J Am Chem Soc 110: 3690
45. Baldwin MJ, Ross PK, Pate JE, Tyeklár Z, Karlin KD, Solomon EI (1991) J Am Chem Soc 113: 8671
46. Tyeklár Z, Jacobson RR, Wei N, Murthy NN, Zubieta J, Karlin KD (1993) J Am Chem Soc 115: 2677
47. Halfen JA, Young VG Jr, Tolman WB (1996) J Am Chem Soc 118: 1920
48. Börzel H, Comba P, Katsichtis C, Kiefer W, Lienke A, Nagel V, Pritzkow H (1999) Chem Eur J 5: 1716
49. Bol JE, Driessen WL, Ho RYN, Maase B, Que L Jr, Reedijk J (1997) Angew Chem Int Ed Engl 36: 998
50. Kitajima N, Koda T, Hashimoto S, Kitagawa T, Moro-oka Y (1988) J Chem Soc, Chem Commun 151
51. Nakamoto K (1997) Infrared and Raman spectra of inorganic and coordination compounds, 5th edn. Wiley, New York
52. Kitajima N, Fujisawa K, Moro-oka Y (1989) J Am Chem Soc 111: 8975
53. Bradley DC, Ghotra JS, Mart FA, Hursthouse MB, Raithby PR (1977) J Chem Soc Dalton Trans 1166
54. Haegele R, Boeyens JCA (1977) J Chem Soc Dalton Trans 648
55. Kitajima N, Koda T, Hashimoto S, Kitagawa T, Moro-oka Y (1991) J Am Chem Soc 113: 5664
56. Sorrell TN, Allen WE, White PS (1995) Inorg Chem 34: 952
57. Lynch WE, Kurtz DM Jr, Wang S, Scott RA (1994) J Am Chem Soc 116: 11030
58. Mahapatra S, Halfen JA, Wilkinson EC, Que L Jr, Tolman WB (1994) J Am Chem Soc 116: 9785
59. Pidcock E, Obias HV, Zhang CX, Karlin KD, Solomon EI (1998) J Am Chem Soc 120: 7841

60. Sanyal I, Karlin KD, Strange RW, Blackburn NJ (1993) J Am Chem Soc 115: 11,259
61. Klein Gebbink RJM, Martens CF, Feiters MC, Karlin KD, Nolte RJM (1997) Chem Commun 389
62. Pidcock E, Obias HV, Abe M, Liang H-C, Karlin KD, Solomon EI (1999) J Am Chem Soc 121: 1299
63. Jung B, Karlin KD, Zuberbühler AD (1996) J Am Chem Soc 118: 3763
64. Halfen JA, Mahapatra S, Wilkinson EC, Kaderli S, Young VG Jr, Que L Jr, Zuberbühler AD, Tolman WB (1996) Science 271: 1397
65. Mahapatra S, Young VG Jr, Kaderli S, Zuberbühler AD, Tolman WB (1997) Angew Chem Int Ed Engl 36: 130
66. DuBois JL, Mukherjee P, Collier AM, Mayer JM, Solomon EI, Hedman B, Stack TDP, Hodgson KO (1997) J Am Chem Soc 119: 8578
67. Mahadevan V, Hou Z, Cole AP, Root DE, Lal TK, Solomon EI, Stack TDP (1997) J Am Chem Soc 119: 11,996
68. Mahapatra S, Halfen JA, Wilkinson EC, Pan G, Wang X, Young VG Jr, Cramer CJ, Que L Jr, Tolman WB (1996) J Am Chem Soc 118: 11,555
69. Zang Y, Dong Y, Que L Jr, Kauffmann K, Münck E (1995) J Am Chem Soc 117: 1169
70. Manchanda R, Brudvig GW, Crabtree RH (1995) Coord Chem Rev 144: 1
71. Hikichi S, Yoshizawa M, Sasakura Y, Akita M, Moro-oka Y (1998) J Am Chem Soc 120: 10567
72. Mahadevan V, DuBois JL, Hedman B, Hodgson KO, Stack TDP (1999) J Am Chem Soc 121: 5583
73. Flock M, Pierloot K (1999) J Phys Chem A 103: 95
74. Estiú GL, Zerner MC (1999) J Am Chem Soc 121: 1893
75. Liu X-Y, Palacios AA, Novoa JJ, Alvarez S (1998) Inorg Chem 37: 1202
76. Bérces A (1997) Inorg Chem 36: 4831
77. Cramer CJ, Smith BA, Tolman WB (1996) J Am Chem Soc 118: 11,283
78. Holland PL, Cramer CJ, Wilkinson EC, Mahapatra S, Rodgers KR, Itoh S, Taki M, Fukuzumi S, Que L Jr, Tolman WB (2000) J Am Chem Soc (in press)
79. Cahoy J, Holland PL, Tolman WB (1999) Inorg Chem 38: 2161
80. Mahapatra S, Kaderli S, Llobet A, Neuhold Y-M, Palanché T, Halfen JA, Young VG Jr, Kaden TA, Que L Jr, Zuberbühler AD, Tolman WB (1997) Inorg Chem 36: 6343
81. Holland PL, Rodgers KR, Tolman WB (1999) Angew Chem Int Ed Engl 38: 1139
82. Itoh S, Taki M, Nakao H, Holland PL, Tolman WB, Que L Jr, Fukuzumi S (2000) Angew Chem Int Ed 39: 398
83. Enomoto M, Aida T (1999) J Am Chem Soc 121: 874
84. Mahapatra S, Halfen JA, Tolman WB (1996) J Am Chem Soc 118: 11,575
85. Obias HV, Lin Y, Murthy NN, Pidcock E, Solomon EI, Ralle M, Blackburn NJ, Neuhold Y-M, Zuberbühler AD, Karlin KD (1998) J Am Chem Soc 120: 12,960
86. Pidcock E, DeBeer S, Obias HV, Hedman B, Hodgson KO, Karlin KD, Solomon EI (1999) J Am Chem Soc 121: 1870
87. Cole AP, Root DE, Mikherjee P, Solomon EI, Stack TDP (1996) Science 273: 1848
88. Root DE, Henson MJ, Machonkin T, Mukherjee P, Stack TDP, Solomon EI (1998) J Am Chem Soc 120: 4982
89. Reim J, Krebs B (1994) Angew Chem Int Ed Engl 33: 1969
90. Stomberg R, Trysberg L, Larking I (1970) Acta Chem Scand 24: 2678
91. Micklitz W, Bott SG, Bentsen JG, Lippard SJ (1989) J Am Chem Soc 111: 372
92. Reim J, Werner R, Haase W, Krebs B (1998) Chem Eur J 4: 289
93. Parish EJ, Kizito SA (1995) In: Paquette LA (ed) Encyclopedia of reagents for organic synthesis, vol 2. Wiley, Chichester, p 1333
94. Ghosh D, Mukherjee R (1998) Inorg Chem 37: 6597
95. Ryan S, Adams H, Fenton DE, Becker M, Schindler S (1998) Inorg Chem 37: 2134
96. Alzuet G, Casella L, Villa ML, Carugo O, Gullotti M (1997) J Chem Soc Dalton Trans 4789
97. Cruse RW, Kaderli S, Karlin KD, Zuberbühler AD (1988) J Am Chem Soc 110: 6882

98. Karlin KD, Nasir MS, Cohen BI, Cruse RW, Kaderli S, Zuberbühler AD (1994) J Am Chem Soc 116: 1324
99. Sanyal I, Mahroof-Tahir M, Nasir MS, Ghosh P, Cohen BI, Gultneh Y, Cruse RW, Farooq A, Karlin KD, Liu S, Zubieta J (1992) Inorg Chem 31: 4322
100. Itoh S, Kondo T, Komatsu M, Ohshiro Y, Li C, Kanehisa N, Kai Y, Fukuzumi S (1995) J Am Chem Soc 117: 4714
101. Itoh S, Nakao H, Berreau L, Kondo T, Komatsu M, Fukuzumi S (1998) J Am Chem Soc 120: 2890
102. Xie D, Gutsche CD (1998) J Org Chem 63: 9270
103. Blain I, Bruno P, Giorgi M, Lojou E, Lexa D, Réglier M (1998) Eur J Inorg Chem 1297
104. Lange SJ, Que L Jr (1998) Curr Opin Chem Biol 2: 159
105. Tommos C, Babcock GT (1998) Acc Chem Res 31: 18
106. Prigge ST, Kolhekar AS, Eipper BA, Mains RE, Amzel LM (1997) Science 278: 1300
107. Reddy KV, Jin S-J, Arora PK, Sfeir DS, Maloney SCF, Urbach FL, Sayre LM (1990) J Am Chem Soc 112: 2332
108. Feiters MC, Klein Gebbink RJM, Solé VA, Nolting H-F, Karlin KD, Nolte RJM (1999) Inorg Chem 38: 6171
109. Klein Gebbink RJM, Martens CF, Kenis PJA, Jansen RJ, Nolting H-F, Solé VA, Feiters MC, Karlin KD, Nolte RJM (1999) Inorg Chem 38: 5755
110. Becker M, Heinemann FW, Schindler S (1999) Chem Eur J 5: 3124
111. Liang H-C, Dahan M, Karlin KD (1999) Curr Opin Chem Biol 3: 168
112. Klein Gebbink RJM, Bosman AW, Feiters MC, Meijer EW, Nolte RJM (1999) Chem Eur J 5: 65
113. Becker M, Schindler S, Karlin KD, Kaden TA, Kaderli S, Palanché T, Zuberbühler AD (1999) Inorg Chem 38: 1989
114. Henson MJ, Mukherjee P, Root DE, Stack TDP, Solomon EI (1999) J Am Chem Soc 121: 10332
115. Kodera M, Katayama K, Tachi Y, Kano K, Hirota S, Fujinami S, Suzuki M (1999) J Am Chem Soc 121: 11006

Rhenium-Oxo and Rhenium-Peroxo Complexes in Catalytic Oxidations

Fritz E. Kühn[1], Wolfgang A. Herrmann[2]

Anorganisch-Chemisches Institut der Technischen Universität München, Lichtenbergstraße 4, 85747 Garching b. München, Germany

[1] *E-mail: fritz.kuehn@ch.tum.de*

[2] *E-mail: wolfgang.herrmann@ch.tum.de*

Rhenium oxo complexes have been known for about 70 years. However, until recently they and their organometallic derivatives have been widely regarded as curiosities. The situation has changed drastically since the successful high yield synthesis of methyltrioxorhenium(VII) (usually abbreviated as MTO) in the late 1980s and its bisperoxo complex about 5 years later. Since then numerous applications of MTO and other related rhenium oxo and peroxo complexes have been found, most notably in oxidation catalysis. A comprehensive review of the most important rhenium oxo and peroxo complexes and their applications in oxidation catalysis is presented here.

Keywords: Methyltrioxorhenium, Nucleophilic additions, Oxidation catalysis, Peroxo complexes, Rhenium oxides

Structure and Bonding, Vol. 97

1 Introduction

Rhenium oxo complexes were described shortly after the element rhenium had been discovered by Noddack and Tacke. Structures of rhenium oxides are also well known [1]. However, the entire class of organorhenium oxides and organorhenium peroxides is less than 30 years old [2]. For a considerable part of this time these complexes have been widely regarded as mere curiosities. This picture changed drastically during the last decade. Today not only an amazing number of these complexes are known and are easy to synthesize, but some also have very interesting applications in both catalysis and material sciences. Especially in the catalytic field the use of organorhenium(VII) oxides is in rapid progress and therefore prototypal for organometal oxides in general. The detailed examination of organorhenium oxides and peroxides has also contributed to a renaissance of the interest in the rhenium oxides and their potential applications in oxidation catalysis [3].

2 Synthesis of Rhenium Oxo Derivatives

2.1 Inorganic Rhenium Oxo Complexes

The key compounds of Re high oxidation state compounds are perrhenates and dirhenium heptaoxide (Re_2O_7). Perrhenates and perrhenic acid, formally $HReO_4$ ($pK_a = -1.25$), can easily be prepared by oxidation of Re metal with H_2O_2. Re_2O_7 is usually prepared by burning the metal in an atmosphere of oxygen at 400 °C [4, 5]. Re_2O_7 belongs to a small group of binary metal oxides which are soluble in certain organic solvents, other examples being Mn_2O_7, Tc_2O_7, RuO_4, and OsO_4. While these oxides keep strictly molecular structures in the crystal [6], Re_2O_7 exhibits an infinite polymeric solid state structure like most insoluble metal oxides, e.g., MoO_3 and WO_3. However, the solid state structure of Re_2O_7 is unique: an equal number of nearly regular ReO_4

tetrahedra and of (highly distorted) ReO_6 octahedra share corners through oxygen bridges as drawn in **I** [7]. Important in this context is that the bridging Re—O—bonds (172.5–216.0 pm) are longer than the terminal (165.0–174.2 pm) ones. Also, the four different oxygens bound to a tetrahedrally coordinated Re-center differ from each other by only about 10 pm in length (168–180 pm) while the octahedrally coordinated Re atoms form three longer (205–216 pm) and three significantly shorter (165–175 pm) Re—O bonds.

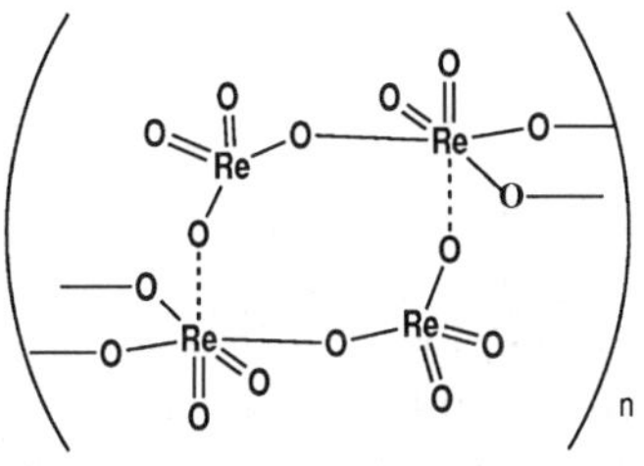

I

The solubilization of Re_2O_7 in organic solvents has provided some conflicting reports. At a certain stage, several studies described Re_2O_7 as being insoluble in any organic solvent in the strict absence of water [8]. Solubility was then attributed to the formation of soluble perrhenic acid from residual water in the solvent. Re_2O_7 dissolves freely in water to give perrhenic acid, which can be crystallized out of these solutions as the bis(aqua) adduct of Re_2O_7, i.e., $Re_2O_7 \cdot (H_2O)_2$ [9, 10]. Both water molecules are coordinated to the same rhenium atom which becomes octahedral. The other Re atom remains tetrahedral and the Re—O—Re bond is nearly linear (bond angle = −179.3°). Careful studies on the hydrolysis of Re_2O_7 have, however, shown that the first product of the interaction of Re_2O_7 with water, in 1,4-dioxane, can actually be considered as a dimer of $ReO_3(OH)$, namely $Re_2O_6(OH)_2 \cdot 3$(1,4-dioxane) which adds water to give the bis-water adduct [11]. This very sensitive compound is stabilized in the crystal by coordination of one solvent molecule to the Re atom and by hydrogen bonding of the other 1,4-dioxane molecules to the bridging OH groups [12]. Two recent structures of supra molecular adducts of $Re_2O_7 \cdot 2(H_2O)$ with 1,4-dioxane [13] and diglyme [14] have been reported. In both cases there is no coordination of the dioxane or diglyme molecules to the Re atoms. Instead, they are associated to the coordinated H_2O molecules by hydrogen bonds.

A reinvestigation of this solubility problem was recently carried out in order to establish the true nature of the species existing in the solutions of Re_2O_7 used as the starting material for the synthesis of many other Re(VII) oxide derivatives, including the tetra(peroxo) complex of Re_2O_7 (see Sect. 3.2) and methyltrioxorhenium (see Sect. 2.2). If bulk Re_2O_7 is dissolved in aprotic, organic donor solvents such as CH_3CN, THF, or DME, solvent adducts of Re_2O_7 are obtained, $Re_2O_7 \cdot (Solv)_2$ (Solv = solvent), which were shown by X-ray crystallography to be structurally similar to $Re_2O_7 \cdot 2H_2O$ [14, 15].

Therefore, it is clear that the species existing in solutions of Re_2O_7 are solvates of the type $Re_2O_7 \cdot (Solv)_2$ and $Re_2O_7 \cdot$ (1,4-dioxane) could also be isolated in the absence of water. The polymeric solid state structure of Re_2O_7 breaks down at the expense of the newly formed Re-(Solv) bonds. As experimentally observed, non-donating or weakly donating solvents are unable to dissolve Re_2O_7.

2.2
Organorhenium Oxides

The first organorhenium(VII) oxides, namely $(CH_3)_3ReO_2$ and $((CH_3)_3CCH_2)_3ReO_2$ were prepared 1975 by Mertis and Wilkinson [16]. In 1979 Beattie and Jones reported on the preparation of CH_3ReO_3 (MTO) [17]. A few other complexes were published during the following years but the breakthrough for catalytic applications came in 1987 when Herrmann et al. developed an efficient synthesis for MTO, starting from dirhenium heptaoxide and tetramethyltin (Eq. 1). This synthetic pathway made it possible to conduct the synthesis of MTO within a few hours and in amounts up to 10 g [18].

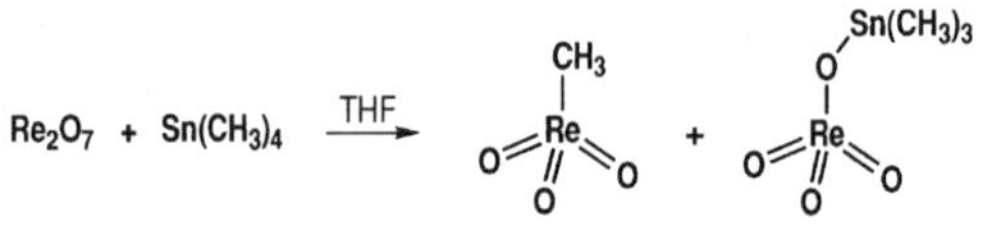

(1)

The method was further improved in 1992 when it was proven possible to reach nearly quantitative Re yields by the addition of carboxylic acid anhydrides to the reaction mixture and additionally by utilizing the trimethylstannyl perrhenate, formed as by-product in the original synthesis (see Eq. 1) [19, 20]. Recently the latter method has been extended for other perrhenates as starting materials so that the moisture sensitive Re_2O_7 can be replaced by more convenient-to-handle starting materials [21]. This synthetic progress was accompanied by the discovery of a plethora of derivatives and catalytic applications of organorhenium(VII) oxides (for general types see formulae **II–IV**). The use of these complexes in catalysis, especially in oxidation catalysis, however, is still very much dominated by MTO itself [22–31].

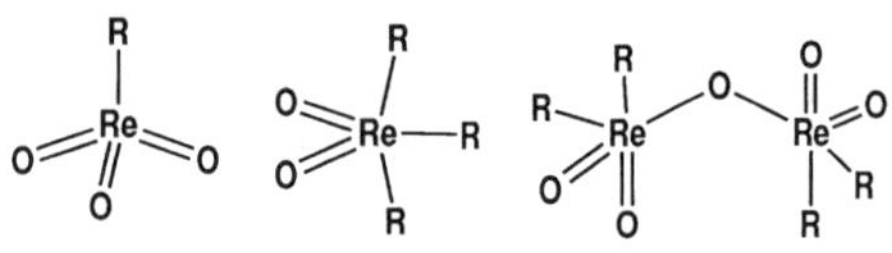

II (Re^{VII}) III (Re^{VII}) IV (Re^{VII})

3
Synthesis of Rhenium Peroxo Complexes

3.1
Organorhenium Peroxo Complexes

Probably the most thoroughly examined field of applications of organorhenium(VII) oxides is the oxidation catalysis. The catalytic activity of MTO and some of its derivatives in the oxidation of olefins was noticed soon after these complexes were accessible in higher amounts [32, 33]. However, the breakthrough in the understanding of the role of organorhenium(VII) oxides of type **II** in oxidation catalysis was the isolation and characterization of the reaction product of MTO with excess H_2O_2 by Herrmann et al. According to Eq. (2) a bisperoxo complex of stoichiometry $(CH_3)Re(O_2)_2O$ (**3**) is formed:

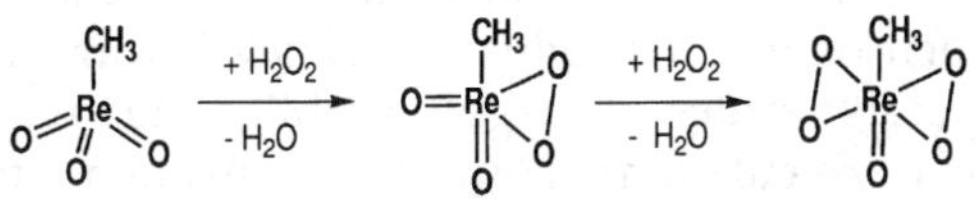

1 **2** **3** (2)

In the solid state it is isolated as an adduct with a donor ligand L (L = H_2O, **3a**; L = $-O{=}P(N(CH_3)_2)_3$, **3b**) [34, 35], which is lost in the gas phase. The structures of **3** (electron diffraction) [23], **3a** and **3b** (X-ray diffraction) were determined; the structure of ligand free complex (**3**) is known from the gas phase. Complex **3** has a trigonal bipyramidal structure, with significant shorter Re—O and Re—C bond distances than in **3a,b**. This is very likely due to the increased Lewis acidity of the free complex **3**. If each peroxo group is considered as a coordinative unit, the molecular structure of complexes **3a,b** can also be described as trigonal bipyramidal. The adduct **3a** melts at 56 °C and can be sublimed at room temperature in oil pump vacuum whereas **3b** melts at 65 °C and decomposes at ca. 75 °C. Both **3a** and **3b** are explosive.

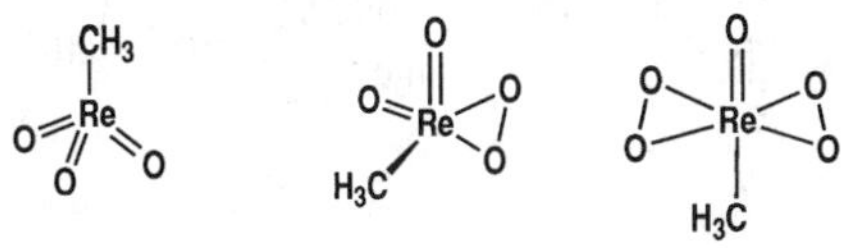

1 **2** **3**

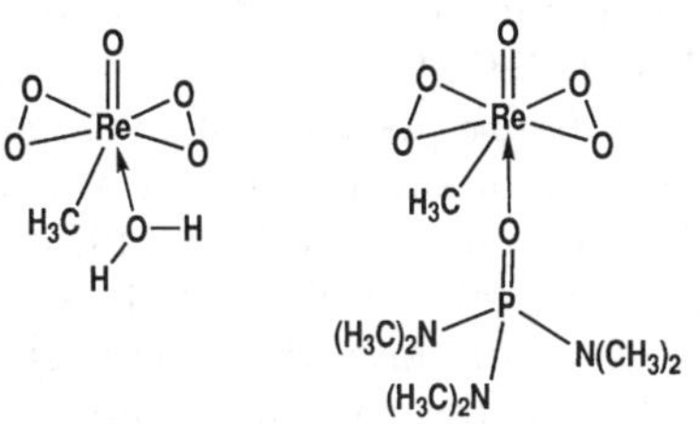

3a **3b**

^{17}O-NMR experiments revealed that **3a,b** exchange their ligand L rapidly in solution, specially in coordinating solvents [23, 24, 34]. The terminal oxygen of **3a** is observed at $\delta(^{17}O) = 762$ ppm, the peroxo groups at $\delta(^{17}O) = 422$ and 362 ppm, respectively, in **3b** the terminal oxygen is observed at $\delta(^{17}O) = 774$ ppm. Only the terminal oxygen atom is involved in an oxygen exchange with water [34, 35].

Experiments with the isolated bis(peroxo)complex **3a** have shown that it is an active species in oxidation catalysis, e.g., in the oxidation of olefins [35]. In situ experiments show that the reaction of MTO with one equivalent of H_2O_2 leads to a monoperoxo complex **2** according to Eq. (2) [35–37]. It has to be noted that the monoperoxo complex has never been isolated. All obtained results on this species are derived from in situ experiments, where **2** exists only in equilibria with MTO and **3**. Complex **2** is also catalytically active in oxidation processes as discussed below. Kinetic experiments conducted by Espenson indicate that the rate constants for the transformation of most substrates into their oxidation products by catalysis with the compounds **2** and **3** are of a comparable order of magnitude [37]. This result is supported by density functional calculations [38]. The transition states in the olefin epoxidation process starting from **2** and **3a** are not different enough in energy to exclude totally one of these two catalysts from the catalytic process. The transition states with the lowest energy levels are depicted as **V** and **VI**. Taking into account energies of hydration and peroxidation as well as energies of activation, it was concluded that olefin epoxidation catalyzed by **1**/H_2O_2 occurs as direct oxygen transfer via transition states of spiro structure, likely involving both mono- and bisperoxo complexes depending on the reaction conditions. Given the rather small differences among the various lower lying transition states, entropy effects may have a major influence on the selection among possibly catalytically active complexes.

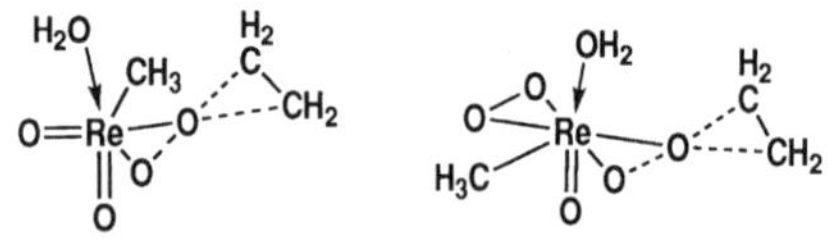

V VI

The decomposition of **1**, **2**, and **3** in solution was also investigated [39–42]. Concerning complex **1**, the full kinetic pH profile for the base-promoted decomposition to CH_4 and ReO_4^- was examined. Spectroscopic and kinetic data give evidence for mono- and dihydroxo complexes of formulae $CH_3ReO_3(OH^-)$ and $CH_3ReO_3(OH^-)_2$ prior to and responsible for the decomposition process. In the presence of hydrogen peroxide complexes **2** and **3** decompose to methanol and perrhenate at a rate that is dependent on $[H_2O_2]$ and $[H_3O]^+$. The complex peroxide and pH dependencies are explained by two possible pathways: attack of either hydroxide on **2** or HO_2^- on MTO. The bisperoxo complex **3** decomposes much more slowly to yield O_2 and MTO [39]. Mechanistic studies in the gas phase show that high valent rhenium compounds, e.g., derivative **3** and high valent chromium compounds, are oxygen transfer reagents while iron derivatives usually catalyze the decomposition of H_2O_2 [43]. The activation parameters for the coordination of H_2O_2 to MTO have also been determined. They indicate a mechanism involving a nucleophilic attack. The protons lost in converting H_2O_2 to a coordinated O_2^{2-} ligand are transferred to one oxide oxygen which remains on the Re center as the aqua ligand L. The rate of this reaction is not pH-dependent [44]. The photochemical and photophysical behavior of **3** has also been closely examined [45].

3.2 Inorganic Rhenium Peroxo Complexes

The rhenium oxides Re_2O_7 and ReO_3 react with hydrogen peroxide solutions yielding peroxo complexes that efficiently catalyze the oxidation of olefins, aromatics, and certain organometallic compounds. In contrast, related oxides of molybdenum, e.g., MoO_3 and vanadium, e.g., V_2O_5 do not activate H_2O_2 under comparable conditions. The active rhenium peroxo complex was isolated from the system Re_2O_7/H_2O_2 by Herrmann et al.: the crystalline orange-red compound of formula $H_4Re_2O_{13}$ (**4**) was also characterized by single crystal X-ray analysis. Its structure resembles a peroxo perrhenic acid. It is structurally defined as two corner-sharing pentagonal bipyramids with apical oxo and aquo ligands; the equatorial positions are occupied by the bridging oxygen and by two η^2-peroxo groups per rhenium. In contrast to the compounds **3a** and **3b**, complex **4** decomposes hydrolytically during the catalytic cycle and can thus not compete with **3** in terms of catalytic activity in oxidation reactions involving H_2O_2 in aqueous systems. Hydrolysis yields perrhenic acid ($Re_2O_7 \cdot 2H_2O$) [13]. It has been shown by Sharpless and coworkers that the use of bis(trimethylsilyl)peroxide $[(Me_3Si)O]_2$ avoids the decomposition of Re_2O_7 (or its tetraperoxocomplex) during the catalytic oxidation [58]. However, it is completely unclear whether Re_2O_7 or $HReO_4$ (which has also been claimed to be an active catalyst with $[(Me_3Si)O]_2$ as oxygen source) react under these conditions to form $Me_3SiOReO_3$, $Me_3SiOOReO_3$, or other related peroxide-containing complexes. $Me_3SiOReO_3$ is known to be formed easily both from Re_2O_7 and ReO_4^- [47]; $Re^{VI}O_3$, which was also found to be catalytically active both with H_2O_2 and $[(Me_3Si)O]_2$ is easily oxidized to "$HReO_4$". It cannot be excluded that the inorganic rhenium oxides Re_2O_7, ReO_3, $HReO_4$, and $(CH_3)_3SiOReO_3$ form the same active species in the presence of $[(Me_3Si)O]_2$.

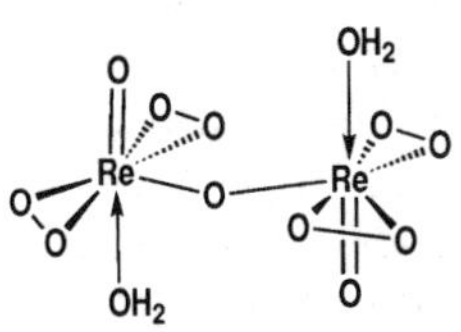

4

4 Catalytic Applications

4.1 Oxidation of Alkenes

One of the most intensively examined catalytic processes using organorhenium(VII) oxides, especially MTO, as catalysts is olefin epoxidation [32–39, 46, 48–68]. Usually less than 85 wt% of H_2O_2 is applied and MTO is typically used in concentrations of 0.2–0.6 mol%. Turnover numbers more than 2000 [mol/mol catalyst] and turnover frequencies of ca. 1500 [mol/mol catalyst/h] can be achieved.

The catalytic system MTO/H_2O_2 is already active below room temperature, e.g., at −30 °C. The reactions between **3** and alkenes are about one order of magnitude faster in semiaqueous solvents (e.g., 85% H_2O_2) than in methanol. The rate constants for the reaction of **3** with aliphatic alkenes correlate closely with the number of alkyl groups on the olefinic carbons. Theoretical calculations support these results [53]. The reactions become significantly slower when electron-attracting groups such as −OH, −CO, −Cl, and −CN are present. Interestingly, the MTO/H_2O_2 system has also been successfully employed for the oxidation of [60]fullerene to 1,2-epoxy[60]fullerene [54] and of pentacyclo[$6.6.0^{2,6}.0^{3,13}.0^{10,14}$]tetradeca-4,11-diene (PCTD) to its exo,exo-diepoxide [55].

Based on Eq. (2), two catalytic pathways may be described corresponding to the concentration of the hydrogen peroxide used. If 85% hydrogen peroxide is used, the described equilibrium is on the right side of Eq. (2) and only compound **3** seems to be responsible for the epoxidation activity (Scheme 1).

When a solution with 30 wt% or less H_2O_2 is used the monoperoxo complex **2** is responsible for the epoxidation process as well and a second catalytic cycle is involved as shown in Scheme 2.

For both cycles a concerted mechanism is suggested in which the electron-rich double bond of the alkene attacks a peroxidic oxygen of **3**, according to Schemes 1 and 2. It has been deduced from experimental data that the system might show a spiro arrangement [32, 36, 48, 54]. The selectivity towards epoxides can be enhanced by the addition of Lewis O- or N-bases such as quinuclidine, pyridine, pyrazole, or 2,2′-bipyridine to the system [36, 46, 57–68]. Lewis acids catalyze ring opening reactions and diol formation. These reactions are suppressed after the addition of Lewis bases. An excess of aromatic N-base ligands, especially under two phase conditions, leads to accelerated reactions [46, 53, 57, 59–68]. Several functionalized epoxides can be prepared in high yields by this method [59, 60]. The activity of the peroxo complexes in these cases depends on the Lewis bases, the redox stability of the ligands, and the excess of the ligands. Density functional calculations show that pyridine and pyrazole as ligands lead to significantly stabilized intermediates and moderate energies of the transition states in olefin epoxidation. This ultimately causes an acceleration of the epoxidation reaction. Non-aromatic nitrogen bases were found to reduce the catalytic performance. The frontier

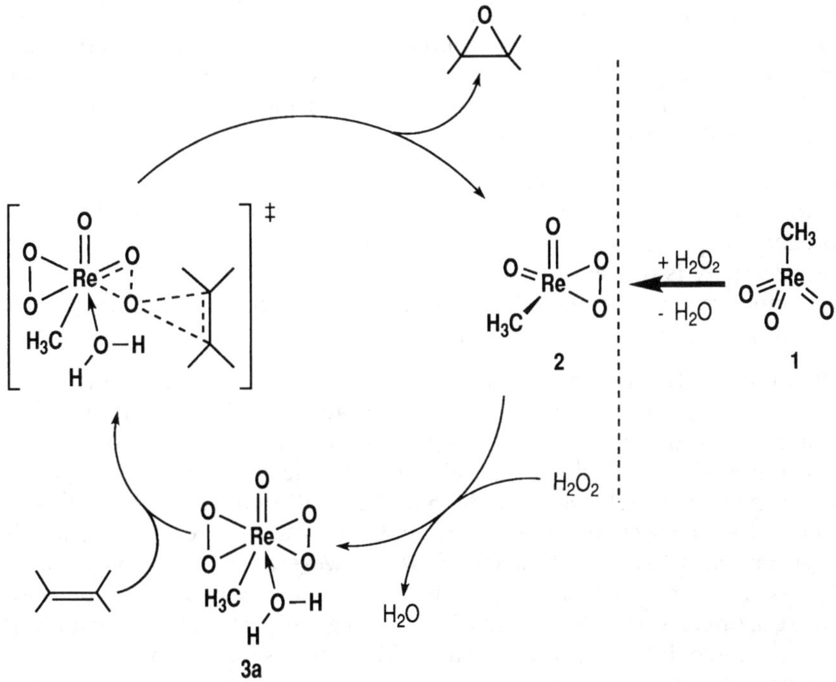

Scheme 1.

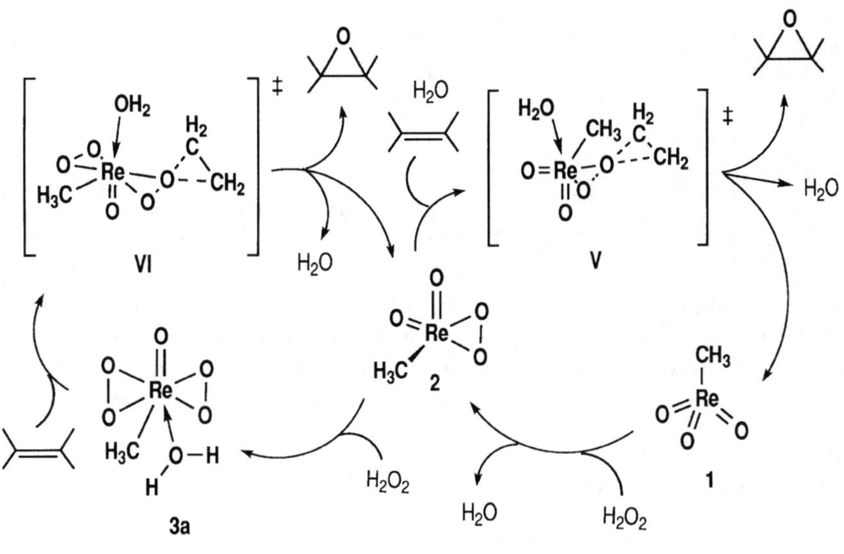

Scheme 2.

orbital interaction between the olefin HOMO π(C—C) and the orbitals with σ^*(O—O) character in the LUMO group of the Re-peroxo moiety controls the olefin epoxidation [53]. In comparison to the standard system for epoxidation which uses *m*-chloroperoxybenzoic acid as oxidizing agent, the MTO/H_2O_2/ aromatic Lewis base-system displays several advantages:

a) it is safer but equal in price
b) since it avoids the opening of the epoxide ring it can be applied in a broader scope
c) its selectivity is higher
d) it is more reactive, requires less solvent, the product workup is easier, and the only byproduct formed is water

Another possibility for enhancing the selectivity towards epoxides is the use of the urea-H_2O_2 adduct. It enables the oxidation to be carried out in water-free solutions, thus avoiding formation of any diols and other side reactions. In the case of the oxidation of chiral allyl alcohols (see Sect. 4.3) high diastereoselectivities have been achieved [50, 54, 55]. The ability to transform olefins to epoxides diastereoselectively seems to indicate that the reaction proceeds through a peracid-like transition state (**VII**). However, a drawback of the urea-H_2O_2 system is the insolubility of the polymeric complex. Electron-poor olefins resist oxidation to the point of competing with catalyst deactivation. It has to be noted in this context that MTO catalyzes the formation of 1,3-dioxolanes in good yields from the reaction of epoxides with aldehydes and ketones (Eq. 3):

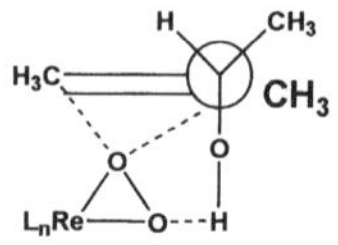

VII

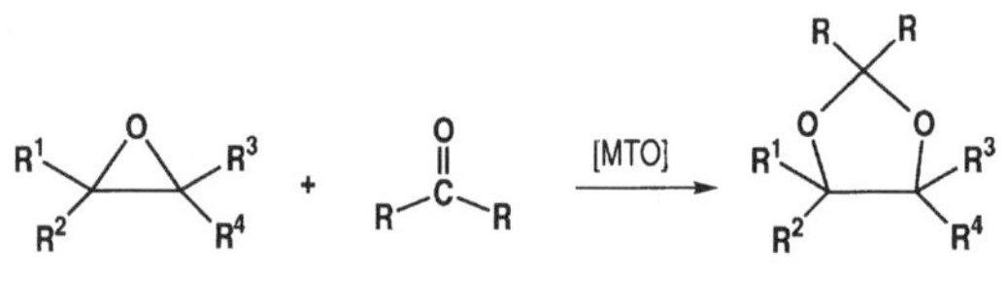

(3)

Lewis base adducts of some derivatives of MTO, e.g., (cyclopropyl)trioxorhenium(VII) have shown even higher activity in the olefin epoxidation than MTO itself [36]. However, the somewhat more difficult synthesis and especially the reduced stability of these complexes has prevented a widespread use up to now.

4.2 Oxidation of Conjugated Dienes

Conjugated dienes are oxidized to epoxides (or diols, if water is present) with the MTO/H_2O_2 system [68]. Using urea/H_2O_2 avoids the subsequent epoxide ring opening. Electron-rich and conjugated dienes are more easily oxidized than electron-poor dienes and dienes with isolated double bonds. According to kinetic measurements complex **2** plays no important role as catalyst in this case. Compound **3** acts as active species [68].

4.3 Epoxidation of Allylic Alcohols and 1,3-Transposition of Allylic Alcohols

Allylic alcohols are epoxidized to the epoxy alcohols by hydrogen peroxide in the presence of MTO [69–71]. Provided acid is not added, the product is mostly epoxide, accompanied by small amounts of triol that results from acid-catalyzed ring opening of the epoxide. With added acid, only the triol is present. According to kinetic data only the bisperoxo complex **3**, but not the monoperoxo complex **2** acts as a catalyst. For this reaction it has been suggested that an O atom of one peroxo group engages in oxygen transfer, in a transition state that is stabilized by hydrogen bonding between the allylic OH group and an O atom of the other peroxo group. Calculations to support this view have been performed [69]. The steric effects of allylic substituents in the epoxidations with MTO/urea/H_2O_2 have been assessed for a series of 3-alkyl-substituted cyclohexenes. The *trans* selectivity increases with the size of the substituent and the diastereoselectivities follow linear correlations with the steric substituent constants of Charton or Taft. It is concluded from these examinations that the MTO/urea/H_2O_2 oxidant favors a quasi-equatorial arrangement of the hydroxy group for effective hydrogen bonding [72].

Furthermore, MTO catalyzes the 1,3-transposition of allylic alcohols [73]. This reaction does not require the presence of peroxides or peroxo complexes. Theoretical investigations on the allylic rearrangement have been performed [74]. A possible mechanism is shown in Scheme 3.

4.4 Oxidation of Alcohols and Ethers

Primary and secondary alcohols are oxidized using the MTO/H_2O_2 catalyst system to aldehydes and ketones, respectively [75–77]. The dominant and reactive form of the catalyst is compound **3**. A mechanism featuring hydride abstraction has been proposed [66]. The addition of a catalytic quantity of bromide ions, such as HBr or NaBr, significantly enhances the reaction rate [76]. The bromide is oxidized to the hypobromide ion, BrO^- which combines with additional bromide to give bromine. Bromine oxidizes the alcohols to aldehydes and ketones. The system MTO/H_2O_2/HBr/TEMPO (=2,2,6,6-tetramethyl-1-piperidinyloxyl) catalyzes the selective oxidation of terminal alcohols to the corresponding aldehydes with excellent selectivities and yields. The system allows the oxidation of alcohols either selectively to aldehydes or to the corresponding acids depending on the reaction parameters. This technique is applicable especially to the oxidation of carbohydrates [77].

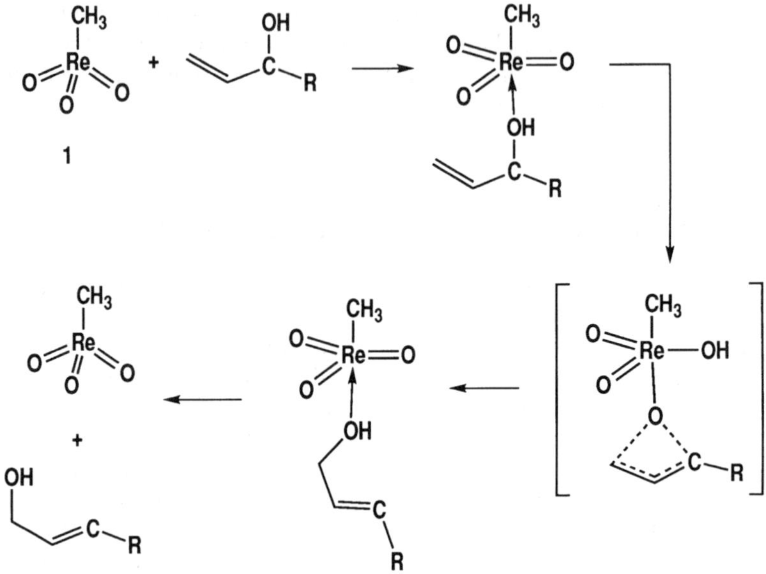

Scheme 3.

The MTO hydrogen peroxide system transforms furans in yields usually $\gg$70% to enediones. Substituted pyranones are obtained from furans with hydroxymethyl groups at the 2-position. The yields in this case are >75%. Acetonitrile as solvent leads to the fastest reactions although the work-up is reported to be easier in CH_2Cl_2 [78]. Silyl enol ethers are oxidized to α-hydroxy ketones by MTO/H_2O_2 with successive disilylation with KF (Eq. 4) [79]. However, for this reaction to be successful, acetic acid and pyridine have to be present. It is assumed that pyridine prevents the formation of hydrolysis products while acetic acid is necessary to prevent base-induced MTO decomposition. Yields are usually >90%. In the case of conjugated systems, the yields are significantly lower:

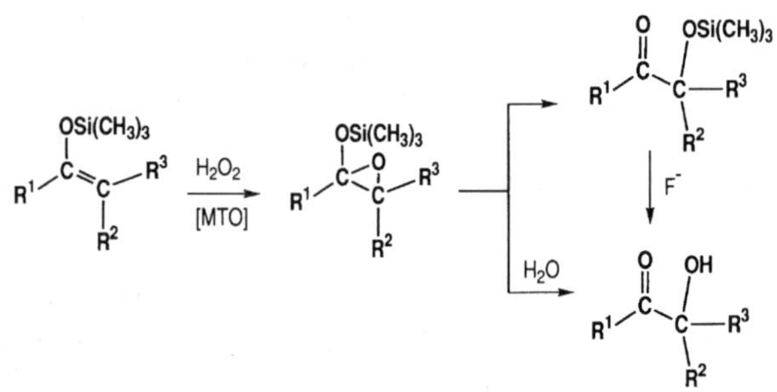

(4)

4.5 Oxidation of Alkynes

Internal alkynes yield carboxylic acids and α-diketones when oxidized with the MTO/H_2O_2 system (Eq. 5) [80]. Rearrangement products were observed only for aliphatic alkynes. Terminal alkynes give carboxylic acids, derivatives thereof, and α-keto acids as the major products. The yields of these products vary with the solvent used [80]. The mechanisms leading to the products observed remain highly speculative to date. Several possibilities are still under discussion.

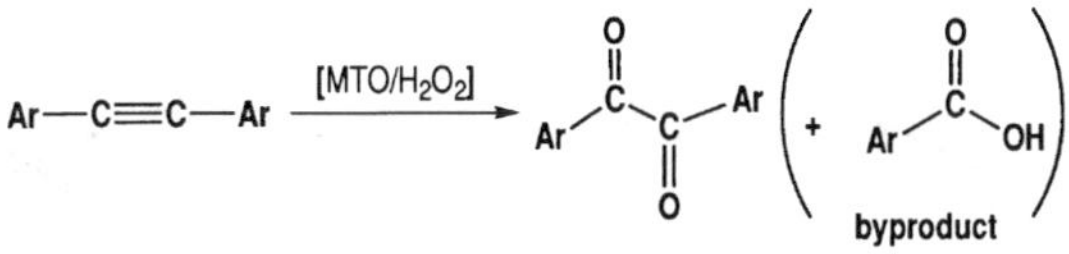

(5)

4.6 Oxidation of Cyclic β-Diketones

The MTO/H_2O_2 system furthermore catalyzes the oxidation of cyclic β-diketones to carboxylic acids (Eq. 6) [81]. Conversions are usually above 85% and the product selectivity is nearly quantitative. The reaction is performed in a 1:1 water acetonitrile solution at room temperature. It has been assumed that enolic forms which exist in solution are first epoxidized. After a rearrangement step the C—C bond is cleaved and an oxygen inserted. Then an α-diketone intermediate forms which is finally oxidized to the carboxylic acid [81].

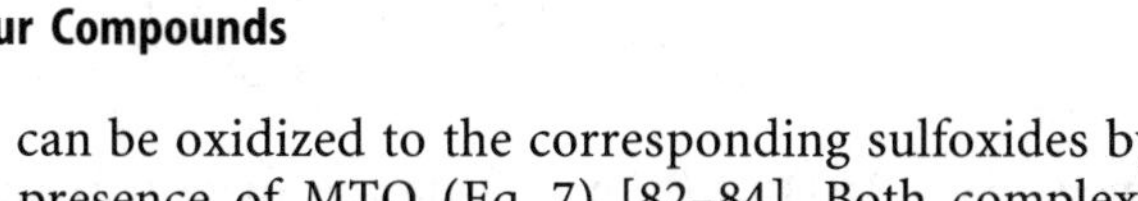

(6)

4.7 Oxidation of Sulfur Compounds

Organic sulfides can be oxidized to the corresponding sulfoxides by hydrogen peroxide in the presence of MTO (Eq. 7) [82–84]. Both complexes **2** and **3** seem to be active in this reaction but kinetic results indicate that **2** might be more active than **3**. The kinetic results also point to a mechanism that involves the nucleophilic attack of the sulfur atom on a coordinated peroxide oxygen since electron donating substituents have accelerating effects. Using ethanol as solvent, the MTO/H_2O_2 system can be used to oxidize dialkyl, diaryl and alkyl aryl sulfides to sulfoxides (R_2S:H_2O_2 = 1:1.1) or sulfones (R_2S:H_2O_2 = 1:2.2) with excellent yields and selectivity even in the presence of oxidatively sensitive functions on the sulfide side chain [82–89]; increased sulfone yields can also be achieved by higher catalyst concentrations. Functional groups in the side chain of the sulfide, e.g., carbon double bonds, are not affected under the reaction conditions: the sulfur atom is selectively oxidized [82]. MTO can also be used in the oxidation of sulfides with the water free system urea-hydrogen peroxide as oxidant in acetonitrile. Partial enantioselective oxidation of methylphenyl sulfide was observed with MTO in the presence of a (+)-camphor derived pyrazine carboxylic acid [83]. Trichloro oxorhenium phosphane complexes, e.g., mer,*trans*-$Re(O)Cl_3(PPh_3)_2$, also catalyze the oxidation of sulfides to sulfoxides. These complexes do not so easily overoxidize the sulfides to sulfones as does the MTO-containing system.

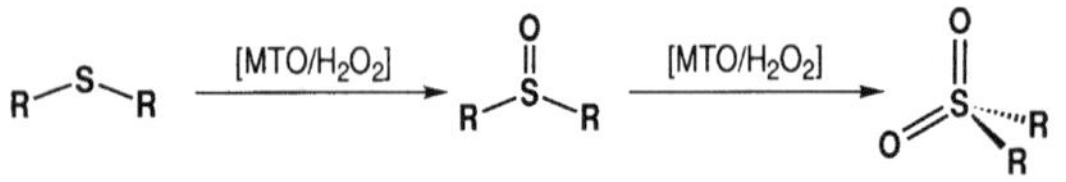

(7)

The mild selectivity of MTO/H_2O_2 has also been demonstrated in the oxidation of thioether Fischer carbene complexes to sulfoxides (Eq. 8)[85]:

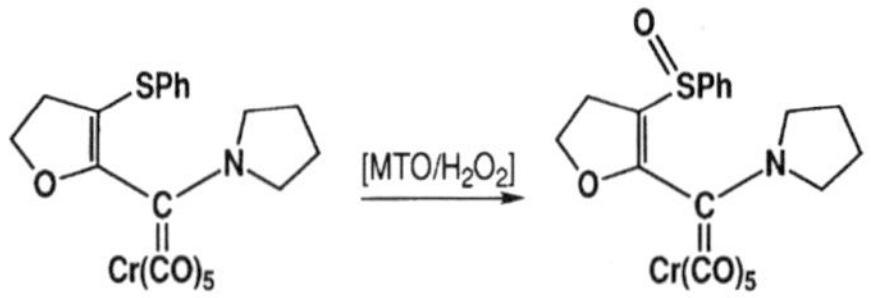

(8)

Oxidation of thiophene and its derivatives has been achieved with this system. However, the rate constants are two to four orders of magnitude below those reported for the "aliphatic" sulfides where the S atom is not part of an aromatic heterocyclic ring [86]. The delocalization of electrons from the sulfur into the aromatic ring causes a decreased reactivity of the sulfur towards the electron-deficient peroxide group of the metal center. MTO is able to catalyze the transformation of thiols to disulfides with DMSO (Eq. 9). The mechanism of this reaction is not completely clear. It is assumed that Re(V) intermediates may be involved [88].

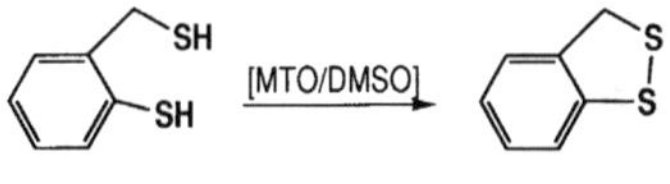

(9)

Re(V) catalysts, e.g., $Re(O)Cl_3(PPh_3)_2$, catalyze oxygen scrambling between sulfoxides (most preferably diphenyl sulfoxide) and sulfides (Eq. 10). Nearly quantitative yields are reached and the catalytic system is comparatively stable against moisture and oxygen [89].

(10)

Oxidation studies of coordinated thiolates were carried out on the model Co(III) complex $[(en)_2Co(SCH_2CH_2NH_2)]^{2+}$. The thiolato complex is first oxidized to a sulfenato complex, $[(en)_2Co(S(O)CH_2CH_2NH_2)]^{2+}$, which is then more slowly oxidized to the sulfinato complex $[(en)_2Co(S(O)_2CH_2CH_2NH_2)]^{2+}$. The second step is ca. 1500 times slower than the first. From these kinetic examinations, **2** seems to be the catalytically active species which is attacked by the nucleophilic S atom [87].

4.8
Oxidation of Anilines and Amines

The MTO/H_2O_2 system also catalyzes the oxidation of anilines (Eq. 11). The major product of this oxidation is nitrosobenzene. For 4-substituted *N,N*-dimethylanilines, the *N*-oxide is the only oxidation product. Electron-withdrawing substituents inhibit the reaction.

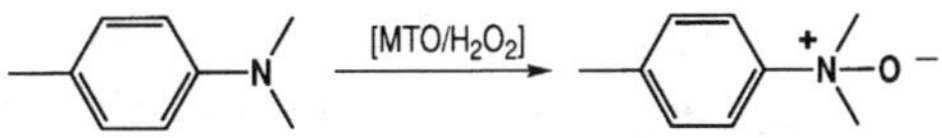

(11)

Kinetic results suggest that both compounds **2** and **3** are involved in the oxidation process [90]. It has been suggested that the rate-controlling step is the nucleophilic attack of the nitrogen lone pair electrons of the anilines on a peroxidic oxygen of the catalyst. Electron-donating groups attached to the nitrogen atom of aniline increase the rate constant. In the case of $ArNH_2$ derivatives, the oxidation proceeds ca. 50 times faster than without catalyst [90]. In general, the reactions are facile and give high yields at or below room temperature [91]. Furthermore, a broad variety of aromatic and aliphatic secondary amines can be oxidized to the corresponding *N*-oxides [92–95]. The amines are converted to the corresponding hydroxylamines before conversion to the nitrones in very good yields [92–95]. The hydroxylamine formation is rate controlling [93]. Both H_2O_2 and the urea-hydrogen peroxide complex can be used together with MTO. Benzylamines are selectively oxidized to oximes. Primary amines are oxidized to nitro compounds by MTO catalysis (Eq. 12) [96, 97]:

$$RNH_2 \xrightarrow{[MTO/H_2O_2]} RNO_2 \qquad (12)$$

It is noteworthy in this context that the *N*-oxides also form adducts with MTO [56]. While aliphatic *N*-oxides form temperature-sensitive adducts with MTO and are inactive in the catalytic epoxidation of olefins, aromatic *N*-oxide adducts of MTO are both stable and catalytically active. In some cases their selectivity in olefin epoxidation is even higher than with *N*-base adducts of MTO (see Sect. 4.1).

4.9
Oxidative Conversion of *N,N*-Dimethylhydrazones to Nitriles

N,N-Dimethylhydrazones of aldehydes react with hydrogen peroxide in the presence of catalytic amounts of MTO to give in high yield the corresponding nitriles. Two different sets of conditions have been applied to perform this transformation. The reaction is either carried out adding dropwise the *N,N*-dimethylhydrazone in ethanol to a solution of 1.5% of MTO and 2 eq. of 35% H_2O_2 in EtOH at −50 °C [98] or in an acetonitrile-acetic acid-pyridine mixture (94.5:5:0.5) at room temperature with a 30% solution of H_2O_2 in water in the presence of MTO [99]. The pyridine is applied to suppress hydrolysis and the acetic acid is used to prevent deactivation of MTO by the basic hydrazones, according to the authors. The nitrile yields are reported in most cases considerably higher then 90%. The reaction starts with the oxidation of the *N,N*-dimethylhydrazone, and proceeds presumably through an intermediate shown in Eq. (13) which undergoes a Cope-type elimination to yield the nitrile [99, 100]:

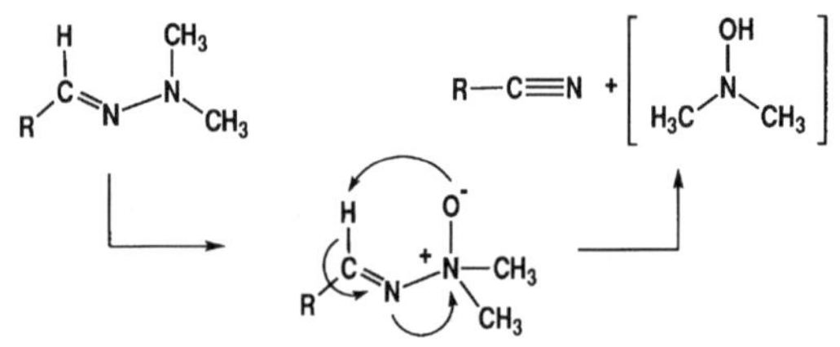

(13)

4.10
Oxidation of Phosphines, Arsines, and Stibines

Tertiary phosphines, triaryl arsines, and triaryl stibines are converted to their oxides, R_3EO (E = P, As, Sb) by MTO/H_2O_2. Kinetic studies lead to the assumption that **2** and **3** have similar catalytic activities in all cases. The kinetic data support a mechanism involving nucleophilic attack of the substrate at the rhenium peroxides. The proposed catalytic cycle is given in Scheme 4 [101].

In the absence of peroxides MTO also catalyzes the oxidation of tertiary phosphines to phosphine oxides [102, 103]. The oxygen sources in this case are sulfoxides which are deoxygenated. It is assumed that MTO forms a sulfoxide adduct first which is then deoxygenated by triphenylphosphane under formation of MTO and sulfide. Another mechanism, suggesting reduction of MTO to methyldioxorhenium by the phosphine and reoxidation by the sulfoxide, seems to be less likely. The oxygen transfer from sulfoxides to phosphines is also catalyzed by several Re(V) species, e.g., mer,*trans*-$Re(O)Cl_3(PPh_3)_2$ and fac,*cis*-$Re(O)Cl_3(CNCMe_3)_2$ [104–106]. Mechanistic studies concerning these transfer reactions have been performed, showing that both ligand-centered and metal-centered oxygen-transfer reactions occur depending on the particular Re(V) catalyst used [105, 106].

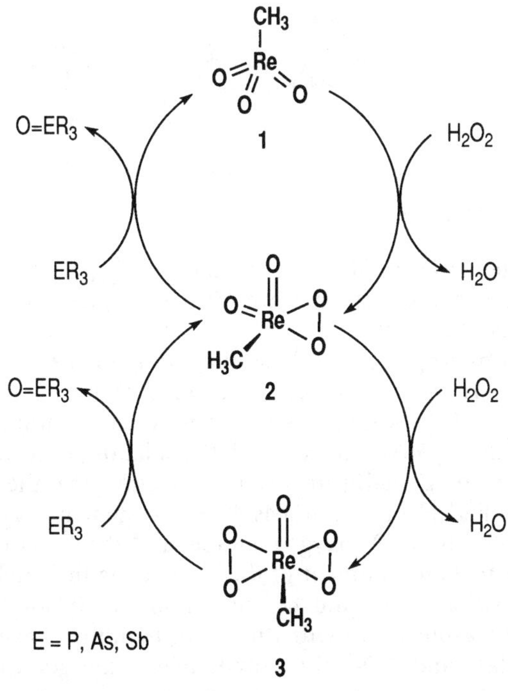

Scheme 4.

4.11
Oxidation of C—H Bonds

MTO/H_2O_2 also catalyzes the insertion of oxygen into a variety of activated and inactivated C—H bonds with yields varying from good to excellent. Alcohols or ketones are formed as shown in Eqs. (14) and (15). In the case of tertiary substrates, alcohols are obtained as the products. Suitable substrates proved that the reaction is stereo specific with retention of the configuration. Alcohols, e.g., ethanol and *t*-butanol, or acetonitrile are used as solvents. The reaction temperatures range between 40 °C and 80 °C and nearly quantitative yields have been obtained in several cases. However, the reaction times are generally longer than those used for most epoxidations, ranging between 10 h and 72 h [107]. The reaction can be accelerated by the addition of pyrazine-2-carboxylic acid, and the total yield is also increased [108].

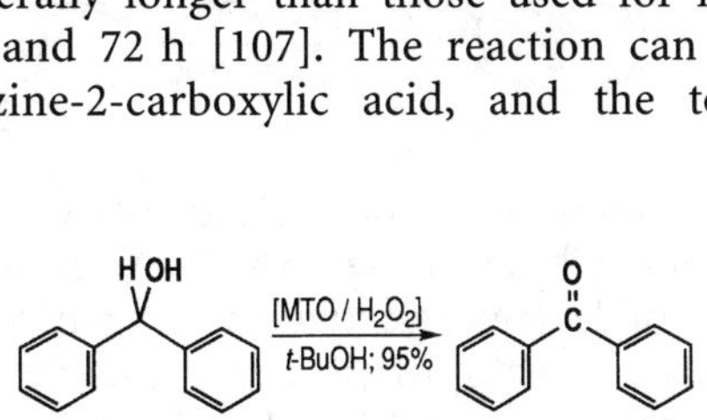

(14)

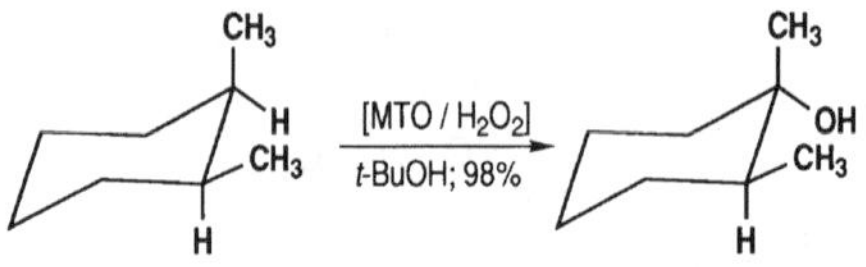

(15)

4.12 Oxidation of Si—H Bonds

The catalyst system MTO/H_2O_2 also catalyzes oxygen atom insertion into Si—H bonds. Silanols and siloxanes are formed as products, the latter being the main products [107, 109].

When the urea/hydrogen peroxide adduct (UHP) is used as oxygen source instead of aqueous H_2O_2, MTO catalyzes the oxidation of silanes to silanols in high conversions (>95% within 18 h at room temperature) and excellent selectivities (usually ≫90%) in favor of the silanol (only tiny amounts of disiloxane are formed, if at all) (Eq. 16). It is assumed that the Si—H oxidation takes place in the helical urea channels. The urea matrix serves as a host for the silane substrate, the H_2O_2 oxygen source, and the MTO metal catalyst as guest. In the confined environment of these channels the catalyst is stabilized against decomposition while the condensation of silanol to disiloxane is avoided for steric reasons [109]. Other matrices than urea have been examined too, namely amylase and SiO_2, but neither match the good results obtained with urea. In both cases the conversions are lower and the amount of disiloxane obtained is considerably higher [109].

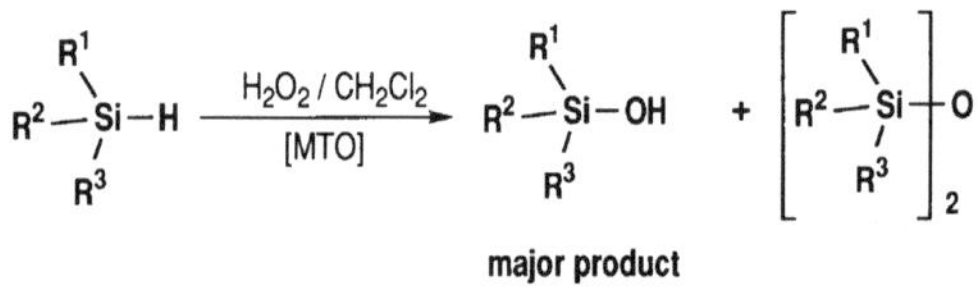

(16)

4.13 Oxidation of Aromatic Compounds

Arenes are oxidized to *p*-benzoquinones by hydrogen peroxide with MTO as catalyst [110–118]. Noteworthy is the high regioselectivity, most notably in the industrially interesting synthesis of vitamin K_3 (see Eq. 17):

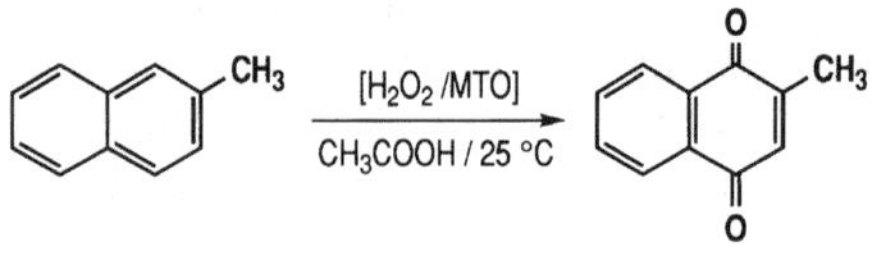

(17)

Since water is an inhibitor, concentrated (85 wt%) H_2O_2 is preferred. Alternatively, commercially available 35% H_2O_2 in acetic anhydride can be employed; a considerable regioselectivity is obtained with this system. The conversion is higher for electron-rich arenes (nearly 100%) and selectivities of more than 85% have been reached [110]. Biphenylene can be oxidized with the MTO system in chloroform affording an *o*-quinone product (83% conversion) [111]. Hydroxy substituted arenes can be oxidized by aqueous hydrogen peroxide (85 wt%) in acetic acid to afford the corresponding *p*-quinonones in isolated yields of up to 80% [114]. It has been shown that using a mixture of acetic acid and acetic anhydride further improves the product yield [117, 118]. Instead of acetic acid, HBF_4 in EtOH can be used too [114, 115]. Anisol was also found to undergo selective oxidation with the MTO/H_2O_2 system to yield *o*- and *p*-methoxyphenols. There is no need to use a solvent in this case. Bidentate Lewis base adducts of halogeno rhenium(VII) oxides also catalyze the arene oxidation in the presence of H_2O_2 [116].

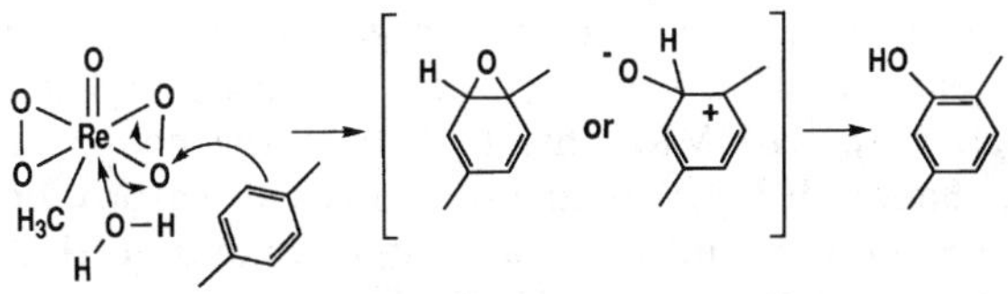

3a

(18)

Concerning the mechanism of the arene oxidation it is assumed that the first reaction step proceeds according to Eq. (18) to form a phenol. The phenols and methoxy benzenes, being more electron-rich than arenes, are more reactive with the peroxorhenium compounds. Therefore, the subsequent steps in the oxidation occur more rapidly than the first [118].

Benzaldehydes with hydroxyl or methoxy substituents in ortho or para positions are oxidized to the corresponding phenols (carboxylic acids are formed as by product) in good yields [113] The yield is temperature and solvent dependent (Eq. 19):

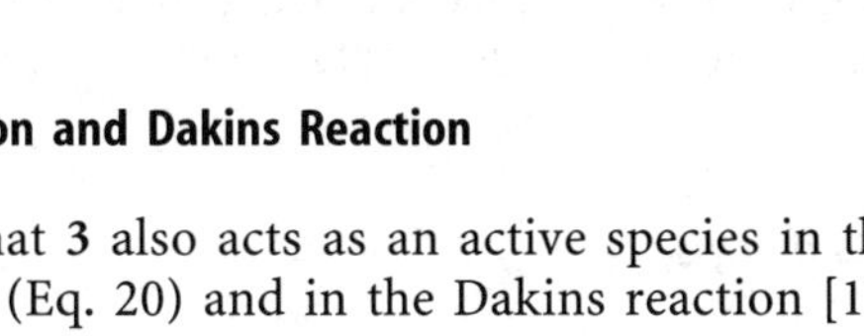

(19)

4.14 Baeyer-Villiger Oxidation and Dakins Reaction

It has been shown that **3** also acts as an active species in the Baeyer-Villiger oxidation of ketones (Eq. 20) and in the Dakins reaction [119, 120].

(20)

It is somewhat surprising that the MTO/H_2O_2 system presents this activity since these oxidations involve nucleophilic attack at the carbonyl group which contrasts with all the preceding examples where the substrates attacked the electrophilic Re-peroxo complexes, e.g., in the olefin epoxidation. Nevertheless, **3** reacts stoichiometrically with cyclobutanone in the absence of H_2O_2. This reversed behavior may be due to substrate binding to rhenium. The unsymmetric geometry of compound **3a**, displaying a polarity within the peroxo ligands might also be responsible for the observed behavior.

The reaction was found to be strongly solvent dependent by means of the mechanistic probe for oxygen transfer reactions, thiantrene-5-oxide [81, 119]. Donor solvents such as acetonitrile seem to enhance the nucleophilicity of the peroxo groups. It has been suggested that the double bond of the enol form (the major tautomer) attacks a peroxo oxygen of the peroxorhenium complex **2** or **3**. This reaction affords a 2-hydroxy-1,3-dicarbonyl intermediate which can be detected by ^{1}H-NMR. This hydroxy intermediate is susceptible to cleavage via Baeyer-Villiger oxidation to yield carboxylic acids as final products. Low H_2O_2 concentrations are sufficient and no H_2O_2 decomposition is observed at temperatures up to 70 °C. This is an advantage of the catalytic MTO/H_2O_2 system over the known transition-metal Baeyer-Villiger catalysts containing V, Mo, Mn, or Os. However, the nucleophilic character of the peroxidic atoms in **3** is not as pronounced as in Pt or Ir peroxo complexes that react with CO_2 or SO_2 to give isolable cycloaddition products [119]. In the case of **3** turnover frequencies of 18000 [mol/mol cat. per hour] are obtained for cyclobutanone, but in other cases turnover numbers of ca. 100 are usual [120]. Cycloketones can be converted into lactones even below room temperature (15 °C) by diluted hydrogen peroxide (10 wt%).

4.15 Oxidation of Halide Ions

Another application of the MTO/H_2O_2 system is the catalytic oxidation of chloride and bromide ions in acidic aqueous solutions. The chloride oxidation steps are three to four orders of magnitude slower than the corresponding bromine oxidation steps. Both compounds **2** and **3** have been shown to be active catalysts in these processes. In both cases the catalyzed reactions were about 10^5 times faster than the uncatalyzed reactions under similar conditions. In a first step HOX is formed, then HOX reacts with X^- to form X_2. When H_2O_2 is used in excess the reaction yields O_2 [121, 122].

4.16 Oxidation of Metal Carbonyls

MTO and some of its derivatives, e.g., $CpReO_3$ and $EtReO_3$, catalyze the oxidation of metal carbonyls to metal oxides with H_2O_2 [13, 123, 124]. This reaction runs at room temperature and yields of up to 90% are obtained. However, only organometal carbonyls with oxidation resistant organic groups can be oxidized, e.g., (pentamethylcyclopentadienyl) tricarbonyl rhenium(I)

(Eq. 21) [13]. In all other cases the organic ligand is also oxidized, leading to decomposition of the product complex [124].

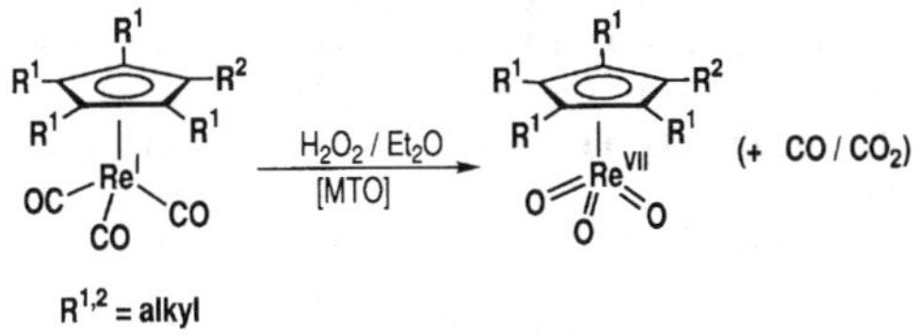

(21)

5 Conclusions

Rhenium(VII) oxides have found numerous applications in catalysis during the recent years. MTO, a water soluble, air and temperature stable metal alkyl has become the most versatile oxidation catalyst in organic chemistry. Its outstanding reactivity in oxidation catalysis is due to the highly Lewis acidic and sterically unsaturated rhenium(VII) center and the thermally and chemically very stable rhenium-carbon bond. Inorganic rhenium oxides, e.g., Re_2O_7, and their peroxo complexes are far less versatile in water-containing systems due to their pronounced moisture sensitivity, but can be applied in water free oxidation systems, e.g., with $(Me_3SiO)_2$. However, in spite of a rich body of MTO chemistry and applications available now, this area is far from being mature. This is clearly seen by the still improving number of papers published each year on new applications of MTO. Several derivatives of MTO, e.g., $CpReO_3$, are also stable enough to be used as catalysts but have not been examined in detail concerning their catalytic abilities. We particularly believe that organometal oxides and derivatives in rhenium's neighborhood of the periodic table will still improve the catalytic strength of this class of compounds.

6 References

1. Rouschias G (1974) Chem Rev 74: 531
2. Herrmann WA, Kühn FE (1997) Acc Chem Res 30: 169
3. Warwel S, Rüsch M, Sojka M (1991) J Chem Soc Chem Commun 1587
4. Noddack J, Noddack W (1929) Z Anorg Allg Chem 181: 1
5. Smith WT, Line LE, Bell WA (1952) J Am Chem Soc 74: 4964
6. Wells AF (1984) Structural inorganic chemistry, 5th edn. Clarendon, Oxford, pp 509–550
7. Krebs B, Müller A, Beyer HH (1968) J Chem Soc Chem Commun 263
8. Krebs B, Müller A (1968) Z Naturforsch 23b: 415
9. Beyer H, Glemser O, Krebs B, Wagner G (1970) Z Anorg Allg Chem 376: 87
10. Beyer H, Glemser O, Krebs B (1968) Angew Chem Int Ed Engl 7: 295
11. Fischer D, Krebs B (1982) Z Anorg Allg Chem 491: 73
12. Krebs B, Fischer D, Paul G (1996) Z Anorg Allg Chem 622: 448
13. Herrmann WA, Correia JDG, Kühn FE, Artus GRJ, Romão CC (1996) Chem J Eur 2: 168

14. Kiprof P, Herrmann WA, Kühn FE, Scherer W, Kleine M, Elison M, Rypdal K, Volden HV, Gundersen S, Haaland A (1992) Bull Soc Chim Fr 129: 655
15. Herrmann WA, Roesky PW, Kühn FE, Elison M, Scherer W, Artus GRJ, Romão CC, Lopes AD, Basset JM (1995) Inorg Chem 34: 4701
16. Mertis K, Wilkinson G (1976) J Chem Soc Dalton Trans 1488
17. Beattie IR, Jones PJ (1979) Inorg Chem 18: 2318
18. Herrmann WA, Kuchler JG, Felixberger JK, Herdtweck E, Wagner W (1988) Angew Chem Int Ed Engl 27: 394
19. Herrmann WA, Kühn FE, Fischer RW, Thiel WR, Romão CC (1992) Inorg Chem 31: 4431
20. Herrmann WA, Thiel WR, Kühn FE, Fischer RW, Kleine M, Herdtweck E, Scherer W, Mink J (1993) Inorg Chem 32: 5188
21. Herrmann WA, Kratzer R, Fischer RW (1997) Angew Chem Int Ed Engl 36: 2652
22. Herrmann WA (1988) Angew Chem Int Ed Engl 27: 394
23. Herrmann WA (1995) J Organomet Chem 500: 149
24. Kühn FE, Fischer RW, Herrmann WA (1999) Chem i u Zeit 33: 192
25. Espenson JH (1995) Chemical kinetics and reaction mechanisms. McGraw-Hill, New York
26. Romão CC, Kühn FE, Herrmann WA (1997) Chem Rev 97: 3197
27. Cusumaro JA (1992) CHEMTECH 482
28. Schmidt B (1997) J Prakt Chem 339: 439
29. Espenson JH, Abu-Omar MM (1997) ACS Adv Chem 253: 3507
30. Gable KP (1997) Adv Organomet Chem 41: 127
31. Brown SN, Mayer JM (1996) J Am Chem Soc 118: 12,119
32. Herrmann WA, Fischer RW, Marz DW (1991) Angew Chem Int Ed Engl 30: 1638
33. Yamazaki S, Espenson JH, Huston P (1993) Inorg Chem 32: 4683
34. Herrmann WA, Correia JDG, Artus GRJ, Fischer RW, Romão CC (1996) J Organomet Chem 520: 139
35. Herrmann WA, Fischer RW, Scherer W, Rauch MU (1993) Angew Chem Int Ed Engl 32: 1157
36. Herrmann WA, Fischer RW, Rauch MU, Scherer W (1994) J Mol Catal 86: 243
37. Espenson JH (1999) Chem Commun 479
38. Gisdakis P, Antonczak S, Köstlmeier S, Herrmann WA, Rösch N (1998) Angew Chem Int Ed Engl 37: 2211
39. Abu-Omar MM, Hansen PJ, Espenson JH (1996) J Am Chem Soc 118: 4966
40. Laurenczy G, Lukács F, Roulet R, Herrmann WA, Fischer RW (1996) Organometallics 15: 848
41. Espenson JH, Tan H, Mollah S, Houk RS, Eager MD (1998) Inorg Chem 37: 4621
42. Brittingham KA, Espenson JH (1999) Inorg Chem 38: 744
43. Schröder D, Fiedler A, Herrmann WA, Schwarz H (1995) Angew Chem Int Ed Engl 34: 2636
44. Pestovsky O, v. Eldik R, Huston P, Espenson JH (1995) J Chem Soc Dalton Trans 133
45. Hatzopoulos I, Brauer HD, Geisberger M, Herrmann WA (1996) J Organomet Chem 520: 201
46. Al-Ajlouni AM, Espenson JH (1995) J Am Chem Soc 117: 9243
47. Herrmann WA (1993) In: Herrmann WA (ed) Organic peroxygen chemistry, vol 164. Springer, Berlin Heidelberg New York, p 130
48. Boehlow TR, Spilling CD (1996) Tetrahedron Lett 37: 2717
49. Murray RW, Singh M, Williams BL, Moncrieff HM (1995) Tetrahedron Lett 36: 2437
50. Al-Ajlouni AM, Espenson JH (1996) J Org Chem 61: 3969
51. Kühn FE, Santos AM, Roesky PW, Herdtweck E, Scherer W, Gisdakis P, Yudanov IB, Di Valentin C, Rösch N (1999) Chem Eur J 5: 3603
52. Adam W, Mitchell CM (1996) Angew Chem Int Ed Engl 35: 533
53. Suzuki T, Iida K, Wada K, Kondo T, Mitsudo T (1999) Tetrahedron Lett 40: 2997
54. Murray RW, Iyanar K (1997) Tetrahedron Lett 38: 335

55. Herrmann WA, Kühn FE, Mattner MR, Artus GRJ, Geisberger M, Correia JDG (1997) J Organomet Chem 538: 203
56. Herrmann WA, Correia JDG, Rauch MU, Artus GRJ, Kühn FE (1997) J Mol Catal 118: 33
57. Rudolph J, Redding KL, Chiang JP, Sharpless KB (1997) J Am Chem Soc 119: 6189
58. Yudin AK, Sharpless KB (1997) J Am Chem Soc 119: 11,536
59. Copéret C, Adolfsson H, Sharpless KB (1997) Chem Commun 1565
60. Herrmann WA, Wojtczak WA, Artus GRJ, Kühn FE, Mattner MR (1997) Inorg Chem 35: 465
61. Herrmann WA, Ding H, Kratzer R, Kühn FE, Haider JJ, Fischer RW (1997) J Organomet Chem 97: 3197
62. Nakajima M, Sasaki Y, Iwamoto H, Hashimoto S (1998) Tetrahedron Lett 39: 87
63. Herrmann WA, Kratzer RM, Ding H, Thiel WR, Glas H (1998) J Organomet Chem 555: 293
64. Gansäuer A (1997) Angew Chem Int Ed Engl 36: 2591
65. Rudler H, Gregorio JR, Denise B, Brégeault JM, Deloffre A (1998) J Mol Catal 133: 255
66. Wang WD, Espenson JH (1998) J Am Chem Soc 120: 11,335
67. Villa de PAL, De Vos DE, Montes de C C, Jacobs PA (1998) Tetrahedron Lett 39: 8521
68. Tan H, Espenson JH (1998) Inorg Chem 37: 467
69. Tetzlaff HR, Espenson JH (1999) Inorg Chem 38: 881
70. Adam W, Corma A, Martínez A, Mitchell CM, Reddy TI, Renz M, Smerz AK (1997) J Mol Catal 117: 357
71. Zhu Z, Espenson JH (1996) J Org Chem 61: 324
72. Adam W, Mitchell CM, Saha-Möller CR (1999) Eur J Org Chem 785
73. Jacob J, Espenson JH, Jensen JH, Gordon MS (1998) Organometallics 17: 1835
74. Bellemin-Laponnaz S, LeNy JP, Dedieu A (1999) Chem Eur J 5: 57
75. Zauche TH, Espenson J (1998) Inorg Chem 37: 6827
76. Espenson JH, Zhu Z, Zauche TH (1999) J Org Chem 64: 1191
77. Herrmann WA, Zoller JP, Fischer RW (1999) J Organomet Chem 581: 404
78. Finlay J, McKervey A, Gunaratne HQN (1998) Tetrahedron Lett 5651
79. Stancovic S, Espenson JH (1998) J Org Chem 63: 4129
80. Zhu Z, Espenson JH (1995) J Org Chem 60: 7728
81. Abu-Omar MM, Espenson JH (1996) Organometallics 15: 3543
82. Yamazaki S (1996) Bull Chem Soc Jpn 69: 2955
83. Gunaratne HQN, McKervey MA, Feutren S, Finlay J, Boyd J (1998) Tetrahedron Lett 39: 5655
84. Adam W, Mitchell CM, Saha-Möller CR (1994) Tetrahedron 50: 13,121
85. Beddoes RL, Painter JE, Quayle P, Patel P (1996) Tetrahedron Lett 37: 9385
86. Brown KN, Espenson JH (1996) Inorg Chem 35: 7211
87. Huston P, Espenson JH, Bakac A (1993) Inorg Chem 32: 4517
88. Jacob J, Guzei IA, Espenson JH (1999) Inorg Chem 38: 1040
89. Arterbum JB, Nelson SL (1998) J Org Chem 61: 2260
90. Zhu Z, Espenson JH (1995) J Org Chem 60: 1326
91. Vassell KA, Espenson JH (1994) Inorg Chem 33: 5491
92. Goti A, Nannnelli L (1996) Tetrahdron Lett 37: 6025
93. Zauche TH, Espenson JH (1997) Inorg Chem 36: 5257
94. Murray RW, Iyanar K, Chen I, Wearing T (1996) J Org Chem 61: 8099
95. Yamazaki S (1997) Bull Chem Soc Jpn 70: 877
96. Murray RW, Iyanar K, Chen I, Wearing T (1996) Tetrahedron Lett 37: 809
97. Murray RW, Ianar K (1996) J Org Chem 61: 8099
98. Rudler H, Denise B (1998) Chem Commun 2145
99. Stancovic S, Espenson JH (1998) Chem Commun 1579
100. Zhu Z, Espenson JH (1997) J Mol Catal 121: 139
101. Abu-Omar MM, Espenson JH (1995) J Am Chem Soc 117: 272
102. Zhu Z, Espenson JH (1995) J Mol Catal 103: 87

103. Abu-Omar MM, Appelman EH, Espenson JH (1996) Inorg Chem 35: 7751
104. Abu-Omar MM, Khan SI (1998) Inorg Chem 37: 4979
105. Bryan JC, Stenkamp RE, Tulip TH, Mayer JM (1987) Inorg Chem 26: 2283
106. Rybek WK, Zagiczek AJ (1992) Coord Chem 26: 79
107. Murray RW, Iyanar K, Chen L, Wearing JT (1995) Tetrahedron Lett 36: 6415
108. Schuchardt U, Mandelli D, Shul'pin GB (1996) Tetrahedron Lett 37: 6487
109. Adam W, Mitchell CM, Saha-Möller CR, Weichold O (1999) J Am Chem Soc 121: 2097
110. Adam W, Herrmann WA, Lin J, Saha-Möller CR, Fischer RW, Correia JDG (1994) Angew Chem Int Ed Engl 33: 2475
111. Adam W, Balci M, Kilic H (1998) J Org Chem 63: 8544
112. Karasevich EI, Nikitin AV, Rubailo VL (1994) Kinet Katal 35: 810
113. Yamazaki S (1995) Chem Lett 127
114. Adam W, Herrmann WA, Lin J, Saha-Möller CR (1994) J Org Chem 59: 8281
115. Adam W, Herrmann WA, Saha-Möller CR, Shimizu M (1995) J Mol Catal 97: 15
116. Kühn FE, Haider JJ, Herdtweck E, Herrmann WA, Lopes AD, Pillinger M, Romão CC (1998) Inorg Chim Acta 279: 44
117. Herrmann WA, Haider JJ, Fischer RW (1999) J Mol Cat 270: 55
118. Jacob J, Espenson JH (1998) Inorg Chim Acta 270: 55
119. Herrmann WA, Fischer RW, Correia JDG (1994) J Mol Catal 94: 213
120. Fischer RW (1994) Ph D Thesis, Technical University Munich
121. Espenson JH, Pestovsky O, Huston P, Staudt S (1994) J Am Chem Soc 116: 2869
122. Hansen PJ, Espenson JH (1995) Inorg Chem 34: 5389
123. Thiel WR, Fischer RW, Herrmann WA (1993) J Organomet Chem 459: C9
124. Herrman WA, Geisberger MR, Kühn FE, Artus GRJ, Herdtweck E (1997) Z Anorg Allg Chem 623: 1229

Structure, Reactivity, and Selectivity of Metal-Peroxo Complexes Versus Dioxiranes

Waldemar Adam, Catherine M. Mitchell, Chantu R. Saha-Möller, Oliver Weichold

Institut für Organische Chemie der Universität Würzburg, Am Hubland, D-97074 Würzburg, Germany
E-mail: adam@chemie.uni-wuerzburg.de; http://www-organik.chemie.uni-wuerzburg.de

Dioxiranes, MTO/H_2O_2, and MTO/UHP are becoming increasingly important as efficient and selective oxidants in organic chemistry and numerous publications have appeared on each of them. Here we present a comprehensive review, comparing the chemistry of these reagents with particular emphasis on the oxidation of π bonds, heteroatoms, and σ bonds. We will discuss structural aspects and on this basis, will outline the reactivity and selectivity of these oxidants, along with the mechanistic aspects of the oxygen-transfer process.

Keywords: Peroxo complexes, Dioxiranes, Oxygen transfer, Reactivity, Chemoselectivity, Regioselectivity, Diastereoselectivity, Enantioselectivity

List of Abbreviations

$^{1,2}A$	1,2-allylic strain
$^{1,3}A$	1,3-allylic strain
config.	configuration
convn	conversion
Cp*	pentamethylcyclopentadienyl
Dec	decyl
DMD	dimethyldioxirane
Hept	heptyl
mal	malonyl
MTO	methyltrioxorhenium
NaY	Na-exchanged Y-zeolite with faujasite (FAU) structure
Np	naphthyl
O	oxidant
pic	picolinyl
quin	quinoline
SOSO	thianthrene 5,10-dioxide
SSO	thianthrene 5-oxide
SSO_2	thianthrene 5,5-dioxide
t	time
T	temperature
TFA	trifluoroacetone
TFD	methyl(trifluoromethyl)dioxirane
TON	turnover number
UHP	urea-hydrogen-peroxide adduct

1 Introduction

Oxygen-transfer reactions are of increasing interest to both organic and inorganic chemists and have been the subject of an intensifying partnership [1]. Due to structural similarities between 'organic' and 'organometallic' oxidants, the study of common reactions and mechanisms, and also their differences, is essential for a better understanding of this burgeoning research field, the goal being the development of new and more efficient oxygen-transferring reagents. Of particular relevance for oxyfunctionalization reactions are those reagents which allow mild reaction conditions and lead to the desired product in both high yield and selectivity. Catalytic reaction conditions are desirable, but should not be enforced at the price of selectivity.

A wide range of both organic and organometallic oxygen-transferring agents has been studied to date with respect to their reactivity and selectivity in oxyfunctionalizations. Different types of metal-based oxidants have been scrutinised, i.e., oxo, superoxo, peroxo, peroxy, perhydrate complexes (Fig. 1), and numerous reports have appeared on oxygen-transfer reactions with such oxidants. We will limit our review to the reactivity and selectivity of rhenium peroxo complexes derived from methyltrioxorhenium (MTO) and H_2O_2, but

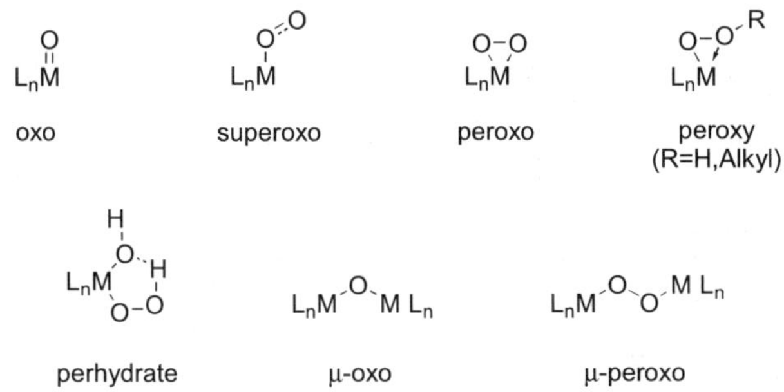

Fig. 1. Types of metal complexes used for oxygen-transfer reactions

for convenience an overview of selected examples of oxygen-transfer reactions for other transition-metal η-peroxo complexes is also given in Table 1. Most of the examples use stoichiometric amounts of the complexes and disadvantages such as poor selectivity, double-bond cleavage, and *cis/trans* isomerization are common problems. Nonetheless, Table 1 demonstrates that peroxo complexes are known for most transition metals, which are capable of transferring oxygen. Hydrogen peroxide usually serves as the oxygen source for early transition metals and molecular oxygen for the late ones. Among these, the rhenium diperoxo complex $CH_3ReO(O_2)_2 \cdot H_2O$ stands out in view of its high reactivity and few side-reactions compared to the other peroxo complexes in Table 1. Indeed, none of the other peroxo complexes competes with the rhenium peroxo complex $CH_3ReO(O_2)_2 \cdot H_2O$ in terms of the broad scope of oxyfunctionalisation (cf. Scheme 1) and its catalytic efficacy.

The structural similarity between metal peroxo complexes and the organic dioxiranes is made evident in Fig. 2, and the goal of this essay is, therefore, a comparison of the prototypes of these two classes of peroxides, namely the catalytic rhenium-based MTO/H_2O_2 oxidant (rhenium peroxo complexes have been proposed as the active oxygen-transferring agent [13]) and the stoichiometric dimethyldioxirane (DMD) and methyl(trifluoromethyl)dioxirane (TFD). We shall focus on the work from our research group, since we have for some time employed these peroxides in the selective synthesis of oxyfunctionalized products and in the mechanistic elucidation of the oxygen-transfer processes on a comparative basis. After an initial brief discussion of the structural parameters of the rhenium peroxo species and of the most commonly utilized dioxiranes, a comparison of their reactivity and selectivity shall be made, and finally the mechanistic features of these oxygen-transferring agents will be highlighted.

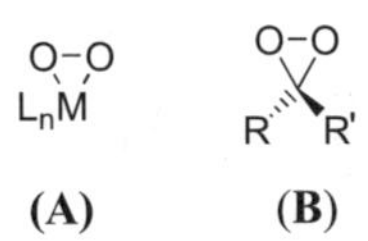

Fig. 2A,B. Structural similarity between: **A** metal-peroxo complexes; **B** dioxiranes

Table 1. Transition-metal peroxo complexes used in oxygen-transfer reactions

Peroxo complex	Oxygen source	Reaction type/preferred substrates	Drawbacks	Stoichiometric or catalytic?	Refs.
$Ti(O_2)(OEt)_2(H_2O)_2$	H_2O_2	– epoxidation of cyclohexene – hydroxylation of phenol	poor selectivity		[2]
$Zr(O_2)(mal)(quin)$[a]	H_2O_2	– oxidation of PPh_3		stoichiometric	[3]
$VO(O_2)(pic)(H_2O)_2$[b]	H_2O_2	– epoxidation – hydroxylation of arenes – hydroxylation of alkanes	*cis/trans* isomerisation double-bond cleavage mixture of alcohols and ketones	stoichiometric	[4]
$Nb(Cp)_2(O_2)$ Cl	H_2O_2	– oxidation of phosphines		stoichiometric	[5]
$CrO(O_2)_2L$ (L = pyridine, HMPA)	H_2O_2	– oxidation of alcohols to carbonyl compounds – oxidation of hydroquinones to quinones – sulfoxidation – *N*-oxidation	double-bond cleavage	stoichiometric	[6]
$MoO(O_2)_2(H_2O)(HMPA)$	H_2O_2	– epoxidation – oxidation of alcohols to carbonyl compounds – phosphine oxidation	double-bond cleavage	stoichiometric	[7]
$WO(O_2)_2(H_2O)(HMPA)$	H_2O_2	– epoxidation – oxidation of alcohols to carbonyl compounds		stoichiometric	[1b]
$CH_3ReO(O_2)_2(H_2O)$	H_2O_2	cf. Scheme 1		catalytic	
$Ru(O_2)Cl(NO)(PPh_3)_2$	O_2	– oxidation of phosphines		stoichiometric	[8]
$Co_2(CN)_4(PMe_2Ph)_5(O_2)$	O_2	– oxidation of phosphines		stoichiometric	[9]
$[Rh(O_2)(AsMe_2Ph)_4](ClO_4)$	O_2	– co-oxidation of terminal olefins to methyl ketones and PPh_3 to $OPPh_3$		stoichiometric	[10]
$Ni(O_2)(t\text{-}BuNC)_2$	O_2	– oxidation of phosphines – oxidation of isocyanides to isocyanates		stoichiometric	[11]
$Pt(O_2)(PPh_3)_2$	O_2	– oxidation of phosphines		stoichiometric	[12]

[a] Hmal = malonic acid; quin = quinoline.
[b] Hpic = picolinic acid.

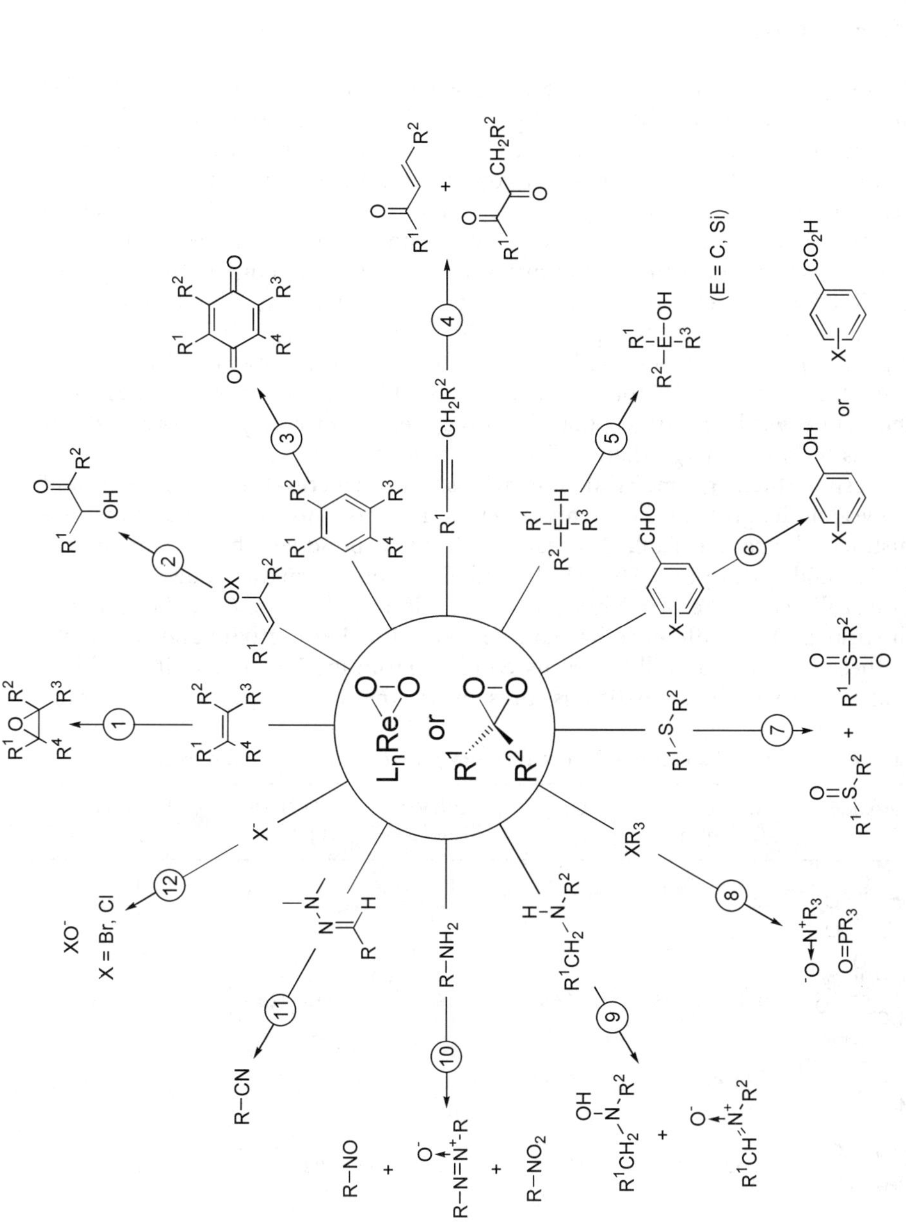

Scheme 1. Major oxidative transformations of MTO/H_2O_2 and dioxiranes

2
Structural Aspects

Although the dioxirane structure was postulated for the first time in 1899 (as an intermediate in the Baeyer-Villiger oxidation of menthone by Caro's acid), convincing evidence for its existence appeared only in the late 1970s through to the mid 1980s by means of microwave and NMR spectroscopy [14]. The year 1985 marks the crowning achievement of the isolation of DMD as acetone solution [15], in 1994 the first crystalline dioxirane, i.e., dimesityldioxirane [16], was prepared. Structural information of the rhenium diperoxo complexes $CH_3ReO(O_2)_2 \cdot L$ (L=H_2O, 4-*tert*-butylpyridine *N*-oxide or HMPA) was made available by X-ray crystallography of the adduct of $CH_3ReO(O_2)_2 \cdot H_2O$ with diglyme [17] and corroborated by NMR and IR spectroscopy [18]. For both the rhenium mono- and diperoxo complexes [19] and some dioxiranes [20], theoretical work confirmed that the three-membered-ring peroxo-type structure is of lower energy than its isomeric acyclic form.

Despite their geometrical similarities, the structural parameters (bond angles and lengths) of the rhenium peroxo functionality and of dioxiranes are distinct. The data in Table 2 reveal that the O—O bond lengths are comparable (147.4 and 151.6 pm). The M—O and C—O bond lengths, however, differ drastically since the M—O bond is ca. 30% longer than the C—O bond in the dioxiranes. This difference has consequences for the reactivity and selectivity of these oxidants, as will be discussed later. From the UV spectra it is evident that a peroxo functionality is present in the MTO/H_2O_2 and dioxirane

Table 2. Spectral and structural data of rhenium peroxo complexes and dioxiranes

Peroxide	M-O or C-O [pm]	O—O [pm]	UV/Vis [nm]	^{17}O-NMR [ppm][a]	References
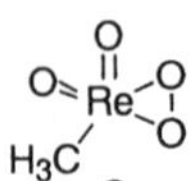	–	–	310	–	[17]
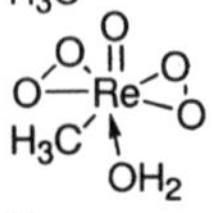	190.4; 191.8[b]	147.4[b]	364	363; 422	[17]
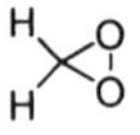	138.8[c]	151.6[c]	–	250	[14a,b]
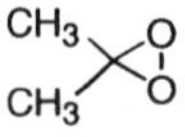	141.7[d]	152.1[d]	335	302	[14b, 21]
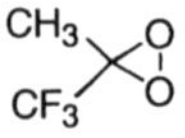	–	–	347	297	[14c]

[a] Values only given for the peroxo oxygen atoms.
[b] Determined by X-ray crystallography.
[c] Determined by microwave spectroscopy.
[d] Calculated by MP2/6-31G*.

oxidants. The absorption bands for both the mono- and diperoxo complexes (310 nm and 364 nm respectively) as well as for DMD (335 nm) and TFD (347 nm) are located in the near UV, which results in a yellow colour for these species. The ^{17}O-NMR data (which are characteristic for mono- and diperoxo complexes) constitute further evidence for the peroxo ring in the rhenium oxidant [22].

3
Reaction Types and Reactivity

Numerous oxidative transformations have been performed with both MTO-based oxidants and dioxiranes (Scheme 1), which are known to be highly active oxygen-transferring species.

These reactions include the oxidation of π bonds, σ bonds and heteroatoms. Depending on the reaction conditions, high yields and selectivities have been achieved for most of these reaction types with both oxidants. Although MTO/H_2O_2 has been claimed as one of the most active catalytic oxidation systems [18], compared to DMD, it is less reactive; however, DMD is used stoichiometrically and MTO catalytically. The choice of solvent and the appropriate additive may improve the reactivity and selectivity for the MTO catalyst.

The MTO catalyst only efficiently activates hydrogen peroxide for oxidation reactions, but not other oxygen sources such as molecular oxygen, alkyl hydroperoxides, and alkyl peroxides [23]. Also, $Me_3SiOOSiMe_3$ has been employed, but presumably adventitious moisture hydrolyses the sensitive silyl peroxide into H_2O_2 and $Me_3SiOSiMe_3$ [24] and, thus, the real oxygen source is again H_2O_2.

For the preparation of dioxiranes, potassium persulfate (commercial names Caroate, Curox, or Oxone) has proved to be the only practical oxygen donor; even then, most of the persulfate is decomposed into molecular oxygen and the yield of the isolated dioxirane is low (ca. 5% for DMD) [25]. In the in-situ generation [26] the yield of oxidation product may be quite high, especially if the substrate is very reactive. Rigorous pH control is necessary, as the dioxirane is otherwise decomposed to the ketone and molecular oxygen [27]. Although dioxiranes are usually used stoichiometrically, catalytic variants by in-situ generation have recently been developed for asymmetric oxidations [28].

Although the oxidants MTO/H_2O_2 and DMD show similar reaction behaviour, a number of differences are, nonetheless, observed: Some oxidations are only possible with MTO/H_2O_2 or with DMD; also, differences in their oxidative efficacy have been noted. A major difference in their reaction behaviour derives from the Lewis acidity of the MTO/H_2O_2 oxidation system, which leads to competing acid-catalysed reactions. Such a handicap may now be overcome by using additives [29]. By contrast, oxidations with isolated DMD are conducted under neutral conditions, which is highly beneficial in the synthesis of acid-sensitive compounds. Of the two most commonly used dioxiranes, DMD and TFD, the latter is both more reactive *and* more selective than DMD [30].

3.1
Oxidation of π Bonds (Olefins, Alkynes, Arenes)

The catalytic activity of MTO was first discovered for the epoxidation of olefins [31]. Initial results showed drastically different yields of epoxide, which depended on the substrate. Thus, turnover numbers (TON) of 100–200 were usually obtained; however, in some cases the TON may be up to 2000 [18]. The reactivity of the MTO catalyst depends on several factors: the polarity of the solvent, the concentration of the catalyst, and the type of additive. The best reaction conditions to date are the use of 0.1 mol% MTO with 5 mol% of pyrazole as additive in trifluoroethanol as polar solvent, which display TONs of up to 2500 [32].

Apart from the high reactivity observed for the MTO catalyst, good selectivity may be achieved by careful control of the reaction conditions. The epoxidation of styrene serves as an example (Table 3). Thus, while MTO/ ca. 10% H_2O_2 shows reasonable reactivity (60% conversion after 3 h at 30 °C), the initial epoxide does not persist under the reaction conditions and subsequent Lewis-acid-catalyzed reactions lead to secondary products as a result of epoxide-ring opening, Hock cleavage, and pinacol rearrangement. These undesired transformations may be avoided by the addition of an appropriate additive [29], e.g. urea, pyridine, 3-cyanopyridine, or pyrazole. Although the presence of all of these additives affords styrene oxide in excellent product selectivity, the efficiency is strongly dependent on the type and the amount of additive. An increase of the amount of additive reduces the reactivity of the oxidant, while a decrease results in poorer selectivity.

Table 3. Reactivity and selectivity of styrene oxidation by selected MTO-based oxidants

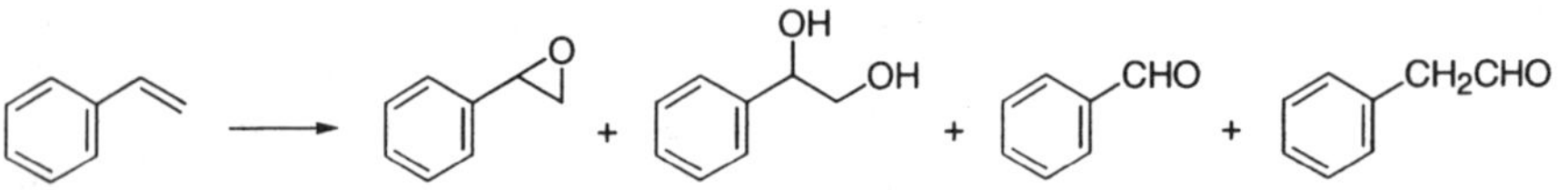

Oxidant[a]	t (h)	Convn (%)	Product distribution (%) epoxide	1,2-diol	aldehyde 1	aldehyde 2	Refs.
MTO/H_2O_2 (1:100)	3	60	–[b]	37	41	22	[18]
MTO/UHP (1:100:100)	19	46	≥95	–	–	–	[29a]
MTO/H_2O_2/pyridine (0.5:200:12)	5	70	≥99	–	–	–	[29b]
MTO/H_2O_2/3-cyano-pyridine (0.5:200:10)	5	≥99[c]	≥99	–	–	–	[33]
MTO/H_2O_2/pyrazole (0.5:200:12)	5	≥99	≥99	–	–	–	[29b]

[a] All entries in CH_2Cl_2 except entry 1 (in *t*-BuOH); the ratios of catalyst to oxygen donor and/or additive is given in parentheses, with the amount of substrate being normalised to 100%.

[b] No epoxide was isolated.

[c] 85% Yield of isolated epoxide.

Two rhenium peroxo complexes [$CH_3ReO_2(O_2)$ and $CH_3ReO(O_2)_2 \cdot H_2O$] are generated during the catalytic oxidation process and both are reported to be catalytically active. Remarkably, they possess similar reactivities towards most substrates, which include olefins [13]. Kinetic data are available for a series of alkyl-substituted styrenes, for which the rate constants of the two peroxo complexes are comparable. As a comparison, the rates of the DMD epoxidation with these substrates are similar [34], but TFD is generally more reactive by a factor of ca. 100 [30a]. DMD and TFD, which are the most frequently used dioxiranes, are usually employed stoichiometrically and in isolated form.

The reactivity of C=C bonds towards epoxidation by the MTO-based oxidant and dioxiranes depends on several factors, most prominently the degree and pattern of substitution at the double bond, and the electronic and steric nature of these substituents. Perusal of the literature reveals that the orders of reactivity given in Fig. 3 apply for the MTO/H_2O_2 and DMD oxidants. The mechanistic aspects of these reactivities shall be discussed in Sect. 4. These reactivity orders are expected for electrophilic oxidants, for which electron-rich double bonds, i.e., double bonds with electron-donating substituents and/or a higher degree of substitution, are more prone to oxidation. Electronic effects of the substituents follow the order OR > R > Ar > COR > sCO_2R, but steric properties operate as well. Thus, *cis* olefins are usually oxidized faster than *trans* olefins, except in the case of *cis-* and *trans*-cyclooctene, for which the latter reacts faster by a factor of 100 in the epoxidation with DMD [35].

While the MTO/UHP (UHP = urea-hydrogen-peroxide adduct) oxidation system and DMD show much similarity in their epoxidation of olefins (Table 4), some differences should, nonetheless, be noted. In most cases, the yields for DMD are slightly higher than with MTO/UHP; however, double bonds with electron-withdrawing groups (e.g., 2-cyclohexenone) are efficiently epoxidized by DMD (convn 83%), while with MTO/UHP only marginal conversion is observed (convn $\leq$ 5%). Additionally, hydrolytic cleavage of the epoxide occurs for particularly vulnerable epoxides (e.g., α-methylstyrene) in

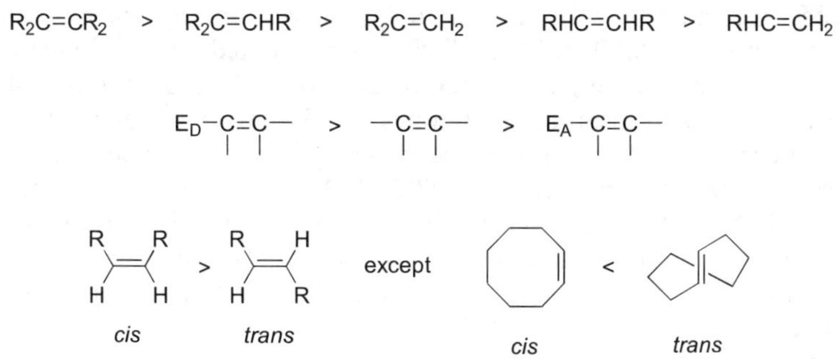

Fig. 3. Epoxidation reactivity of MTO/H_2O_2 and DMD oxidants as a function of double-bond substitution

Table 4. Oxidation of selected olefins by MTO/UHP and DMD

Olefin	MTO/UHP[a] Product ratio (%)			DMD Product ratio (%)		
	epoxide	:	diol	epoxide	:	diol
	≥95	:	5	≥95[b]	:	5
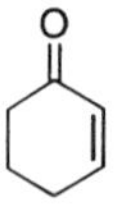	≥95	:	5	≥95[c]	:	5
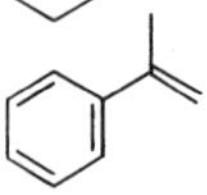	68	:	32	92[c]	:	8
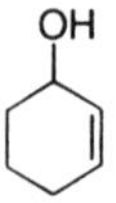	≥95[d]	:	5[e]	70[f]	:	30[e]
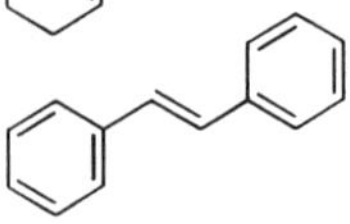	≥95	:	5	≥95[g]	:	5

[a] See [29a].
[b] See [34].
[c] See [23b].
[d] Diastereomeric ratio (dr) = 81:19 (*cis/trans*); see [36].
[e] Minor product is 2-cyclohexenone.
[f] dr = 47:53 (*cis/trans*); see [37].
[g] See [15].

the case of MTO/UHP. In contrast, the chemoselectivity is poorer for DMD, as demonstrated in the oxidation of 2-cyclohexenol, from which a mixture of epoxide and enone (CH oxidation) is obtained with DMD, but only the epoxide for MTO/UHP. Fortunately, *cis/trans* isomerization is not a problem for either oxidant.

Alkynes are much less reactive than the corresponding alkenes, and are initially oxidized by DMD and MTO/[O] to a postulated oxirene intermediate, which subsequently undergoes either further oxidation or a 1,2-H shift to α-diketones and α,β-unsaturated ketones as the main products (Table 5). Smaller amounts of other products occur through oxidation/rearrangement or oxidative cleavage. The reactivity of MTO/H_2O_2 towards alkynes is somewhat inferior to that of DMD; however, by far the most reactive is TFD, which gives excellent conversion of the alkynes within minutes. The product distributions obtained with DMD and TFD are similar, but for MTO/H_2O_2 more α-diketone is produced.

The oxidation of arenes is still more difficult than alkynes. As expected, the oxidation of benzenes is substantially harder to achieve than that of polycyclic arenes. DMD oxidises arenes even under the standard reaction conditions (i.e.

Table 5. Oxidation of alkynes by MTO/H_2O_2 and DMD

Alkyne	Oxidant	t	Convn (%)	Products (%)[a]						Refs.
Ph—≡—Ph	MTO/H_2O_2	2 days	69	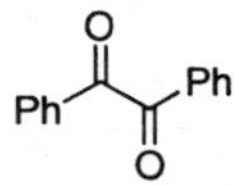	(57)	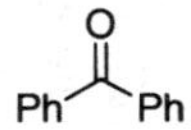	(0)			[38]
	DMD	8 h	50	"	(15)	"	(29)			[39]
	TFD	6 min	82	"	(25)	"	(49)			[39]
n-Pr—≡—n-Pr	MTO/H_2O_2	2 days	≥97	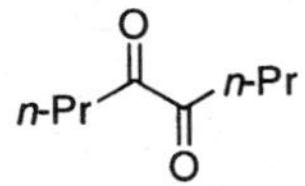	(43)	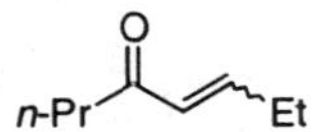	(11)	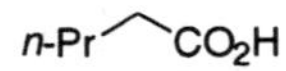	(24)	[38]
n-Hept–C—≡—C–n-Hept	DMD	8 h	>96	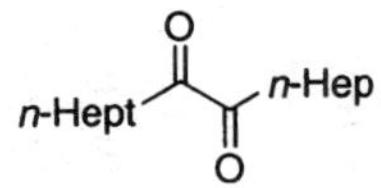	(18)	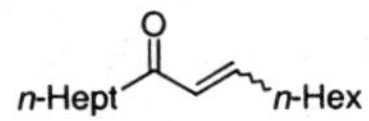	(78)			[39]
	TFD	2 min	>96	"	(19)	"	(77)			[39]

[a] The yields are given in parentheses

0–20 °C, acetone as solvent), whereas the reactions with MTO/H_2O_2 are performed in glacial acetic acid [40] and usually at elevated temperatures. The role of the acetic acid in the case of MTO/H_2O_2 seems to be one of increasing the electrophilicity of the peroxo complex by protonation of a peroxy oxygen atom [18].

The MTO-based oxidant and the dioxiranes show different product selectivity in the oxidation of arenes (Scheme 2), although epoxides are initially formed. In the case of DMD, these are isolable [44], while for the MTO oxidant the epoxides do not persist under the reaction conditions and rearrange in-situ to the hydroquinones. These are subsequently oxidized further to the corresponding quinones. Thus, $MTO/H_2O_2/AcOH$ leads predominantly to the *para*-quinones. With biphenylene the *ortho*-quinone was obtained, while phenanthrene is overoxidized to diphenyl-2,2′-dicarboxylic acid. The formation of undesired side-products from reactions of epoxide intermediates may be a problem (Scheme 3).

The epoxides may undergo ring-opening with H_2O_2 to form the β-hydroxy hydroperoxide, which give water-soluble products after further oxidation. For the 3,5-di-*tert*-butyl derivative, the corresponding muconic anhydride is formed, while for the 2,6-dimethyl case a [4 + 2] cycloadduct is observed as side-product.

In the oxidation of arenes, DMD gives epoxides in the majority of cases; exceptions are anthracene and the electron-rich 1,2,3-trimethoxy-5-methylbenzene, which give the respective *para*-quinones. TFD oxidizes 2-methylnaphthalene to the diepoxide (the same product as with DMD) in 66% yield after only 1 h at 0 °C [44], but phenanthrene is transformed to the corresponding *ortho*-quinone (9,10-quinone) in ca. 75% yield [14c]. Similar behavior is observed in the oxidation of phenols with all three oxidants, i.e., a mixture of hydroquinones and quinones (although dioxiranes have a greater propensity towards the formation of *ortho*-quinones than does $MTO/H_2O_2/AcOH$) [40b, 46]. However, in the oxidation of catechol with TFD, oxidative cleavage of the benzene ring occurs and *Z,Z*-muconic acid is obtained in high yield [47].

Some of the quinones are of industrial interest. For example, 2-methyl-1,4-naphthoquinone (Vitamin K_3) is produced on an industrial scale as a supplement for animal feed [40a], while 2,3-dimethoxy-5-methyl-*p*-benzoquinone is a key intermediate in the synthesis of coenzyme Q [40c].

3.2 Oxidation of Heteroatoms (S, N)

The oxidation of sulfides with MTO/H_2O_2 and DMD leads to the corresponding sulfoxides in high yield and excellent selectivity. If an excess of oxidant is used, the reaction may be driven towards the sulfone as the major product (Scheme 4). In the case of sulfoxidations with the MTO-based oxidant, both the monoperoxo rhenium complex $CH_3ReO_2(O_2)$ and the diperoxo complex $CH_3ReO(O_2)_2 \cdot H_2O$ are claimed to be catalytically active; however, the rate of the oxygen transfer for the monoperoxo $CH_3ReO_2(O_2)$ is greater by approx-

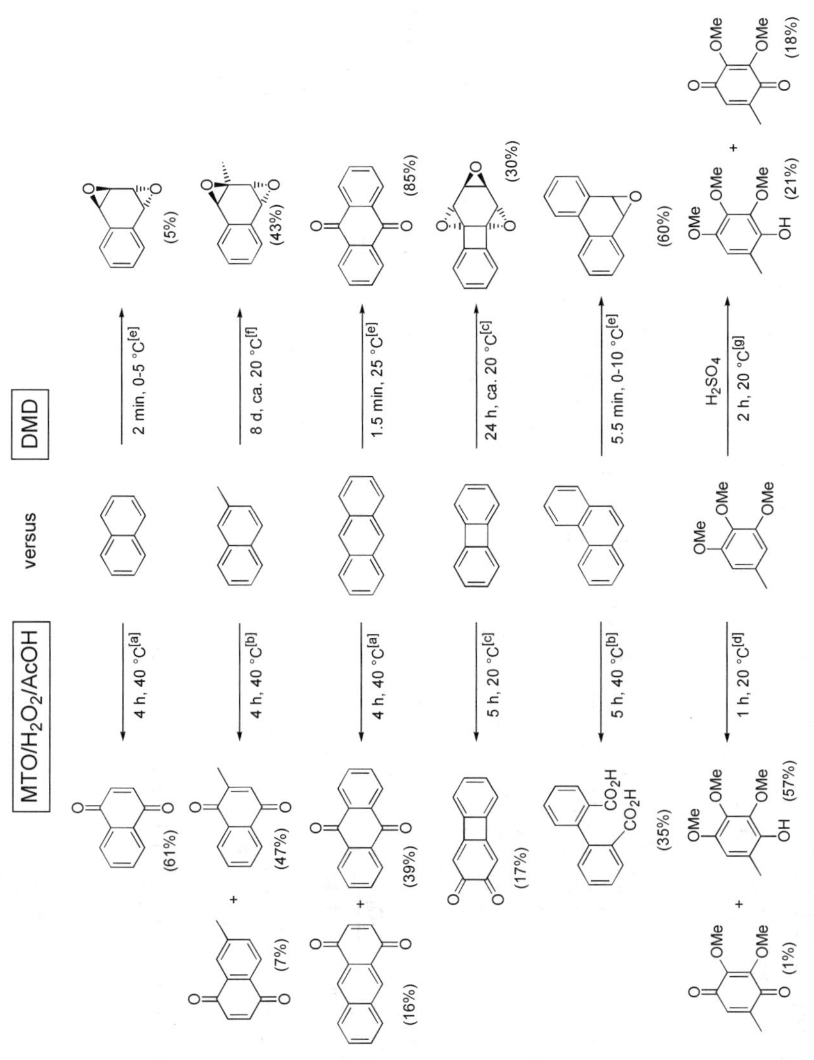

[a] See ref. [41]. [b] See ref. [40a]. [c] See ref. [42]. [d] See ref. [40c]. [e] See ref. [43]. [f] See ref. [44]. [g] See ref. [45]

Scheme 2. Oxidation of arenes by MTO/H_2O_2 versus DMD

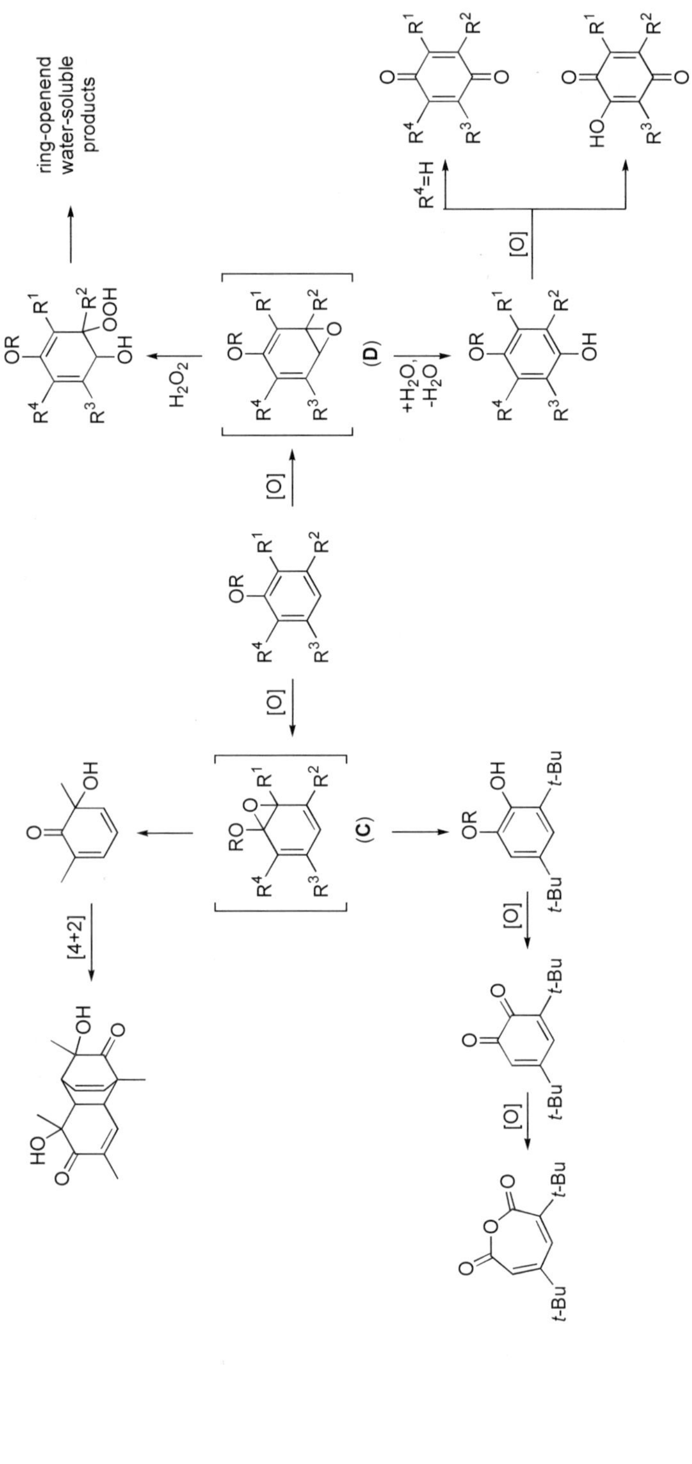

Scheme 3. MTO-catalysed oxidation of phenols and methoxybenzenes

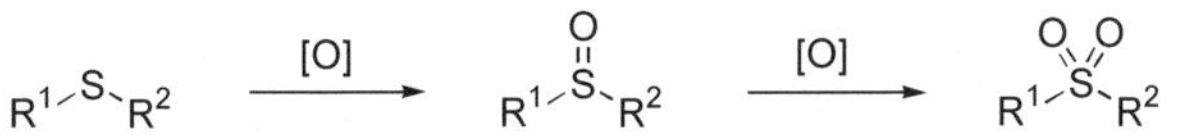 (1)

Scheme 4. Oxidation of sulfides to sulfoxides and sulfones

imately one order of magnitude than for the diperoxo rhenium complex $CH_3ReO(O_2)_2 \cdot H_2O$ [13] (the oxidation of thiophenes is an exception since the diperoxo complex is more active towards these substrates [48]). Kinetic data for the oxidation of *para*-substituted thioanisoles by DMD and the monoperoxo rhenium complex $CH_3ReO_2(O_2)$ show the reactivity of both to be similar [49]. The rate constants follow linear Hammett relationships with reaction constants (ρ) of −0.98 for MTO/H_2O_2, −1.00 for DMD, and −0.34 for TFD. Thus, the selectivity (sulfoxide vs sulfone) order for these electrophilic oxidants is usually TFD < MTO/H_2O_2 < DMD.

Thiophenes are oxidized with both MTO/H_2O_2 and DMD to the corresponding sulfones through sulfoxide intermediates [48, 50]. In the oxidation with DMD, the thiophene 1-oxides could be kinetically stabilized by bulky substituents at the 3- and 4-positions and were isolated in crystalline form [51]. Generally these sulfoxides are difficult to prepare with either oxidant, as the oxidation step from the sulfoxides to the sulfones proceeds at a faster rate than the reaction of the thiophenes to the 1,1-dioxide and the sulfoxides. In addition, thiophene 1-oxide readily dimerises to Diels-Alder products.

MTO/H_2O_2 and DMD both oxidize the sulfur in the ligand sphere of organometallic complexes [13, 52]. In fact, the (thiolato)cobalt(III) complex $(en)_2Co(SCH_2CH_2NH_2)(ClO_4)_2$ was one of the first sulfur-containing compounds to be oxidized by the MTO system [13].

In the oxidation of nitrogen compounds by MTO/H_2O_2 and DMD, numerous products are possible. Thus, from primary amines, the corresponding nitro and nitroso compounds and oximes have been obtained, along with azo and azoxy compounds and the nitroso dimers (Scheme 5). In contrast, secondary amines (Scheme 6) are oxidized to hydroxylamines, nitrones, and in some cases nitroxides. Tertiary amines give the corresponding *N*-oxides cleanly.

The product distribution in the MTO-catalyzed oxidation of primary amines depends on the substituents on the carbon atom α to the amine functionality and on how much of the oxygen source (H_2O_2) is used. In general, nitro compounds are only obtained when the α-carbon atoms are tertiary, as shown for 1-aminoadamantane in Table 6. Instead of azo compounds, further oxidation to the corresponding azoxy compounds is observed under these reaction conditions, (cf. Scheme 5). Thus, the oxidation of aniline initially leads to the corresponding hydroxylamine, which is oxidised further to $PhN(OH)_2$, followed by the elimination of H_2O to give the nitroso benzene. The latter may then either be oxidized to the nitrobenzene or undergo condensation with the aniline to produce azobenzene, which is oxidized further to azoxybenzene [63]. For benzylic amines, the oxime is selectively produced, with none of the azoxy compound or nitroso dimers [55]. This is in contrast to alkylamines (e.g., *n*-decylamine), for which a mixture of

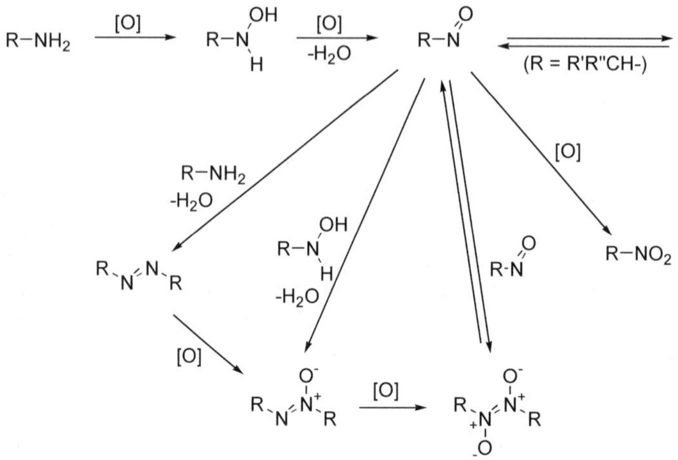

Scheme 5. Oxidation of primary amines with MTO/H_2O_2

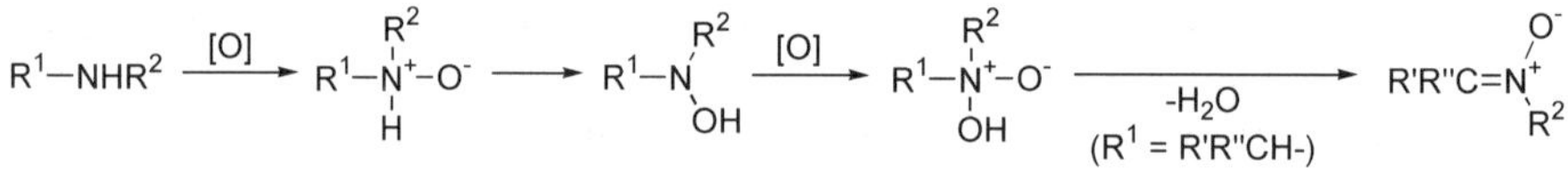

Scheme 6. Oxidation of secondary amines with MTO/H_2O_2

all three of these products is obtained; however, the product ratio depends on the solvent [55].

The product distribution in the oxidation of primary amines with DMD differs from that of MTO/H_2O_2. Thus, due to the greater oxidation power of DMD, the nitro compounds are generally the preferred products, as illustrated for 1-aminoadamantane, aniline, and *n*-butylamine in Table 6. Indeed, it has been shown that both phenylhydroxylamine and nitrosobenzene are readily oxidized to nitrobenzene by DMD [54]. For a few selected substrates nitroso compounds may be isolated, when a sub-stoichiometric amount of the dioxirane is used [64]. These nitroso products readily transform to the corresponding dimers, but do not tautomerize to the oximes. The oxidation of glycosyl amines by DMD to the corresponding hydroxylamines and oximes was reported, without further reaction to the nitro sugars [65].

In the case of secondary amines, e.g., *N*-*tert*-butyl benzylamine, 1.0 equiv. of DMD cleanly afforded the hydroxylamines [56], while 2.0 equiv. led selectively to the nitrones [57]. MTO/H_2O_2 (UHP may also be used) [66] gave the nitrone less selectively in the case of dibenzylamine, and a larger excess (4.0 equiv.) of H_2O_2 was needed to convert all the hydroxylamine intermediate. For the MTO oxidant, the solvent also plays an important role in regard to the product distribution of hydroxylamine vs nitrone and, thus, much more nitrone is obtained when the reaction is conducted in protic solvents instead of chlorinated ones [55, 66a]. The oxidation of unsymmetrical secondary amines may give a mixture of regioisomeric nitrones [55].

Secondary amines with tertiary alkyl groups are oxidised by both MTO/H_2O_2 and DMD in excellent yields to the corresponding nitroxides [53, 58, 67], which are used as spin labels in biochemical and medicinal investigations [68]. In this transformation, the amine is initially oxidized to the hydroxylamine, followed by further oxidation to the hydroxylamine *N*-oxide. This latter product is expected to cleave homolytically to give the nitroxide [58].

In contrast to the complex oxidation of primary and secondary amines, the conversion of tertiary amines is quite straightforward and leads only to *N*-oxides in excellent yields, e.g., quinoline and isoquinoline in Table 6. A whole range of differently functionalized tertiary aromatic amines has successfully been converted to the corresponding *N*-oxides with MTO/H_2O_2 and DMD [59, 60]. The *m*- and *p*-substituted pyridine derivatives are readily oxidized by MTO/H_2O_2 to the corresponding *N*-oxides in the presence of only 0.2–0.5 mol% MTO, regardless of the electronic nature of the substituent. In contrast, *o*-substituted pyridines are sluggish substrates towards oxidation with MTO/H_2O_2 for steric reasons and require higher amounts of catalyst (5 mol% MTO) for conversion.

When DMD or TFD are used as oxidant for the transformation of tertiary aromatic amines, the in-situ-generated *N*-oxides are partially deoxygenated with the release of singlet oxygen ($^1O_2$) [69]. Even when a large excess of dioxirane is used, full conversion cannot be reached for some substrates. Especially the more nucleophilic heteroaromatic *N*-oxides are deoxygenated at comparable or even higher rates than the rate of the amine oxidation. Two of the most notable examples for such deoxygenation are 4-dimethylaminopyridine-*N*-oxide (Scheme 7) and 2′,3′,5′-triacetyladenosine-N^1-oxide [70]. Such a deoxygenation process is not known for the MTO-based oxidants.

Another notable oxyfunctionalization of nitrogen compounds is the oxidation of aldehyde *N*,*N*-dimethylhydrazones to nitriles and *N*-methylene-

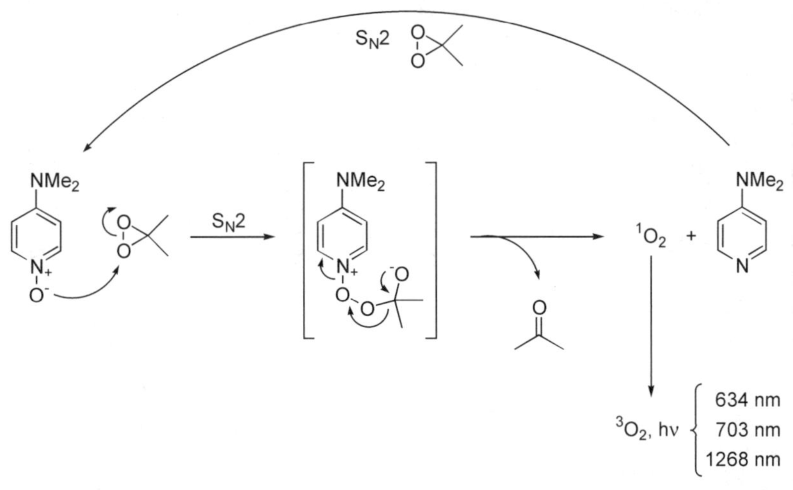

Scheme 7. Formation of singlet oxygen in the deoxygenation of heteroarene *N*-oxides by DMD

Table 6. Oxidation of organonitrogen compounds by MTO/H_2O_2 and DMD

Substrate	Oxidant	T (°C)	t (h)	Products			Refs.
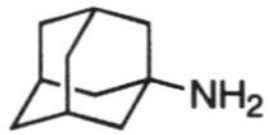				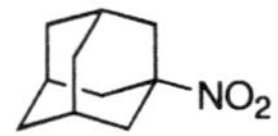			
	MTO/H_2O_2	60	2	88			[53]
	DMD	20	0.5	95			[54]
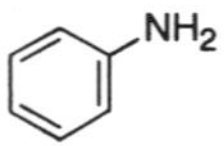				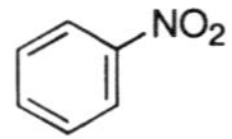	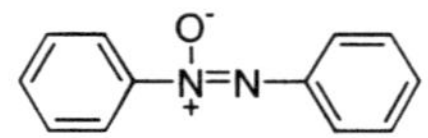		
	MTO/H_2O_2	20	2	70	30		[53]
	DMD	20	0.5	97	–		[54]
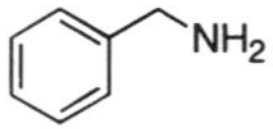				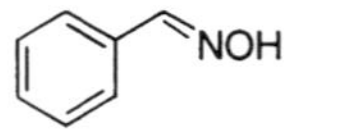			
	MTO/H_2O_2	20	0.5	88			[55]
$CH_3(CH_2)_9$-NH_2				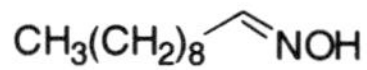	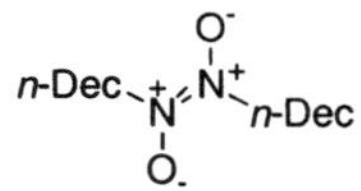	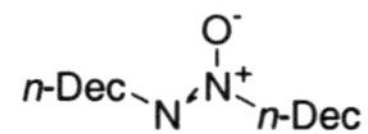	
	MTO/H_2O_2	20	0.5	38	11	47	[55]
$CH_3(CH_2)_3$-NH_2				$CH_3(CH_2)_3$-NH_2			
	DMD	20	0.5	84			[54]
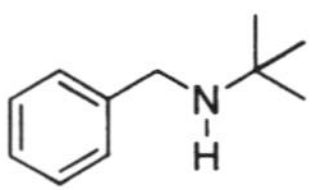				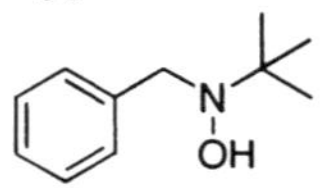	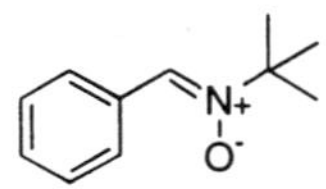		
	DMD (1.0 equiv)	0	0.25	99	–		[56]
	DMD (2.0 equiv)	0	0.15	–	96		[57]
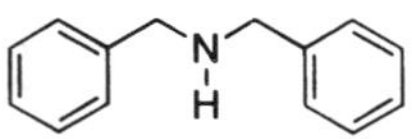				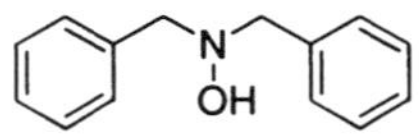	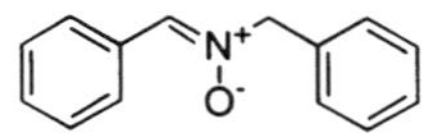		

	MTO/H_2O_2 (1.0 equiv)	0	1.0	66	6	[55]
	MTO/H_2O_2 (4.0 equiv)	0	1.0	–	91	[55]
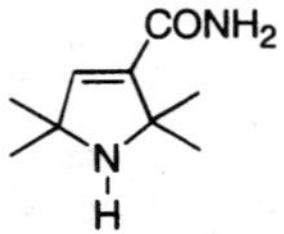				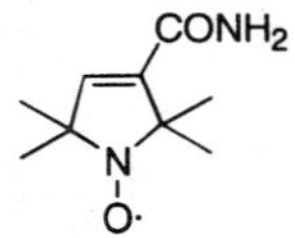		
	MTO/H_2O_2	0	1	100[a]		[53]
	DMD	0	0.5	98[a]		[58]
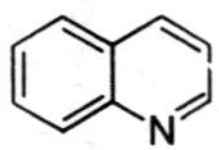				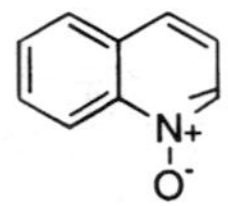		
	MTO/H_2O_2	24	3	98		[59]
	DMD	0	2	98		[60]
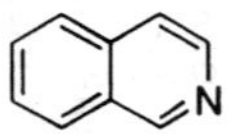				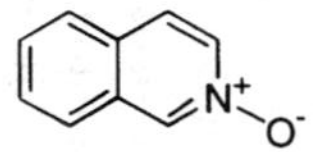		
	MTO/H_2O_2	24	15	92		[59]
	DMD	25	2	92		[60]
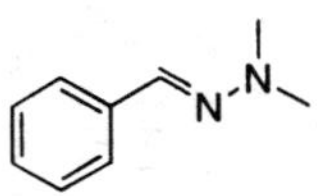				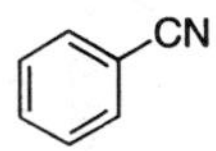		
	MTO/H_2O_2	$-50 \rightarrow 20$	1	93		[61]
	DMD	0	0.05	97		[62]
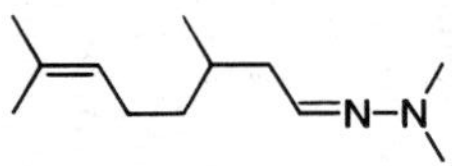				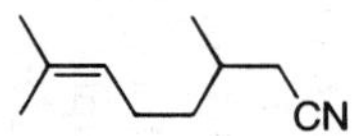		
	MTO/H_2O_2	$-50 \rightarrow 20$	1	96[a]		[61]
	DMD	0	0.05	97[a]		[62]

[a] No epoxide was formed.

N-methyl *N*-oxide in excellent yields with both MTO/H_2O_2 and DMD [61, 62]. The nitriles do not react further and other functional groups (e.g., double bonds) do not affect the transformation.

3.3
Oxidation of σ Bonds (C—H, Si—H)

Probably the most marked difference in the reactivity of the rhenium peroxo oxidants and dioxiranes lies in their behaviour towards oxygen insertion into C—H bonds. Little has been reported on the MTO catalysis for such a transformation, as under the normally used reaction conditions it is several orders of magnitude slower than epoxidations and does not take place [49b]. Much more drastic conditions are necessary, i.e., the use of large amounts of MTO and a large excess of the oxygen source H_2O_2, heating, and high concentrations, in order to achieve C—H insertions (C—H to C—OH and C—OH to C=O) with the rhenium catalyst [71]. The reaction proceeds with retention of configuration, as shown for *cis*- and *trans*-1,2-dimethylcyclohexane and *trans*-decalin (Table 7). The remarkable discrimination between equatorial versus axial C—H positions in the oxidation of *cis*- and *trans*-1,2-dimethylcyclohexane should be noted. When several positions are available in the substrate, the C—H insertion usually stops after one oxyfunctionalization, the multiple oxidation of adamantane is an exception. The selectivity order of the C—H bonds in unactivated alkanes is tertiary > secondary > primary, which is typical for electrophilic oxidants [73]. In the oxidation of alkanes with H_2O_2 catalysed by MTO and in the presence of pyrazine-2-carboxylic acid, the formation of hydroperoxide has also been observed besides alcohol and its carbonyl product [75].

The oxidation of alcohols to the corresponding carbonyl products is inefficient. Thus, only a poor yield (19%) is achieved in the synthesis of acetophenone from (1-phenyl)ethanol (cf. Table 7). Furthermore, much lower yields of ketone/aldehyde than of alcohol are usually obtained in the oxidation of cyclohexane, cyclooctane, and *n*-heptane. The yields of the carbonyl compounds may be improved, when the reaction is performed in neat alcohol or in acetic acid with bromide ions added as a co-catalysts [49b].

These poor results for the MTO-based oxidant are in stark contrast to the DMD oxidation of alkanes and alcohols, and in particular the oxidation of these substrates with the highly reactive TFD (up to 700 times more reactive than DMD [73]). TFD is even capable of oxidizing unactivated hydrocarbons under very mild reaction conditions and sometimes in a matter of minutes [30a]. A prime example is the quantitative conversion of cyclohexane to cyclohexanone at −22 °C in 18 min [73]! As expected for these electrophilic oxidants, primary C—H bonds are not appreciably oxidized, and tertiary C—H bonds are preferred over secondary ones [73], although substrates with only secondary C—H bonds may be oxidized to the corresponding ketones in moderate to high yields under relatively mild conditions with TFD. In fact, TFD is so reactive that all of the alcohol intermediate is converted to the carbonyl product. The use of trifluoroacetic anhydride as solvent allows in-situ

Table 7. C—H insertion by MTO/H_2O_2 and dioxiranes

Substrate	Oxidant	T (°C)	t	Convn (%)	Products (%)			Refs.
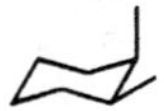					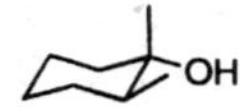			
	MTO/H_2O_2	40	72 h	–	98			[71]
	DMD	22	18 h	–	>95			[72]
	TFD	–22	4 min	–	97			[73]
					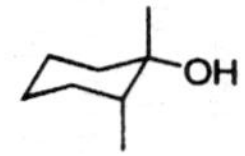	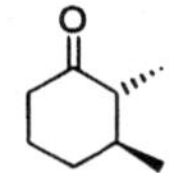	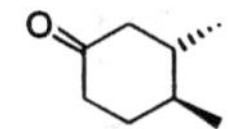	
	MTO/H_2O_2	40	72 h	–	20	–	–	[71]
	DMD	22	18 h	–	45	–	–	[72]
	TFD	–22	10 min	98	90	6	4	[73]
					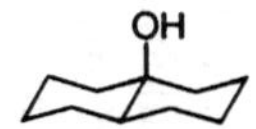	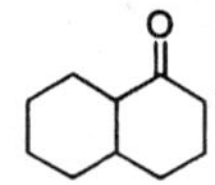	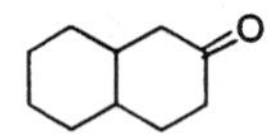	
	MTO/H_2O_2	40	72 h	–	20	–	–	[71]
	DMD	22	17 h	–	20	–	–	[72]
	TFD	–22	11 min	93	66	18	16	[73]
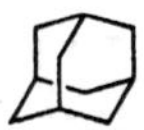					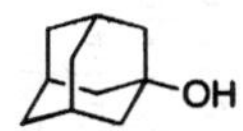	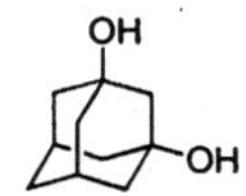	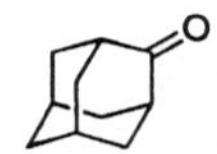	
	MTO/H_2O_2	60	48 h	–	88	6–8	–	[71]
	DMD	22	18 h	–	87	3		[72]
	TFD[a]	–22	1 min	>98	94	5	<1	[73]
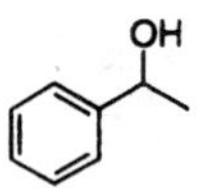					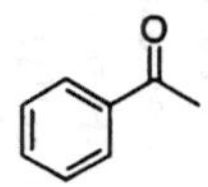	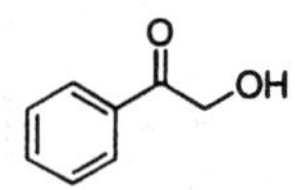		
	MTO/H_2O_2	60	24 h	–	19	–		[71]
	DMD	25	–	–	73	2		[74]

[a] Excess of TFD leads to 1,3,5,7-tetrahydroxyadamantane.

acetylation of the alcohols, which may subsequently be deprotected by hydrolysis and in this way product selectivity in favour of the alcohol may be achieved [76]. In addition to the excellent reactivity of TFD, this oxidant also shows higher selectivity for tertiary over secondary C—H oxyfunctionalization [77] (a notable exception is norbornene [73], which is oxidized preferentially at a secondary carbon atom) and amazing site selectivity for tertiary C—H bonds in steroidal substrates [78]. A remarkable feat of TFD is its ability to perform polyoxyfunctionalisations such as the oxidation of adamantane to adamantan-1,3,5,7-tetraol in 73% yield [77].

The relative rate data for the oxidation of *para*-substituted cumenes was shown to correlate linearly with the Hammett substituent constants (σ) and reveals that the insertion reaction is an electrophilic process with $\rho = -2.76$ [79]. A correlation between the rate constants of a series of substituted adamantanes with the Taft steric substituent constants (σ^*) also gave a straight line with $\rho^* = -1.08$ [79].

That alcohols are the intermediates in the oxidation of alkanes to the corresponding carbonyl compounds was confirmed by the fact that the authentic alcohols were oxidized to the ketones/aldehydes under the reaction conditions used for the oxidation of alkanes. The second step of the reaction (CH—OH to C=O) is approximately 50 times faster than the first one (CH—H to CH—OH), as shown for the oxidation of cyclohexane and cyclohexanol by TFD [73, 80]. Ethers, silyl ethers, and esters derived from secondary alcohols can also be transformed to the ketones, albeit at lower rates than the corresponding alcohols [81]. The rate constants for the oxidation of secondary alcohols (RR′CHOH with R = Ph and R = Me, Et, *n*-Pr, *i*-Pr, *t*-Bu) and their methyl ethers (RR′CHOMe) have been shown to correlate linearly with the steric bulk of the substituents, and a good Hammett correlation was observed in the case of a set of methyl ethers [81].

An important preparative application of the oxidative transformation of alcohols to ketones is the desymmetrization of 1,2- and 1,3-diols to the corresponding keto alcohols [82]. The regioselectivity depends on the electronic effects of the substituents at the carbinol positions [82d]. Thus, electron-donating groups on the phenyl substituent of 1-phenyl-propane-1,2-diol lead to the preferential oxidation of the alcohol functionality next to the aromatic substituent; with electron-accepting substituents the product distribution is reversed. With optically active substrates, the transformation proceeds with retention of configuration at the chiral center adjacent to the one undergoing oxidation to a carbonyl group [82a, 83]. The enantioselective oxidation with chiral dioxiranes leads to optically active α-hydroxy ketones [84].

Oxygen insertion into a Si—H bond (77 kcal/mol) is far easier to achieve than into a C—H (99 kcal/mol) bond, due to the fact that Si—H bonds are substantially weaker than C—H bonds [85]. In addition, for the oxidation of silanes, the net energy gain is far greater because of the formation of strong Si—O bonds. Thus, both the MTO oxidant and dioxiranes are capable of oxidizing silanes to silanols under relatively mild reaction conditions (Table 8). A disadvantage of the MTO-catalyzed oxidation of silanes to silanols compared

Table 8. Oxidation of triorganosilanes

Silane	Oxidant	t	T (°C)	Convn (%)	Product distribution (%) Silanol[a]	Disiloxane	Refs.
$EtMe_2SiH$	MTO/85% H_2O_2	18 h	20	55	≤5	95	[86]
	MTO/UHP	18 h	20	≥95	94	6	[86]
	MTO/85% H_2O_2/NaY	24 h	20	72	≥99	1	[87]
n-$PrMe_2SiH$	MTO/85% H_2O_2	18 h	20	56	≤5	95	[86]
	MTO/UHP	18 h	20	77	82	18	[86]
	DMD	0.5 h	0	63	92	8	[86]
t-$BuMe_2SiH$	MTO/85% H_2O_2	18 h	20	70	≥99	1	[86]
	MTO/UHP	18 h	20	98	≥99	1	[86]
	MTO/85% H_2O_2/NaY	24 h	20	≥99	≥99	1	[87]
	DMD	0.5 h	0	60	≥99	1	[86]
$PhMe_2SiH$	MTO/85% H_2O_2	13 h	20	26	20	80	[86]
	MTO/UHP	13 h	20	87	98	2	[86]
	MTO/85% H_2O_2/NaY	24 h	20	≥99	≥99	1	[87]
	DMD	0.5 h	0	92	≥99	1	[86]
	TFD	<1 min	$-20 \rightarrow 0$	≥98	≥99	1	[85]
Et_3SiH	MTO/85% H_2O_2	24 h	20	54	54	46	[86]
	MTO/UHP	24 h	20	78	≥99	1	[86]
	MTO/85% H_2O_2/NaY	24 h	20	86	99	1	[87]
	TFD	<1 min	$-20 \rightarrow 0$	≥98	≥98	1	[85]
(+)-(α-Np) PhMeSiH	MTO/85% H_2O_2	14 h	20	2	90 (1)	10	[86]
	MTO/UHP	14 h	20	12	96 (91)	4	[86]
	DMD	18 min	$-20 \rightarrow 0$	≥98	≥98 (97)	2	[85]
	TFD	<1 min	$-20 \rightarrow 0$	≥98	≥98 (97)	2	[85]

[a] The ee values of the silanol (+)-(α-Np)PhMeSiOH are given in parentheses.

to DMD is that the condensation of the silanol to the corresponding disiloxanes may take place [86] (cf. the epoxidation of olefins with MTO/H_2O_2, where the epoxides subsequently undergo hydrolysis and cleavage and rearrangement reactions [29a]). Thus, with MTO/85% H_2O_2 a significant amount of condensation of the silanol products occurs, whereas with MTO/UHP and MTO/85% H_2O_2 in the NaY zeolite such secondary reactions are only marginal and excellent yields of the silanols are obtained. A further drawback of the MTO/85% H_2O_2 oxidant is the significant decomposition of the MTO catalyst and the oxygen source H_2O_2 within only 1–2 h, while the MTO/UHP catalyst persists for much longer times. Although MTO/85% H_2O_2 leads to lower conversions of the silane than MTO/UHP, oxidation kinetics with the silane $PhMe_2SiH$ showed that the MTO/85% H_2O_2 oxidant is initially catalytically more active than MTO/UHP, but that the longer lifetime of the MTO/UHP oxidant ensures higher conversions of the silane.

For the silanes listed in Table 8 a drastic improvement of the conversion is observed when MTO/85% H_2O_2 is replaced by MTO/UHP or MTO/85% H_2O_2/NaY; moreover, the silanol is the sole product even for those silanes which are prone to condensation to the disiloxanes (e.g. $EtMe_2SiH$, *n*-$PrMe_2SiH$, $PhMe_2SiH$, and Et_3SiH). A further difference in the behaviour of MTO/85%

H_2O_2 vs MTO/UHP and MTO/85% H_2O_2/NaY manifests itself in the oxidation of the optically active silane (+)-(α-Np)PhMeSiH. The latter is oxidized with retention of configuration by MTO/UHP, while complete racemisation occurs with MTO/85% H_2O_2. Unfortunately, under the reaction conditions employed in Table 8 only modest conversion to the silanol (+)-(α-Np)PhMeSiOH is observed; however, the use of increased amounts of catalyst and UHP leads almost quantitatively to the desired silanol in excellent enantiomeric purity (ee 94%) [86].

Despite the virtues of MTO/UHP and MTO/85% H_2O_2/NaY as catalytic oxidants and their high silanol selectivity, their reactivity lags far behind that of the dioxiranes DMD and TFD. These dioxiranes, which are used stoichiometrically, also yield silanols in excellent yields and selectivity, albeit in a matter of only minutes and at sub-ambient temperatures. For example, the optically active silane (+)-(α-Np)PhMeSiH is quantitatively oxidized at −20 °C to the silanol (+)-(α-Np)PhMeSiOH with both dioxiranes within 18 min and with complete retention of configuration.

MTO/UHP and DMD have been employed in the selective oxidation of metallo-silanes $(L_nM)_x$-$Si(H)_y(R)_z$ to the corresponding silanols $(L_nM)_xSi(OH)_y(R)_z$ with M = Cr, Mo, W, Fe, and Ru [88]. A notable example is given in Fig. 4, in which a tungsten-silanetriol has been prepared in excellent yield from the corresponding tungsten-silane by oxidation with MTO/UHP and DMD [88e,g].

3.4
Other Oxidation Types

Among other reaction types which may be performed with the MTO oxidants and dioxiranes is the oxidation of furans to enediones [89], the oxidation of $Cp^*Re(CO)_3$ to the corresponding rhenium trioxide [31, 90], and the oxidative transformation of phosphines and halides [15, 31, 91]. Also, the catalytic oxidation of arsines and stibines by MTO has been reported [31]. In addition, MTO catalyzes the Baeyer-Villiger oxidation of some selected ketones [31] and the Dakin reaction of substituted benzaldehydes [92].

For DMD, the following oxidations have additionally been reported: aldehydes to the corresponding acids (with minor amounts of the formate in some cases) [15, 93], diazo compounds and phosphorus yields to the

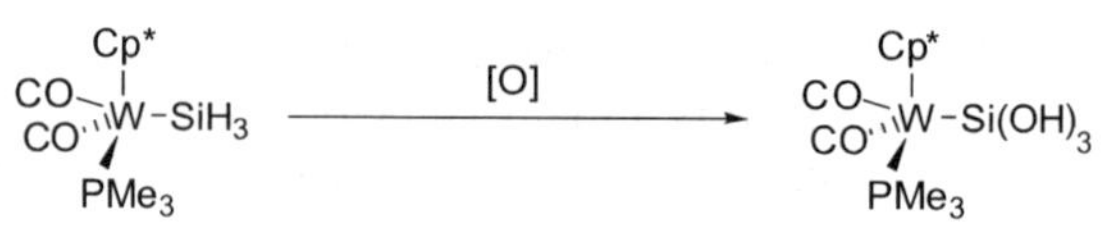

MTO/UHP	94%
DMD	71%

Fig. 4. Oxidation of tungsten silanes (see [88g])

corresponding α-oxo products [94], hydrazines to hydrazine *N*-oxides [95], isocyanates to nitro compounds [96], ketoximes and nitronate anions to their parent ketones [97], and selenophenes to their 1,1-dioxides [98].

4 Mechanistic Considerations

In oxygen-transfer reactions with MTO/H_2O_2, an oxygen atom is transferred from the oxygen source H_2O_2 onto the substrate 'S' by means of a three-membered-ring rhenium-peroxo intermediate (Fig. 5). Two such intermediates are formed during the catalytic cycle, namely the monoperoxo complex **E** and the diperoxo complex **F**, and both are reported to show oxidation activity [13]. Despite the geometrical similarity of DMD and the rhenium peroxo complexes, i.e., the three-membered peroxide ring, the mechanism of the peroxo-bond formation is different. In the case of DMD, the Caroate adds to the carbonyl group of the ketone and subsequently hydrogen sulfate is eliminated, during which the peroxide bond of the Caroate is broken and a new peroxide bond is formed by nucleophilic substitution [27]. In contrast, for the rhenium peroxo complexes, the H_2O_2 nucleophilically adds to one of the oxo-metal bonds to generate a hydroxy-hydroperoxy intermediate, which subsequently eliminates water to the monoperoxo complex **E** [99]. The diperoxo complex **F** is formed in an analogous manner. The initial peroxide bond of H_2O_2 is preserved during the generation of these peroxo complexes **E** and **F** [18].

While the in-situ-generated dioxirane persists even at room temperature for prolonged periods and therewith serves as an efficient oxygen-transfer reagent, appreciable decomposition of the rhenium peroxo complexes is observed under these conditions. The products of this decomposition include methanol

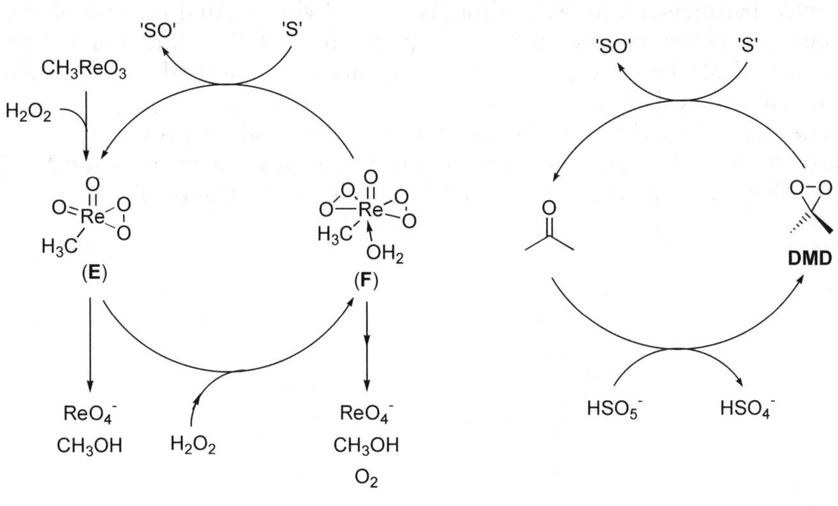

Fig. 5. Catalytic cycles of the oxidations by MTO/H_2O_2 and DMD

and perrhenate [86], whose high acidity causes epoxide-ring opening; additionally, molecular oxygen is liberated from the diperoxo complex **F** [100]. These features diminish the efficiency of MTO as a catalyst for oxygen transfer.

4.1
π-Bond Oxidations

For both MTO/H_2O_2 and DMD, much experimental data has been accumulated for the elucidation of the transition structure of the oxygen transfer in epoxidations. From the reactivity order (cf. Fig. 3) it is apparent that *cis* olefins react faster than *trans* ones, with the *cis/trans* pair of cyclooctene as the only exception, for which the *trans* reacts ca. 100 times faster than the *cis* isomer. While this effect is quite pronounced for DMD ($k_{cis}/k_{trans} \leq 8$ [30a]), it is only marginal for MTO/H_2O_2 (k_{cis}/k_{trans} = 1.3–2.5 [13]). The MTO/H_2O_2 data parallel those for the epoxidation by peracids. Consequently, in view of the greater *cis* reactivity, a *spiro* geometry with partial charge separation is proposed for the DMD transition structure (Fig. 6), which was confirmed by computational studies [101]. An analogous geometry has also been postulated for the MTO/H_2O_2 oxidant, despite the small difference in the epoxidation rates for *cis* and *trans* olefins [13]. While for DMD there is only one possible *spiro* arrangement (structure **G** in Fig. 6), for the peroxo complex of MTO the transition structure with the lowest activation energy and reaction enthalpy comprises an attack of the olefin on the peroxo oxygen opposite to the methyl group and on a S_N2 trajectory with the olefin plane parallel to the oxo ligand (structure **H**) [19]. The fact that the oxygen-transfer process is stereoselective with both oxidants, i.e., *cis* olefins give *cis* epoxides and *trans* olefins the corresponding *trans* epoxides, speaks for a concerted mechanism rather than a diradical one. Although a radical mechanism has been argued for DMD [102], recent experimental data with ultra-fast radical clocks [103] dispose of such claims. Moreover, calculations of the population of the antibonding O—O orbital indicate that the peroxo bond is, although elongated, still a doubly occupied σ-bond [101].

The use of the chiral allylic alcohol with allylic strain provided detailed information on the transition structure for the epoxidation by DMD and MTO/H_2O_2 (Fig. 7) [35, 104]. The 1,3-allylic strain conformationally aligns the

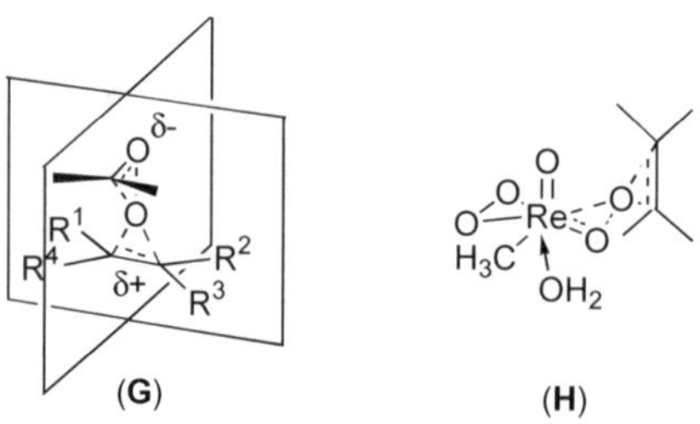

(G) (H)

Fig. 6. Transition structures **G** and **H** for the epoxidation with DMD and MTO/H_2O_2

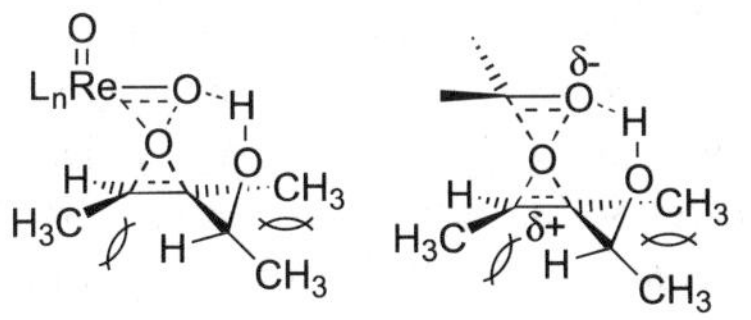

α ca. 130° dr = 80 : 20

α ca. 130° dr = 91 : 9

Fig. 7. Postulated transition states in the epoxidation of allylic alcohols by MTO/H_2O_2 and DMD with their dihedrals angle α and *threo/erythro* diastereoselectivities (see [104])

hydroxy group towards the *threo* face of the chiral allylic alcohol and hydrogen bonding directs both oxidants to this face of the π bond. The hydrogen bonding not only controls the observed *threo* selectivity of the epoxidation, but also enhances the epoxidation rate by compensation of charge separation in the transition state [30a].

The initial step in the oxidation of arenes to their epoxides is similar to that of simple π bonds for both reagents. While for MTO/H_2O_2 the epoxides [40b,c] are postulated on the basis of subsequent products, they may be isolated in the case of DMD (cf. Scheme 2).

4.2 Heteroatom Oxidations

Unlike the epoxidation, only little mechanistic data is available on the oxidation of heteroatoms. For DMD [49a], as well as for MTO/H_2O_2 [49b, 105], reaction rates strongly depend on the electronic nature of the substrate. For example, sulfides are more reactive than the corresponding sulfoxides, such that the oxidation of sulfides may be stopped at the sulfoxide stage by applying a 1:1 stoichiometry of oxidant and substrate. The experimental data for both oxidants are consistent with a nucleophilic attack of the heteroatom lone pair at one of the electrophilic peroxide oxygens along a typical S_N2 trajectory (Fig. 8). Only for substrates with very low oxidation potentials ($E_{ox} \leq 1.60$ eV) was a competing electron-transfer process observed [106].

Unlike for DMD or MTO/H_2O_2, for TDF the sulfone is found as the main oxidation product even when sub-stoichiometric amounts of the oxidant were employed. On the basis of ^{18}O-labeling experiments, a biphilic insertion of the

Nu = R_2S, R_3N, R_3P

Fig. 8. S_N2 transition states for the oxidation of heteroatoms by MTO/H_2O_2 and DMD

sulfide into the peroxide bond of the dioxirane to form a hypervalent sulfurane has been suggested [107a]. However, these results are contrary to those for the oxidation of thianthrene 5-oxide (**SSO**) by TFD, in which mainly the disulfoxides *cis*- and *trans*-**SOSO** have been observed [107b].

The S_N2-type mechanism was established for the oxidation of nitrogen heteroarenes by DMD [60]. The rate constants for the DMD oxidation correlated linearly with those for the alkylation with CH_3I [60], which is a well established, genuine S_N2 process [108].

4.3 σ-Bond Oxidations

Although MTO/H_2O_2 is known to catalyze the oxidation of σ bonds, no mechanistic work has been done on simple alkanes. In contrast, for dioxiranes an oxenoid mechanism has been proposed for the oxidation of the C—H bond (Fig. 9), as early as this unusual reaction was discovered [14c]. This mechanism was experimentally confirmed by the oxidation of cyclopropylcarbinyl radical clocks, for which no products derived from cyclopropane-ring opening were detected [109]. Further definitive evidence against radical intermediates was provided by the oxidation of the enantiomerically enriched (*R*)-2-phenylbutane, which was oxidized with 100% retention by DMD; had the benzyl radical intervened, then racemization should have taken place within picoseconds [103]. The experimental results are backed up by high-level computational investigations [110], which substantiate a concerted oxenoid mechanism for the C—H insertion. All these experimental and theoretical facts unequivocally reject a possible radical-type mechanism [111].

A special case of C—H insertion is the oxidation of alcohols, which is easier to achieve. Two different reaction pathways were detected for the MTO/H_2O_2 oxidation of ^{18}O-labelled alcohols to the corresponding carbonyl products (Fig. 10), of which the minor one involves a concerted oxenoid mechanism, i.e., direct oxygen insertion into the C—H bond to form a *gem* diol. This pathway was confirmed by the detection of labeled water as by-product [49b].

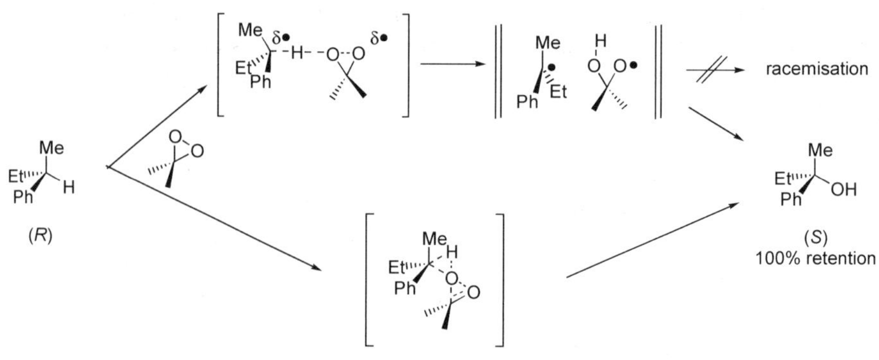

Fig. 9. Mechanism of the C—H insertion of optically active 2-phenylbutane with DMD

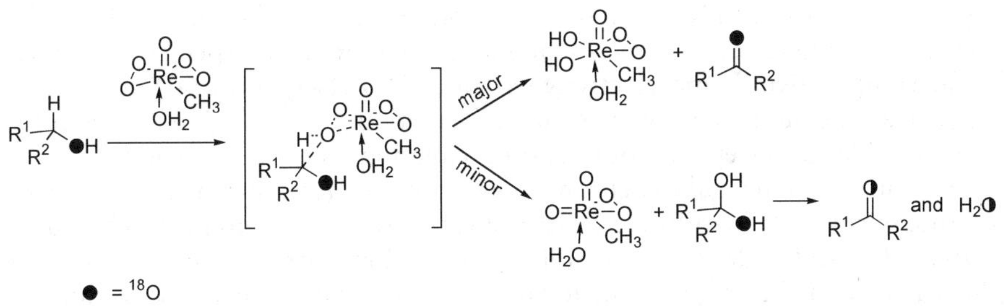

Fig. 10. Dual mechanism of the oxidation of ^{18}O-labeled alcohols by MTO/H_2O_2

Due to the weaker Si—H bond, the insertion is much easier than into C—H bonds. Of mechanistic interest is the oxidation of the chiral (*R*)-(α-Np)PhMeSiH, which was found to proceed with 100% retention for both MTO/UHP and DMD as oxidants [85, 86]. Therefore, a concerted oxenoid mechanism was postulated, with a transition state similar to the one for the oxidation of C—H bonds (Fig. 11).

5
Selectivities

Due to their geometrical similarity, dioxiranes and MTO/H_2O_2 often show similar selectivities in oxygen-transfer reactions, except for some special cases. Although any such differences between the two reagents may be of preparative interest, pronounced similarities in the chemo-, regio-, and diastereoselectivities have so far been useful to elucidate the reaction mechanisms and electronic nature of both oxidants. A comparison between the two systems in terms of enantioselectivity is, however, not possible, as to date no enantioselective oxygen-transfer has been reported for the rhenium-based oxidant, except one example in a patent (cf. Sect. 5.4).

5.1
Chemoselectivity

The peroxo-rhenium complexes and dioxiranes show very similar chemoselectivity in their oxidation reactions. For both the following reactivity order

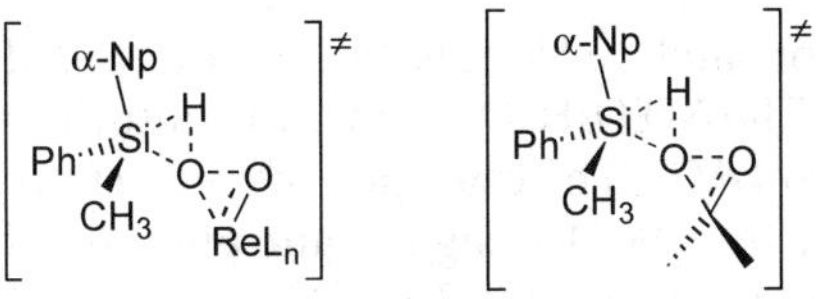

Fig. 11. Proposed transition structures in the oxidation of (+)-(α-Np)PhMeSiH by MTO/UHP and DMD

applies: heteroatom (S,N) oxidation >double-bond epoxidation >C—H insertion [35]. Nonetheless, substrate-dependent differences are expected, as shown by the comparative chemoselectivity data for MTO/H_2O_2 and DMD in Table 9, which features sulfoxidation vs sulfone formation, sulfoxidation vs epoxidation, *N* oxidation vs epoxidation, and epoxidation vs C—H insertion.

For both oxidants, sulfoxidation prevails over sulfone formation, as shown for thioanisol and thianthrene 5-oxide (Table 9). The pronounced chemoselectivity for sulfoxidation over sulfone formation corroborates the electrophilic nature of the oxygen-transferring species. A quantitative assessment of the degree of electrophilicity has been made possible with the mechanistic probe thianthrene 5-oxide (**SSO**), which was developed for this purpose [119]. Thus, **SSO** is oxidized at the sulfide site by electrophilic oxidants to give the *bis*-sulfoxides *cis*- and *trans*-**SOSO** and at the sulfoxide site by nucleophilic ones to give the sulfone **SSO$_2$**. With both the MTO-based oxidants and the dioxiranes, the formation of *cis*- and *trans*-**SOSO** dominates over that of **SSO$_2$** and, thus, the electrophilic character of all these oxidants is unequivocally established. Although sulfoxides are exclusively obtained when only one equivalent of the oxygen source is used, the oxidation may be driven further towards the corresponding sulfones with an excess of dioxirane or H_2O_2 (for MTO). Other data disclose that sulfoxidation is also preferred over epoxidation (Table 9) [114, 115], *N* oxidation [23b, 35], arene oxidation (cf. Table 9), and C—H insertion [105].

In the case of *N* oxidation vs epoxidation, the preference is strongly substrate-dependent. Thus, for the two examples 2,2,5,5-tetramethyl-4-pyrroline-3-carboxamide and 4-(4′-cyclohexenyl)pyridine in Table 9, DMD oxidises the amine functionality and not the C=C bond. In contrast, MTO/H_2O_2 transforms the pyrroline substrate at the ring nitrogen atom to the corresponding nitroxide (double bond is deactivated), while the pyridine derivative is epoxidized. Other examples for which *N* oxidation and not epoxidation takes place are 4-vinylpyridine and 1,2-bis-(2-pyridyl)ethylene [59]. In both cases the double bond is deactivated and, therefore, MTO preferentially catalyzes the *N* oxidation rather than the usual epoxidation.

DMD epoxidation of olefins functionalized with amino groups is possible, if the amino functionality bears an electron-withdrawing group [120] or a protecting group [116]. Protection in the form of a boron-trifluoride adduct prior to the oxidation step with DMD or TFD is particularly attractive, because the free amino functionality of the epoxy product is readily recovered by treatment with dilute $KHCO_3$ solution [116]. The preferential C—H insertion over *N* oxidation with DMD as oxidant may be conducted in an analogous manner [121].

As demonstrated for allylic alcohols (Table 9), epoxidations are preferred over C—H insertions. The MTO/H_2O_2 oxidant is notoriously sluggish for C—H insertions compared to DMD. For example, while MTO/H_2O_2 gives only traces of insertion products, for DMD a significant amount of enone is produced from *cis*- and *trans*-5-(*tert*-butyl)-2-cyclohexen-1-ol (Table 9), which is approximately doubled in the nonpolar 9:1 CCl_4/acetone solvent mixture vs 9:1 MeOH/acetone [118]. Compared to such cyclic substrates, for acyclic allylic

Table 9. Chemoselectivity

Substrate	Product distribution (%)[a]				Refs.
		Sulfoxidation vs sulfone formation			
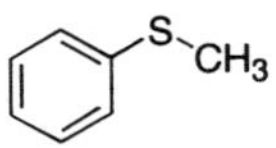		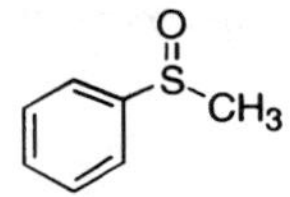	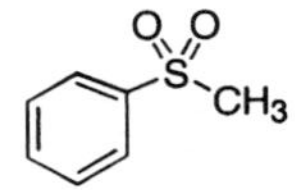		
	MTO/UHP	≥95	: 5		[112]
	MTO/H_2O_2	≥95	: 5		[105]
	DMD	≥95	: 5		[49a]
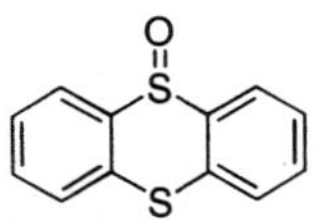		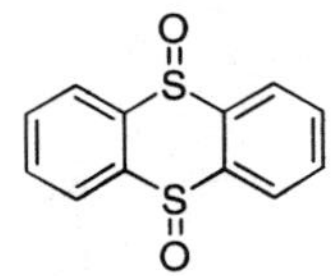	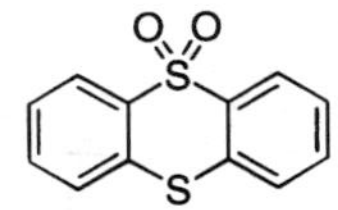	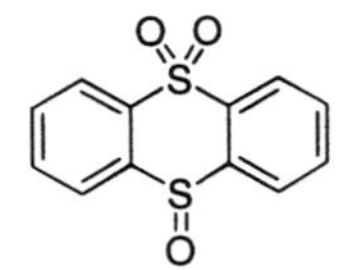	
SSO		**SOSO** (*cis/trans*)	**SSO_2**	**$SOSO_2$**	
	MTO/UHP[b]	93	: 5	: 2	[112]
	MTO/H_2O_2[b]	92	: 3	: 5	[112]
	DMD[c]	88	: 12	: –	[113]
	TFD[d]	89	: 3	: 8	[107]
		sulfoxidation vs epoxidation			
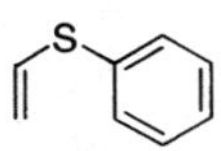		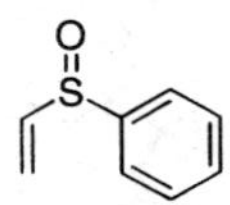	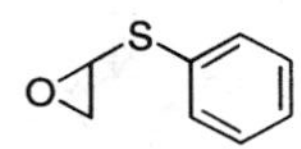		
	MTO/H_2O_2	≥95	: 5		[114]
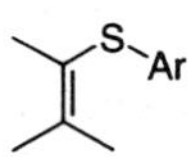		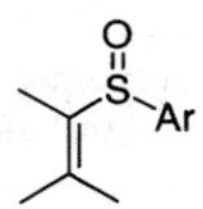	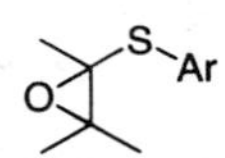		
	DMD	≥95	: 5		[115]

Table 9. (*Contd.*)

Substrate		Product distribution (%)[a]				Refs.
		Sulfoxidation vs sulfone formation				
		N-oxidation vs epoxidation				
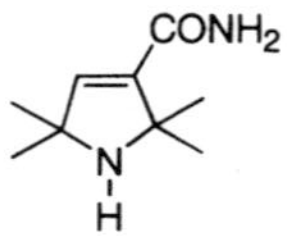		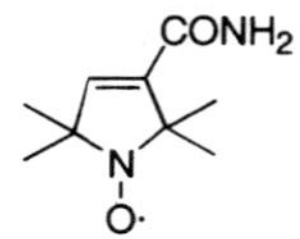		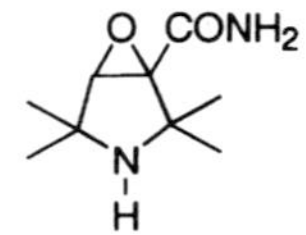		
	MTO/H_2O_2	≥95	:	5		[53]
	DMD	≥95	:	5		[58]
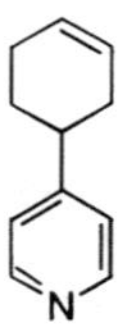		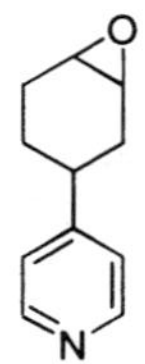		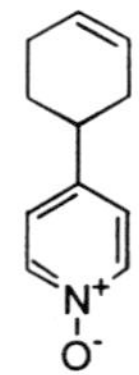		
	MTO/H_2O_2	≥95	:	5		[59]
	DMD	≥5	:	95		[116]
		epoxidation vs C—H insertion				
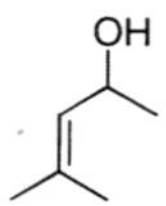		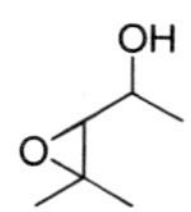		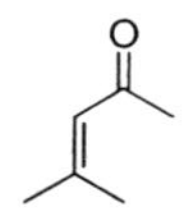		
		(*threo*/*erythro*)				
	MTO/UHP	≥95 (83:17)	:	5	$CDCl_3$	[104]
	$MTO/H_2O_2^e$	≥95 (81:19)	:	5	CH_2Cl_2	[104]
	DMD	≥95 (82:18)	:	5	CCl_4/acetone (9:1)	[117]
	TFD	≥95 (82:18)	:	5	MeOH/acetone (9:1)	[30b]

Substrate	Oxidant	Epoxide		Enone	Solvent	Ref.
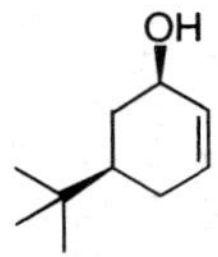		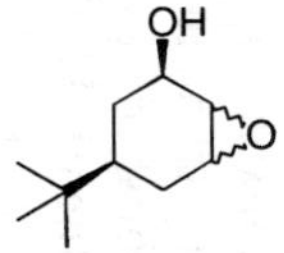		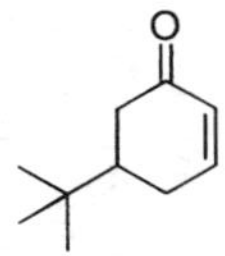		
		(*syn/anti*)				
	MTO/UHP	≥95 (84:16)	:	5	$CDCl_3$	[36]
	$MTO/H_2O_2^e$	91 (80:20)	:	9	$CDCl_3$	[36]
	MTO/UHP	≥95 (50:50)	:	5	CD_3OD	[36]
	DMD	57 (82:18)	:	43	CCl_4/acetone (9:1)	[118]
	DMD	80 (38:62)	:	20	MeOH/acetone (9:1)	[118]
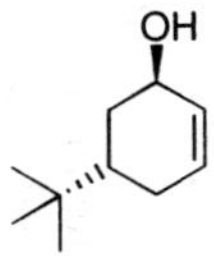		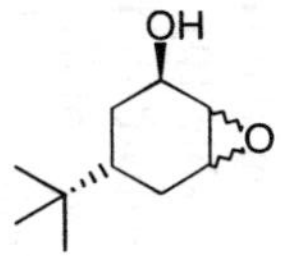		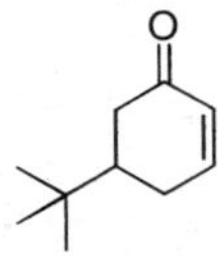		
		(*syn/anti*)				
	MTO/UHP	≥95 (72:28)	:	5	$CDCl_3$	[36]
	$MTO/H_2O_2^e$	83 (72:28)	:	17	$CDCl_3$	[36]
	MTO/UHP	≥95 (15:85)	:	5	CD_3OD	[36]
	DMD	67 (58:42)	:	33	CCl_4/acetone (9:1)	[118]
	DMD	85 (29:71)	:	15	MeOH/acetone (9:1)	[118]

[a] Diastereoselectivities are given in parentheses.
[b] Conducted at ca. 20 °C; X_{SO} = sulfoxide oxidation/total oxidation; calculated according to [113]; $X_{SO} = 0.07$.
[c] T = 0 °C; $X_{SO} = 0.13$.
[d] T = 0 °C; $X_{SO} = 0.10$.
[e] In the presence of 12 mol% pyridine.

alcohols only little enone formation is observed [37]. In the case of cyclic vinyl silanes without an allylic hydroxy group, C—H insertion may also occur; however, the amount of enone formed is significantly less than for the corresponding allylic alcohols due to the lack of assistance by the allylic OH group for oxygen insertion [35, 122]. A comparison of DMD and the more reactive TFD shows that, in contrast to DMD, with TFD both cyclic and acyclic allylic alcohols are chemoselectively oxidized to the corresponding epoxides with no C—H insertion taking place [30b]. Clearly, the scarce data in Table 9 emphasizes the need for more intensive work on the chemoselectivity of oxidants such as DMD and MTO/H_2O_2.

5.2 Regioselectivity

Several examples of regioselectivity in oxidations with the MTO-based oxidants and dioxiranes have been reported. The reaction types include epoxidations, sulfoxidations, *N* oxidations and C-H insertions. Notable examples for regioselective epoxidations are given in Table 10. As expected for electrophilic oxidants (cf. the oxidation of **SSO** in Sect. 4.2), the site of higher nucleophilicity is generally attacked. Thus, in the case of limonene, the more nucleophilic trisubstituted endocyclic double bond is epoxidized more readily than the terminal disubstituted one. For MTO/UHP (95:5) this regiochemical differentiation is slightly higher than for DMD (80:20). As expected, this regioselectivity is switched when the electron-withdrawing carbonyl group is placed in conjugation with the more highly substituted double bond, as demonstrated for carvone (entry 2).

The effect of allylic hydroxy groups on the regioselectivity of epoxidation reactions is illustrated for geraniol (Table 10). This difunctional olefin, in which both double bonds are trisubstituted, has been established as regiochemical probe for assessing the type of association between the oxidant and the allylic alcohol of this substrate, i.e., hydrogen bonding vs metal-alcoholate binding [126]. For the MTO-based oxidants, the unfunctionalized double bond rather than the allylic one is preferentially epoxidized. The ratio of the two regioisomeric epoxides depends only nominally on the solvent, since the shift in the regioselectivity towards the remote double bond is small when the reaction medium is changed from $CDCl_3$ to the protic CD_3OD. This shift in regioselectivity for CD_3OD is due to preferred hydrogen bonding between the substrate and solvent rather than between substrate and oxidant and, thus, the amount of the alkene epoxide is enhanced. No such solvent effect would be expected had metal-alcoholate binding between the MTO catalyst and the allylic alcohol been operating, and the epoxide of the allylic alcohol functionality would have been formed exclusively [127].

A similar, but more pronounced trend is observed for the dioxiranes DMD and TFD as oxidants. In the nonprotic reaction medium the regioselectivity for DMD is 49:51 for geraniol, while in the protic medium a substantial shift to 12:88 in favor of the remote double bond is found [30b]. This effect is smaller

Table 10. Regioselectivity

Substrate	Oxidant	Solvent	Product distribution (%)[a]			Refs.
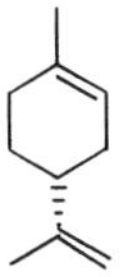			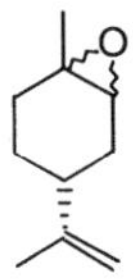		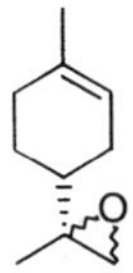	
	MTO/UHP	CH_2Cl_2	≥95	:	5	[123]
	DMD	acetone	80 (33 : 67)	:	20 (50:50)	[124]
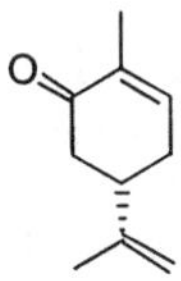			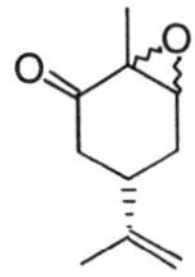		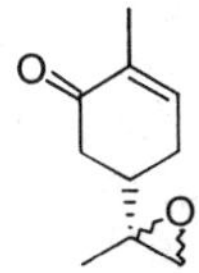	
	MTO/H_2O_2[b]	CH_2Cl_2	≤5	:	95 (50:50)	[125]
	DMD	acetone	≤5	:	95 (45:55)	[124]
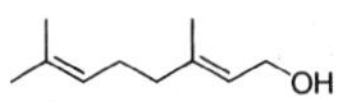			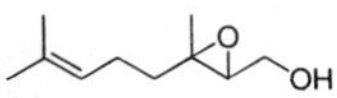		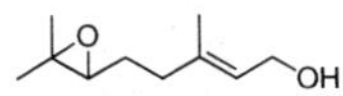	
	MTO/UHP	$CDCl_3$	22	:	78	[104]
	MTO/UHP	CD_3OD	14	:	86	[104]
	MTO/H_2O_2[c]	$CDCl_3$	18	:	82	[104]
	DMD	CCl_4/acetone (9:1)	49	:	51	[30b]
	DMD	MeOH/acetone (9:1)	12	:	88	[30b]
	TFD	CCl_4/TFA (9:1)	43	:	57	[30b]
	TFD	MeOH/TFA (9:1)	27	:	73	[30b]

[a] Diastereoselectivities are given in parentheses.
[b] In the presence of 1.2 mol% bipyridine *N,N'*-dioxide.
[c] In the presence of 12 mol% pyridine.

for TFD than for DMD, which has been ascribed to the greater reactivity of TFD and, hence, its lower sensitivity towards the relative nucleophilicity of the two double bonds. When the methyl or silyl ether derivatives of geraniol were epoxidised by these dioxiranes (data not shown), no solvent effect was found and the ratio of regioisomers is essentially that of geraniol in the protic medium MeOH/acetone (9:1) [30b]. These results, along with the solvent effect observed in the epoxidation of geraniol, provide evidence for intramolecular hydrogen bonding between the allylic hydroxy group and the oxidant in the transition state.

Similar regioselectivity is obtained in the epoxidation of linalool (Scheme 8) with MTO/H_2O_2/bipyridine, for which the unfunctionalised double bond is also preferred over the allylic one. However, for this substrate the major

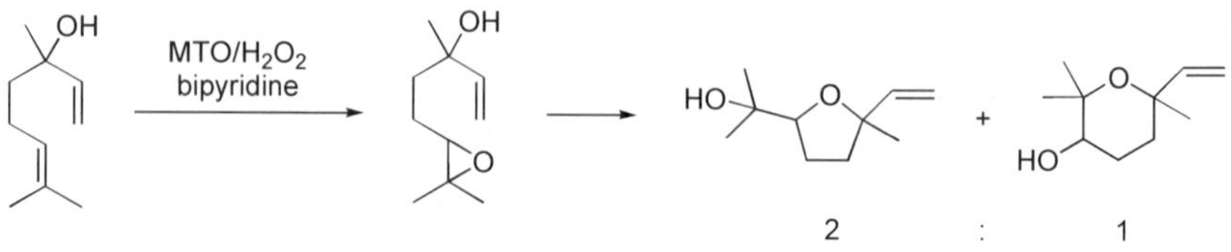

Scheme 8. Oxidation of linalool with MTO/H_2O_2/bipyridine

epoxide could only be isolated in low yield (10%), because it rearranges to a mixture of hydroxyalkyl-substituted oxolanes and hydroxyoxanes under the reaction conditions [128]. Other examples of regioselectivity with dioxiranes include the epoxidation of fluorinated monoterpenes and sesquiterpenes [129], the oxidation of Vitamin D_3 derivatives [78c], and the epoxidation of steroidal 1,4-diene-3-ones [130]. Substrates with similarly reactive double bonds, which would normally be oxidised unselectively, may be regioselectively epoxidised by employing the protection/deprotection strategy [131].

Regioselectivity is also manifested in nitrogen and arene oxidations, C-H insertions, and sulfoxidations. In *N* oxidations, MTO/H_2O_2 preferably attacks the secondary alkyl-substituted amine functionality over the aromatic secondary one in 1,2,3,4-tetrahydro-9*H*-pyrido[3,4-*b*]indole (Scheme 9) [55]. The oxidation of unsymmetrical secondary amines may also give a mixture of regioisomeric nitrones depending on the substituent pattern; however, this has only been observed for *N*-isopropylbenzylamine and *N*-ethylbenzylamine (Scheme 10) [55].

For DMD, the regioselectivity in *N* oxidations has also been thoroughly investigated, as illustrated in Scheme 11 [60, 70, 132]. Due to steric reasons, the *ortho* isomer of dimethylaminopyridine is preferentially oxidized at the aliphatic nitrogen, while in the *para* isomer the more nucleophilic pyridine nitrogen atom is attacked.

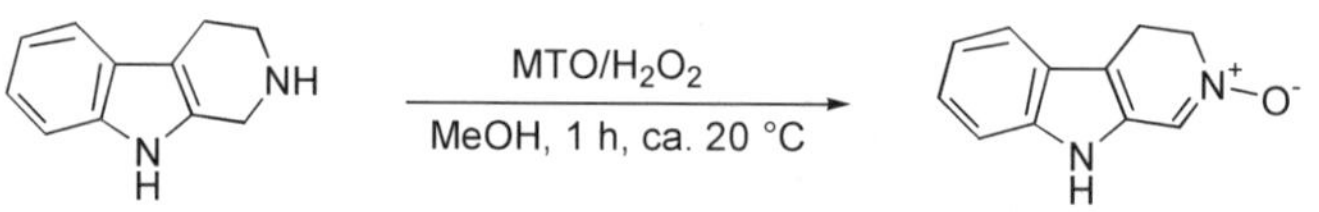

Scheme 9. MTO-catalysed *N* oxidation of 1,2,3,4-tetrahydro-9*H*-pyrido[3,4-*b*]indole

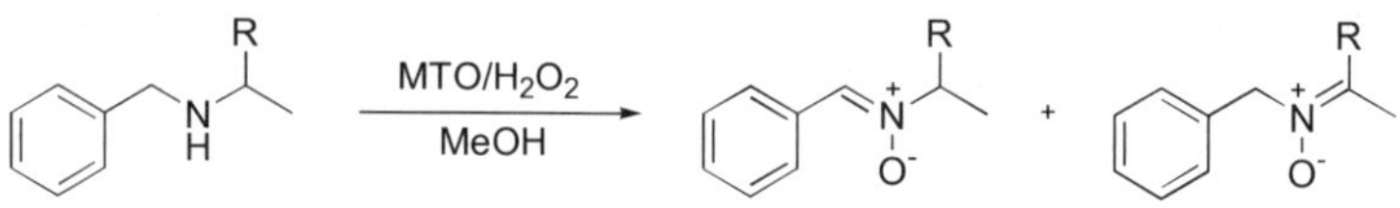

R = H	54	:	46
R = CH_3	87	:	13

Scheme 10. Regioselective *N* oxidation of unsymmetrical secondary amines with MTO/H_2O_2

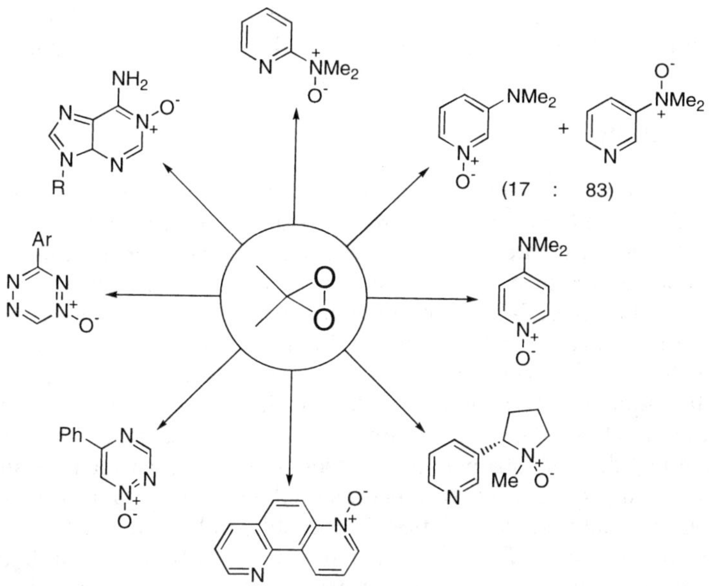

Scheme 11. Regioselective *N* oxidations with DMD

For the *meta* isomer both *N*-oxides are obtained, as expected the dimethylamino oxidized product dominates [133]. Similarly, for nicotine, the oxidation of the nonaromatic nitrogen atom is favored [60]. That 1,7-phenanthroline is selectively oxidized at the N7-position to give the monoxide 1,7-phenanthroline-7-oxide in the reaction with DMD is due to the steric hindrance by the *peri* hydrogen at the C10-position [60]. Triazine and tetrazine *N*-oxides may also be directly, efficiently and in many cases selectively prepared from the corresponding aromatic amines with dioxiranes [132]. In the case of 1,2,4-triazine *N*-oxides, this represents a major improvement over alternative synthetic routes, which often comprise a number of steps or lead to other oxidation products [134]. Similar problems arise for tetrazine *N*-oxides; however, these may be synthesized quantitatively and selectively from the corresponding tetrazines with TFD [132a]. DMD shows the same selectivity in the *N* oxidation of tetrazines, albeit at a drastically reduced reactivity [132a]. The adenosine derivative is oxidized regioselectively in 62% yield to the *N*-oxide depicted in Scheme 11; unfortunately, higher yields cannot be reached, as the *N*-oxide undergoes deoxygenation to the amine and singlet oxygen under the reaction conditions [70].

For C—H insertions, excellent regioselectivity has been reported in dioxirane reactions with steroids [78, 135] and in the oxidation of diols to the corresponding hydroxy ketones [82b, d, 83, 136]. In arene oxidations, the transformation of 2-methylnaphthalene to the regioisomeric 1,4- and 5,8-quinones in ratios of up to 91:9 by MTO/H_2O_2/AcOH deserves special mention in view of the industrial importance of the 1,4 isomer, Vitamin K_3, a widely used animal feed additive [40a, 137]. The sulfoxidation by DMD of a disulfide with an adjacent carbonyl group takes place at the β-sulfur (with respect to the carbonyl group) instead of at the α-sulfide functionality [138].

5.3 Diastereoselectivity

The most prominent example of diastereoselectivity in oxidations with MTO-based oxidants and dioxiranes is the epoxidation of chiral allylic alcohols. Such substrates constitute a useful stereochemical probe for the elucidation of transition structures in oxygen-transfer processes. For this purpose, we have employed a set of methyl- and *tert*-butyl-substituted chiral allylic alcohols with various stoichiometric and catalytic oxidants. In the epoxidation of these allylic alcohols, hydroxy-group directivity operates such that one of the two π faces of the double bond is preferentially attacked to favor one of the product diastereomers [126]. The stereodifferentiation by means of the hydroxy-directing effect depends strongly on the type of bonding between the substrate and the oxidizing species: When metal-alcoholate binding applies (vanadium- and titanium-catalyzed epoxidations), the dihedral angle (C=C—C—O) is generally in the range of 40–90° (Structure **I**, Fig. 12), while for hydrogen-bonding association between the substrate and the oxidant (peracid epoxidations) the angle is between 110-140° (Structure **J**) [126]. In the cases in which an acute dihedral angle applies, 1,2-allylic (^{1,2}A) strain is important and the *erythro* product is favored; in contrast, 1,3-allylic (^{1,3}A) strain dominates for cases with obtuse dihedral angles and the *threo* product prevails. The methyl-substituted chiral allylic alcohols in Table 11 possess either 1,2- or 1,3-allylic strain, or both. Inspection of the diastereochemical ratios of the epoxides discloses that for the oxidants MTO/UHP, MTO/H_2O_2/pyridine, DMD, and TFD *threo* diastereoselectivity (^{1,3}A strain) is observed, except with the first substrate (^{1,2}A strain) and the last substrate (no hydroxy group). This indicates that the electronic nature of the interaction between oxidant and allylic alcohol is one of hydrogen bonding with the dihedral angle in the range of 110–140° (Structures **K** and **L**, Fig. 12). Further evidence comes from the drop in diastereoselectivity, when methanol is used as solvent, which breaks up this intermolecular association (Table 11). Also, the pronounced difference in the diastereomeric ratios obtained for mesitylol and its acetate derivative

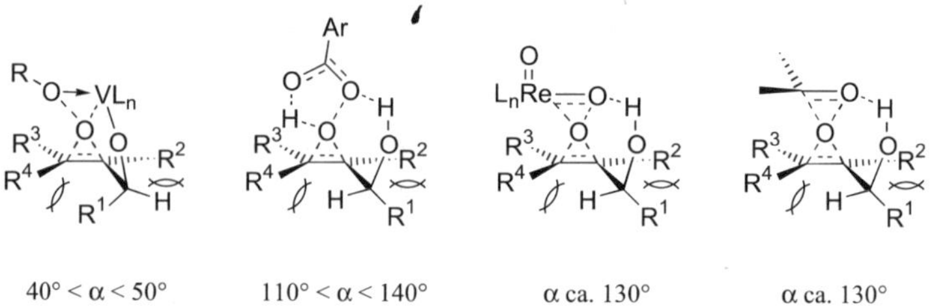

40° < α < 50° (**I**) | 110° < α < 140° (**J**) | α ca. 130° (**K**) | α ca. 130° (**L**)

Fig. 12. Postulated transition states in the epoxidation of allylic alcohols by VO(acac)$_2$/*t*-BuOOH, *m*-CPBA, MTO/H_2O_2, and DMD

Table 11. Diastereoselectivity of the MTO/H_2O_2 and dioxirane epoxidation of chiral allylic alcohols

Substrate	Oxidant	Solvent	dr[a] *threo:erythro*	Refs.
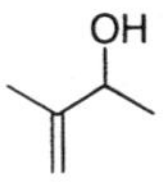	MTO/UHP	$CDCl_3$	50:50	[104]
	MTO/$H_2O_2^b$	CH_2Cl_2	41:59	[104]
	DMD	CCl_4/acetone (9:1)	70:30	[117]
	TFD	CCl_4/TFA (9:1)	62:38	[30b]
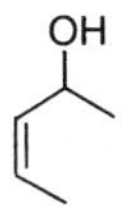	MTO/UHP	$CDCl_3$	82:18	[104]
	MTO/$H_2O_2^b$	CH_2Cl_2	86:14	[104]
	DMD	CCl_4/acetone (9:1)	85:15	[117]
	TFD	CCl_4/TFA (9:1)	88:12	[30b]
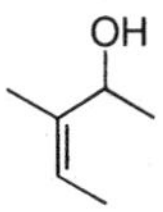	MTO/UHP	$CDCl_3$	91:09	[104]
	MTO/UHP	CD_3OD	39:61	[104]
	MTO/$H_2O_2^b$	CH_2Cl_2	80:20	[104]
	DMD	CCl_4/acetone (9:1)	91:09	[117]
	DMD	MeOH/acetone (9:1)	82:18	[117]
	TFD	CCl_4/TFA (9:1)	90:10	[30b]
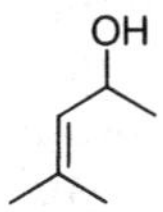	MTO/UHP	$CDCl_3$	83:17	[104]
	MTO/$H_2O_2^b$	CH_2Cl_2	81:19	[104]
	DMD	CCl_4/acetone (9:1)	82:18	[117]
	DMD	MeOH/acetone (9:1)	59:41	[117]
	TFD	CCl_4/TFA (9:1)	82:18	[30b]
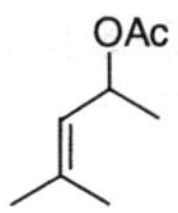	MTO/UHP	$CDCl_3$	34:66	[104]
	DMD	acetone	58:42	[117]

[a] Diastereomeric ratio.
[b] In the presence of 12 mol% pyridine.

(Table 11) substantiates the hydrogen-bonding mechanism. On the basis of these stereochemical data, the transition structures in Fig. 7 have been postulated for the epoxidation of allylic alcohols with the MTO-based oxidants and dioxiranes.

The dihedral angle α (C=C—C—O) in the allylic alcohol substrate dictates whether hydrogen bonding is involved in the transition state or not. Any significant deviation from the preferred angle suppresses the hydroxy-directing effect and steric effects become important. Such competition between electronic (formation of hydrogen bonds) and steric effects has been demonstrated for a set of cyclic allylic alcohols, whose α angle of the preferred ground-state conformation varies systematically from 120° to 199° with increasing ring size [36]. Thus, for 2-cyclopentenol (120°) and 2-cyclohexenol (139°), the MTO-based oxidants show good *cis* selectivity (controlled by hydrogen bonding), while for 2-cycloheptenol (180°) and 2-cyclooctenol (199°) *trans* selectivity (controlled by steric effects) is preferred on account of their larger dihedral angles. More reliable than these conformationally flexible cyclic allylic alcohols for defining the optimal dihedral angle α in the stereoselective oxygen-transfer process are conformationally fixed substrates, the diastereomeric 5-*tert*-butyl derivatives of cyclohexenol (Table 12). The dihedral angles of their ground-state conformations are 110° for the *trans*-configured substrate and 140° for the *cis* one (Fig. 13). The *cis* diastereoselectivity (hydrogen bonding) for the MTO-catalyzed epoxidations is slightly higher for the *cis*-configured cyclic allylic alcohol (Table 12) and a significant solvent effect was observed, in which the *trans* epoxide was enhanced in a protic solvent like methanol. From these data a dihedral angle α of ca. 130° was suggested for MTO/UHP and MTO/30% H_2O_2/pyridine [36]. For DMD, the difference in the *cis/trans* diastereoselectivities is more pronounced (Table 12) and in methanol the *trans* epoxide is favoured [36, 118]. Hydrogen bonding of the allylic alcohol with the protic solvent disrupts the internal hydrogen bonding (*cis* epoxide) and steric effects dominate such that the *trans* epoxide is preferred.

A quantitative measure of the purely steric effects in the epoxidations with MTO/UHP and DMD [36] has been achieved for a series of 3-alkyl-substituted cyclohexenes. For the Me, Et, *i*-Pr, and *t*-Bu derivatives the diastereomeric ratios (*cis/trans*) of the epoxides correlated well with the Taft steric substituent constants (Fig. 14). Thus, with increasing size of the alkyl substituent the *trans* diastereoselectivity increased, i.e., the epoxidation proceeds preferentially from the sterically less hindered side of the double bond. DMD was shown to be more sensitive to steric factors, a fact that is in accordance with other work for DMD on the epoxidation of methylenecyclohexanes [139]. This may be explained by the fact that the C—O bond length in the dioxirane (138.8 pm [14a]) is shorter by ca. 30% than the Re—O ones (190.4 pm and 191.8 pm [17])

Table 12. Diastereoselective epoxidation of conformationally fixed cyclic allylic alcohols

Substrate	Oxidant	Solvent	dr[a] *cis:trans*	Refs.
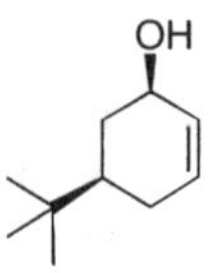	MTO/UHP	$CDCl_3$	84:16	[36]
	MTO/UHP	CD_3OD	50:50	[36]
	MTO/H_2O_2[b]	$CDCl_3$	80:20	[36]
	DMD	CCl_4/acetone (9:1)	82:18	[118]
	DMD	MeOH/acetone (9:1)	38:62	[118]
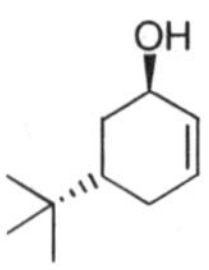	MTO/UHP	$CDCl_3$	72:28	[36]
	MTO/UHP	CD_3OD	15:85	[36]
	MTO/H_2O_2[b]	$CDCl_3$	72:28	[36]
	DMD	CCl_4/acetone (9:1)	58:42	[118]
	DMD	MeOH/acetone (9:1)	29:71	[118]

[a] Diastereoselectivity relative to the hydroxy group.
[b] In the presence of 12 mol% pyridine.

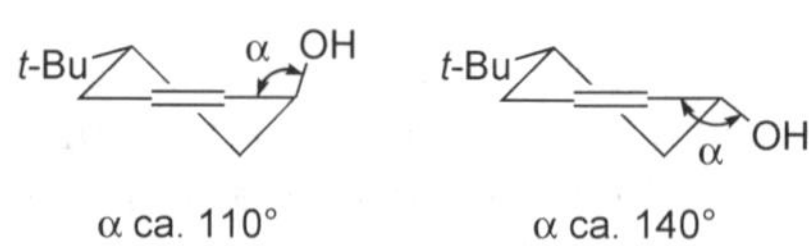

Fig. 13. Preferred ground-state conformations of conformationally fixed cyclohexenols

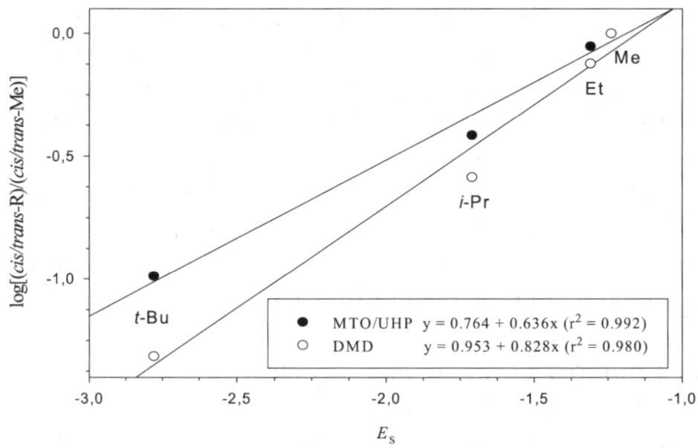

Fig. 14. Diastereoselectivity (*cis/trans*) in the epoxidation of 3-alkyl-substituted cyclohexenes vs Taft's steric substituent constant (E_S)

in its peroxo functionality. A comparison between the diastereomeric ratios of MTO/UHP for 3-methylcyclohexene (*cis/trans* = 49:51) and 2-cyclohexenol (*cis/trans* = 81:19) unequivocally reveals that the hydroxy-directing effect observed for the allylic alcohols is electronic (hydrogen bonding) and not steric in nature.

Other examples of diastereoselectivity in oxidations by MTO-based oxidants and dioxiranes include the epoxidation of steroids [125, 130], glycals [140], and Vitamins D_2 and D_3 and their derivatives [78c, 141], the α hydroxylation of enolates [142], and also the sulfoxidation of chiral ruthenium thioether complexes [143]. The latter opens up a novel stereoselective synthesis of optically active sulfoxides after decomplexation from the ruthenium chiral auxiliary. In the α hydroxylation of enolates derived from chiral ketones, the diastereoselectivity depends on the size of the enolate substituent and its strength of coordination to the enolate oxygen atom, with best results achieved with sterically demanding titanium enolates [142a]. Thus, the diastereoselectivity increases gradually for the small and only weakly coordinated sodium enolates (dr 67:33 in the case of the menthone enolate) to the sterically more demanding silyl derivatives (dr 76:24 for the menthone enolate), and is highest for the sterically most demanding titanium enolates (dr 96:4 for the menthone enolate). Asymmetric oxidation of enolates derived from prochiral ketones to their α-hydroxy derivatives has been achieved, when an optically active titanium fragment is utilized as enolate substituent; the best enantioselectivity (up to 63% ee) was observed for TADDOL as chiral ligand on the titanium atom [144].

5.4
Enantioselectivity

Little is known about enantioselective oxidations with MTO-based oxidants, although several attempts have been made to achieve enantioselectivity in

MTO-catalyzed epoxidations and sulfoxidations [128, 145]. Thus, for the epoxidation of *cis*-β-methylstyrene with MTO/H_2O_2 in the presence of the chiral additive (*S*)-1-(*N*,*N*-dimethylamino)-1-phenylethane, an ee value of 86% was claimed [145b]; however, this method is very limited with respect to the choice of substrate and additive. For *trans*-β-methylstyrene, styrene, and 1-octene the ee values were very small or even negligible (ee up to 13%) when a series of chiral amines was used as ligands. In the MTO-catalyzed sulfoxidation of thioanisole by H_2O_2 with a camphor-derived carboxylic acid as optically active ligand, the corresponding sulfoxide was obtained in 15% ee [145c].

By contrast, a number of highly enantioselective oxidations with dioxiranes have been reported lately. Early studies employed in-situ-generated chiral dioxiranes derived from simple optically active ketones (Fig. 15) with Caroate as oxygen donor, but enantiomeric excesses of only up to 20% were obtained in these epoxidations [146]. A significant improvement was achieved by using achiral dioxiranes in the presence of bovine serum albumin for sulfoxidations [147] or optically active Mn(III)salen catalysts for the epoxidation of chromenes and isoflavones, [148]. Enantiomeric excesses of up to 89% were found for the sulfoxidations and of up to 93% for the epoxidation of chromenes and isoflavones.

A major breakthrough in enantioselective oxidations with in-situ-generated dioxiranes constitutes the use of the chiral fructose-derived ketone with quasi C_2 symmetry depicted in Scheme 12 (first structure). With this ketone, a wide range or *trans*-disubstituted and trisubstituted olefins, which bear functional groups such as acetal, chloro, ester, or silyl ether, may be efficiently and catalytically epoxidized in almost perfect optical purity [149]; however, the ee values for *cis*- and terminally substituted olefins are much lower. For comparison, other recently used chiral ketones are given in Scheme 12, together with the ee values obtained in the epoxidation of *trans*-stilbene. These ketones do not match the efficacy of asymmetric induction that has been demonstrated for the fructose-derived ketone. The general drawback of all these oxidation systems is the fact that the optically active ketones are destroyed by oxidation and, thus, they do not operate truly catalytically.

Other transformations conducted with the fructose-derived ketone are the enantioselective oxidation of silyl enol ethers and *vic*-diols to the corresponding optically active α-hydroxy ketones [84, 154]. For the latter, *meso* diols have been enantiomerically enriched through asymmetrization and *rac* diols through kinetic resolution. Enantiomeric excesses of up to 82% were achieved

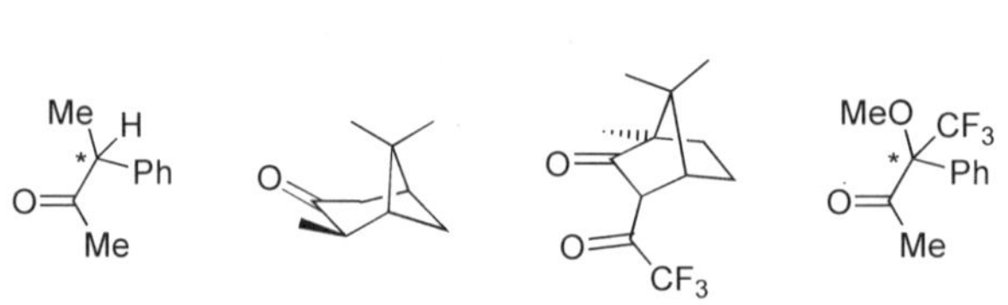

Fig. 15. Simple optically active ketones used in early studies for the in-situ generation of optically active dioxiranes

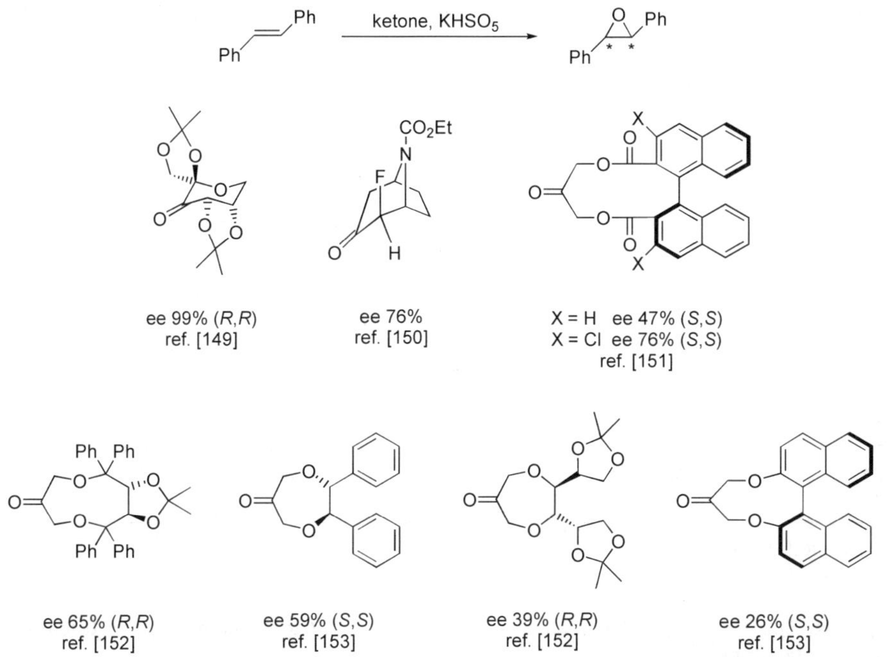

Scheme 12. Enantioselectivity in the epoxidation of *trans*-stilbene by dioxiranes derived from complex optically active ketones

in the oxidation of the silyl enol ethers (Table 13) and of up to 65% for the *meso* and *rac* diols (Table 14). For the silyl enol ethers, bulky substituents both at the silicon atom and *gem* to the enolate substituent promote enantioselectivity. In the case of the *vic* diols, *rac* substrates were shown to work better in regard to enantioselectivity than the corresponding *meso* ones, and acyclic substrates better than cyclic ones.

6 Conclusions

Dioxiranes and MTO-based oxidants have been intensively investigated throughout the last decade. In this review, we have focused on a comprehensive comparison of the versatile chemistry of these two structurally similar oxidants. We have shown that DMD, TFD, MTO/UHP, and MTO/H_2O_2 exhibit similar chemical behavior, although major differences are observed in the case of C—H insertions and arene oxidations, and in the fact that MTO-based oxidations are catalytic in nature, in contrast to most reactions with dioxiranes. By the appropriate choice of solvent and in the presence of appropriate additives, the reactivity and selectivity of the oxygen-transfer process may be enhanced. On the basis of the observed chemo-, regio-, diastereo-, and enantioselectivities, valuable information on the electrophilic

Table 13. Enantioselective oxidation of silyl enol ethers to α-hydoxy ketones by the fructose-derived dioxirane[a]

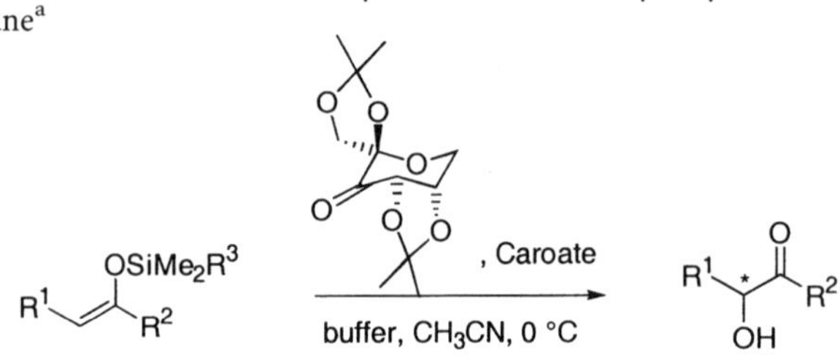

R^1	R^2	R^3	t (h)	Convn (%)	ee (%)	Configuration
Ph	Ph	Me	1.5	36	61	R (−)
Me	Ph	Me	1.5	46	54	R (+)
Me	Ph	*t*-Bu	3.0	92	82	R (+)
Et	Ph	Me	1.5	33	67	R (+)
Ph	Me	Me	1.5	35	18	R (−)
Ph	*t*-Bu	Me	3.0	30	52	S (−)

[a] See Ref. [154].

Table 14. Enantioselective oxidation of *vicinal* diols to α-hydoxy ketones by the fructose-derived dioxirane[a]

Substrate	Time (h)	Convn (%)	Product	ee (%)	Configuration
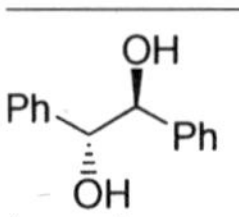 (*meso*)	3.0	89	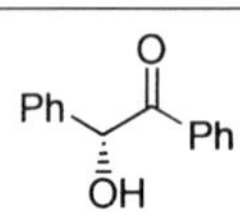	45	*R*-(−)
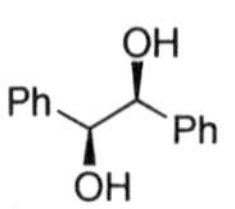 (*rac*)	3.0	51	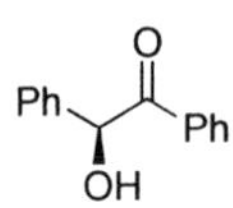	65	*S*-(+)
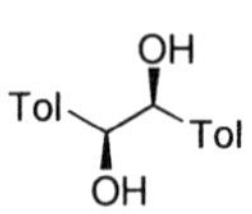 (*rac*)	3.0	12	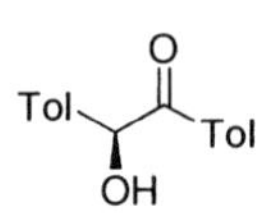	61	*S*-(+)
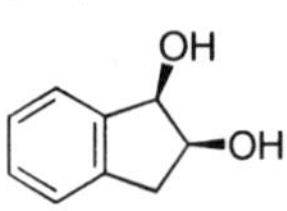	2.0	30	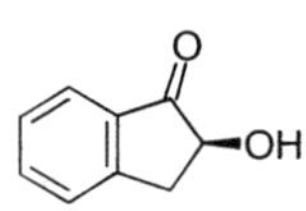	9	*S*-(+)
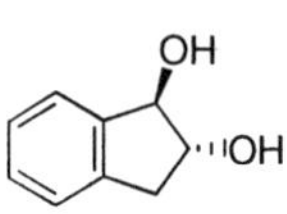	2.0	26	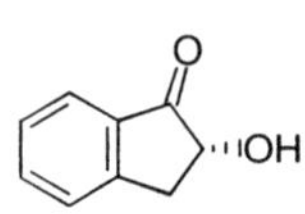	20	*R*-(−)

[a] Ref. [84].

character of the oxygen-transfer step, on the transition structures, and on the likely mechanism has been gained. To date, a major drawback of MTO is that no widely applicable chiral oxidation system has been found; in contrast, for dioxiranes several effective chiral ketones are known, which serve for in-situ asymmetric oxidations. Nonetheless, dioxiranes and the rhenium peroxo oxidants complement each other well and it is expected that they will be even more intensively used for selective oxidations.

7 References

1. a) Mimoun H (1982) Angew Chem, Int Ed Engl 21: 734 b) Mimoun H (1983) Transition-metal peroxides – synthesis and use as oxidizing agents. In: Patai S (ed) The Chemistry of Functional Groups, The Chemistry of Peroxides. Wiley, London, p 463 c) Jørgensen KA (1989) Chem Rev 89: 431 d) Sundermeyer J (1993) Angew Chem, Int Ed Engl 32: 1144 e) Kitajima N, Akita M, Moro-oka Y (1992) Transition Metal Peroxides and Related Compounds. In: Ando W (ed) Organic Peroxides, Wiley, Chichester, p 535 f) Conte V, Di Furia F, Modena G (1992) Transition Metal Catalyzed Oxidation. The Role of Peroxo Complexes. In: Ando W (ed) Organic Peroxides, Wiley, Chichester, p 559
2. Talsi EP, Babushkin DE (1996) J Mol Cat, A: Chemical 106: 179
3. Islam MC, Begum M, Roy HN, Pal SC, Ahmed S (1997) Synth React Inorg Met-Org Chem 27: 17
4. Mimoun H, Daire E, Fischer J, Poskl M, Saussine L, Weiss R (1983) J Am Chem Soc 105: 3101
5. Bkouche-Waksman A, Bois C, Sala-Pala J, Guerchais JE (1980) J Organomet Chem 195: 307
6. a) Fleet GWJ, Little W (1977) Tetrahedron Lett 16: 3749 b) Curci R, Giannattasio S, Sciacovelli O, Troisi L (1984) Tetrahedron 40: 2763 c) Ciminale F, Camporeale M, Mello R, Troisi L, Curci R (1989) J Chem Soc, Perkin Trans 2 417
7. Mimoun H, Sérée de Roch I, Sajus L (1970) Tetrahedron 26: 37
8. Graham BW, Laing KR, O'Connor CJ, Roper WR (1942) J Chem Soc Dalton Trans 1237
9. Halpern J, Goodall BL, Khare GP, Lim HS, Pluth JJ (1975) J Am Chem Soc 97: 2301
10. a) Igersheim F, Mimoun H (1978) J Chem Soc, Chem Commun 559 b) Igersheim F, Mimoun H (1980) Nouveau J Chim 4: 161
11. a) Otsuka S, Nakamura A, Tatsuno S (1969) J Am Chem Soc 91: 6994 b) Otsuka S, Nakamura A, Tatsuno Y, Niki M (1972) J Am Chem Soc 94: 3761
12. Sen A, Halpern J (1977) J Am Chem Soc 99: 8337
13. Gable KP (1997) Adv Organomet Chem 41: 127
14. a) Suenram RD, Lovas FJ (1978) J Am Chem Soc 100: 5117 b) Adam W, Chan Y-Y, Cremer D, Gauss J, Scheutzow D, Schindler M (1987) J Org Chem 52: 2800 c) Mello R, Fiorentino M, Sciacovelli O, Curci R (1988) J Org Chem 53: 3891
15. Murray RW, Jeyaraman R (1985) J Org Chem 50: 2847
16. Kirschfeld A, Muthusamy S, Sander W (1994) Angew Chem Int Ed Engl 33: 2212
17. Herrmann WA, Fischer RW, Scherer W, Rauch MU (1993) Angew Chem Int Ed Engl 32: 1157
18. Fischer RW (1994) PhD thesis, TU München
19. Gisdakis P, Antonczak S, Köstlmeier S, Herrmann WA, Rösch N (1998) Angew Chem Int Ed Engl 37: 2211
20. a) Wadt WR, Goddard WA III (1975) J Am Chem Soc 97: 3004 b) Harding LB, Goddard WR III (1978) J Am Chem Soc 100: 7180 c) Cremer D (1981) J Am Chem Soc 103: 3633
21. a) Cassidei L, Fiorentino M, Mello R, Sciacovelli O, Curci R (1987) J Org Chem 52: 699 b) Cremer D, Schindler M (1987) Chem Phys Lett 133: 293

22. Reynolds MS, Butler A (1996) Inorg Chem 35: 2378
23. a) Smurova LA, Kholopov AB, Nikitin AV, Brin EF, Rubailo VL (1996) Kinet Catal 37: 361 b) Adam W, Mitchell CM unpublished results
24. Yudin AK, Sharpless KB (1997) J Am Chem Soc 119: 11536
25. Adam W, Bialas J, Hadjiarapoglou L (1991) Chem Ber 124: 2377
26. Adam W, Hadjiarapoglou L, Smerz A (1991) Chem Ber 124: 227
27. Murray RW (1989) Chem Rev 89: 1187
28. Adam W, Smerz AK, Zhao C-G (1997) J Prakt Chem 339: 298
29. a) Adam W, Mitchell CM (1996) Angew Chem, Int Ed Engl 35: 533 b) Adolfsson H, Converso A, Sharpless KB (1999) Tetrahedron Lett 40: 3991
30. a) Curci R, Dinoi A, Rubino MF (1995) Pure & Appl Chem 67: 811 b) Adam W, Paredes R, Smerz AK, Veloza LA (1998) Eur J Org Chem 349
31. Herrmann WA, Kühn FE (1997) Acc Chem Res 30: 169
32. van Vliet MCA, Arends IWCE, Sheldon RA (1999) J Chem Soc Chem Commun 821
33. Copéret C, Adolfsson H, Sharpless KB (1997) J Chem Soc Chem Commun 1565
34. Baumstark AL, Vasquez PC (1988) J Org Chem 53: 3437
35. Adam W, Smerz AK (1996) Bull Soc Chim Belg 105: 581
36. Adam W, Mitchell CM, Saha-Möller CR (1999) Eur J Org Chem 785
37. Smerz AK (1996) PhD thesis, Universität Würzburg
38. Zhu Z, Espenson JH (1995) J Org Chem 60: 7728
39. Curci R, Fiorentino M, Fusco C, Mello R, Ballistreri FP, Failla S, Tomaselli GA (1992) Tetrahedron Lett 33: 7929
40. a) Adam W, Herrmann WA, Lin J, Saha-Möller CR, Fischer RW, Correia JDG (1994) Angew Chem, Int Ed Engl 33: 2475 b) Adam W, Herrmann WA, Lin J, Saha-Möller CR (1994) J Org Chem 59: 8281 c) Adam W, Herrmann WA, Saha-Möller CR, Shimizu M (1995) J Mol Cat A: Chemical 97: 15
41. Adam W, Lin J unpublished results
42. Adam W, Balci M, Kilic H (1998) J Org Chem 63: 8544
43. Jeyaraman R, Murray RW (1984) J Am Chem Soc 106: 2462
44. Adam W, Mello R unpublished results
45. Adam W, Shimizu M (1994) Synthesis 560
46. Crandall JK, Zucco M, Kirsch RS, Coppert DM (1991) Tetrahedron Lett 32: 5441
47. Altamura A, Fusco C, D'Accolti L, Mello R, Prencipe T, Curci R (1991) Tetrahedron Lett 32: 5445
48. Brown KN, Espenson JH (1996) Inorg Chem 35: 7211
49. a) Murray RW, Jeyaraman R, Pillay MK (1987) J Org Chem 52: 746 b) Espenson JH (1999) J Chem Soc Chem Commun 479
50. Nakayama J, Nagasawa H, Sugihara Y, Ishii A (1997) J Am Chem Soc 119: 9077
51. Nakayama J, Yu T, Sugihara Y, Ishii A (1997) Chem Lett 499
52. a) Messeguer A (1995) Acros Organics Acta 1: 71 b) Beddoes RL, Painter JE, Quayle P, Patel P (1996) Tetrahedron Lett 37: 9385 c) Schenk WA, Frisch J, Dürr M, Burzlaff N, Stalke D, Fleischer R, Adam W, Prechtl F, Smerz AK (1997) Inorg Chem 36: 2372
53. Murray RW, Iyanar K, Chen J, Wearing JT (1996) Tetrahedron Lett 37: 805
54. Murray RW, Jeyaraman R, Mohan L (1986) Tetrahedron Lett 27: 2335
55. Yamazaki S (1997) Bull Chem Soc Jpn 70: 877
56. Murray RW, Singh M (1989) Synth Commun 19: 3509
57. Murray RW, Singh M (1990) J Org Chem 55: 2954
58. Murray RW, Singh M (1988) Tetrahedron Lett 29: 4677
59. Copéret C, Adolfsson H, Khuong T-AV, Yudin AK, Sharpless KB (1998) J Org Chem 63: 1740
60. Adam W, Golsch D (1993) Angew Chem Int Ed Engl 32: 737
61. Rudler H, Denise B (1998) J Chem Soc Chem Commun 2145
62. Altamura A, D'Accolti L, Detomaso A, Dinoi A, Fiorentino M, Fusco C, Curci R (1998) Tetrahedron Lett 39: 2009
63. Espenson JH, Abu-Omar MM (1997) Adv Chem Ser 253: 99

64. a) Zabrowski DL, Moormann AE, Beck KR Jr (1988) Tetrahedron Lett 29:4501 b) Murray RW, Singh M, Rath N (1996) Tetrahedron Asymmetry 7: 1611
65. Wittman MD, Halcomb RL, Danishefsky SJ (1990) J Org Chem 55: 1981
66. a) Goti A, Nannelli L (1996) Tetrahedron Lett 33: 6025 b) Murray RW, Iyanar K, Chen J, Wearing JT (1996) J Org Chem 61: 8099
67. Murray RW, Iyanar K (1998) Heteroatom Chem 9: 347
68. a) Berliner LJ (ed) (1976) Spin Labeling, Theory and Applications. Academic Press, New York b) Keana JFW (1978) Chem Rev 78: 37 c) Holzman JL (ed) (1984) Spin Labeling in Pharmacology. Academic Press, New York
69. Ferrer M, Sánchez-Baeza F, Messeguer A, Adam W, Golsch D, Görth F, Kiefer W, Nagel V (1998) Eur J Org Chem 2527
70. Adam W, Briviba K, Duschek F, Golsch D, Kiefer W, Sies H (1995) J Chem Soc Chem Commun 1831
71. Murray RW, Iyanar K, Chen J, Wearing JT (1995) Tetrahedron Lett 36: 6415
72. Murray RW, Jeyaraman R, Mohan L (1986) J Am Chem Soc 108: 2470
73. Mello R, Fiorentino M, Fusco C, Curci R (1989) J Am Chem Soc 111: 6749
74. Kovall F, Baumstark AL (1994) Tetrahedron Lett 35: 8751
75. Schuchardt U, Mandelli D, Shul'pin GB (1996) Tetrahedron Lett 37: 6487
76. Asensio G, Mello R, González-Núñez ME, Castellano G, Corral J (1996) Angew Chem Int Ed Engl 35: 217
77. Mello R, Cassidei L, Fiorentino M, Fusco C, Curci R (1990) Tetrahedron Lett 31: 3067
78. a) Bovicelli P, Lupattelli P, Mincione E, Prencipe T, Curci R (1992) J Org Chem 57: 2182 b) Bovicelli P, Lupattelli P, Mincione E, Prencipe T, Curci R (1992) J Org Chem 57: 5052 c) Curci R, Detomaso A, Prencipe T, Carpenter GB (1994) J Am Chem Soc 116: 8112
79. Murray RW, Gu H (1995) J Org Chem 60: 5673
80. Mello R, Cassidei L, Fiorentino M, Fusco C, Hümmer W, Jäger V, Curci R (1991) J Am Chem Soc 113: 2205
81. Baumstark AL, Kovac F, Vasquez PC (1999) Can J Chem 77: 308
82. a) Curci R, D'Accolti L, Detomaso A, Fusco C, Takeuchi K, Ohga Y, Eaton PE, Yip YC (1996) Tetrahedron Lett 34: 4559 b) Bovicelli P, Lupattelli P, Sanetti A, Mincione E (1994) Tetrahedron Lett 35: 8477 c) Bovicelli P, Lupattelli P, Sanetti A, Mincione E (1995) Tetrahedron Lett 36: 3031 d) Bovicelli P, Sanetti A, Lupattelli P (1996) Tetrahedron 52: 10969
83. D'Accolti L, Detomaso A, Fusco C, Rosa A, Curci R (1993) J Org Chem 58: 3600
84. Adam W, Saha-Möller CR, Zhao C-G (1998) Tetrahedron: Asymmetry 9: 4117
85. Adam W, Mello R, Curci R (1990) Angew Chem Int Ed Engl 29: 890
86. Adam W, Mitchell CM, Saha-Möller CR, Weichold O (1999) J Am Chem Soc 121: 2097
87. Adam W, Weichold O unpublished results
88. a) Adam W, Azzena U, Prechtl F, Hindahl K, Malisch W (1992) Chem Ber 125: 1409 b) Malisch W, Hindahl K, Käb H, Reising J, Adam W, Prechtl F (1995) Chem Ber 128: 963 c) Malisch W, Schmitzer S, Lankat R, Neumayer M, Prechtl F, Adam W (1995) Chem Ber 128: 1251 d) Malisch W, Neumayer M, Fey O, Adam W, Schuhmann R (1995) Chem Ber 128: 1257 e) Malisch W, Lankat R, Schmitzer S, Reising J (1995) Inorg Chem 34: 5702 f) Malisch W, Jehle H, Möller S, Saha-Möller CR, Adam W (1998) Eur J Inorg Chem 1585 g) Malisch W, Jehle H, Mitchell CM, Adam W (1998) J Organomet Chem 566: 259
89. a) Adger BM, Barrett C, Brennan J, McKervey MA, Murray RW (1991) J Chem Soc Chem Commun 1553 b) Finlay J, McKervey MA, Gunaratne HQN (1998) Tetrahedron Lett 39: 5651
90. Wolowiec S, Kochi JK (1991) Inorg Chem 30: 1215
91. Montgomery RE (1974) J Am Chem Soc 96: 7820
92. Yamazaki S (1995) Chem Lett 127
93. Baumstark AL, Beeson M, Vasquez PC (1989) Tetrahedron Lett 30: 5567
94. a) Ihmels H, Maggini M, Prato M, Scorrano G (1991) Tetrahedron Lett 32: 6215 b) Saba A (1994) Synth Comm 24: 695 c) Wasserman HH, Baldino CM, Coats SJ (1995) J Org Chem 60: 8231

95. Nelsen SF, Scamehorn RG, De Felippis J, Wang Y (1993) J Org Chem 58: 1657
96. Eaton PE, Wicks GE (1988) J Org Chem 53: 5353
97. a) Olah GA, Liao Q, Lee C-S, Prakash GKS (1993) Synlett 427 b) Adam W, Makosza M, Zhao C-G (1998) Synlett 1335
98. Nakayama J, Matsui T, Sugihara Y, Ishii A, Kumakura S (1996) Chem Lett 269
99. Pestovsky O, van Eldik R, Huston P, Espenson JH (1995) J Chem Soc Dalton Trans 133
100. Hatzopoulos I, Brauer H-D, Geisberger MR, Herrmann WA (1996) J Organomet Chem 520: 201
101. Bach RD, Andrés JL, Owensby AL, Schlegel HB, McDouall JJW (1992) J Am Chem Soc 114: 7207
102. Bravo A, Fontana F, Fronza G, Minisci F (1995) Tetrahedron Lett 38: 6945
103. Adam W, Curci R, D'Accolti L, Dinoi A, Fusco C, Gasparrini F, Kluge R, Paredes R, Schulz M, Smerz AK, Veloza LA, Weinkötz S, Winde R (1997) Chem Eur J 3: 105
104. Adam W, Mitchell CM, Saha-Möller CR (1999) J Org Chem 64: 3699
105. Adam W, Mitchell CM, Saha-Möller CR (1994) Tetrahedron 50: 13121
106. Mello R, Ciminale F, Fiorentino F, Fusco C, Prencipe T, Curci R (1990) Tetrahedron Lett 31: 6097
107. a) Asensio G, Mello R, González-Núñez ME (1996) Tetrahedron Lett 37: 2299 b) Adam W, Golsch D, Görth FC (1996) Chem Eur J 2: 255
108. De La Mare PBD, Swedlund BE (1973) Heterolytic mechanism of substitution involving carbon-halogen bonds. In: Patai S (ed) The Chemistry of Carbon Halogen Bonds, Part I. Wiley, London, p 407
109. a) Vanni R, Garden SJ, Banks JT, Inguld KU (1995) Tetrahedron Lett 36: 7999 b) Simakov PA, Choi SY, Newcomb M (1998) Tetrahedron Lett 39: 8187
110. a) Du X, Houk KN (1998) J Org Chem 63:6480 b) Glukhovtsev MN, Canepa C, Bach RD (1998) J Am Chem Soc 120: 10528
111. Minisci F, Zhao L, Fontana F, Bravo A (1995) Tetrahedron Lett 36: 1697
112. Mitchell CM (1998) PhD thesis, Universität Würzburg
113. Adam W, Golsch D (1994) Chem Ber 127: 1111
114. Yamazaki S (1996) Bull Chem Soc Jpn 69: 2955
115. Adam W, Hadjiarapoglou L (1993) Top Curr Chem 164: 45
116. Ferrer M, Sánchez-Baeza F, Messeguer A, Diez A, Rubiralta M (1995) J Chem Soc Chem Commun 293
117. Adam W, Smerz AK (1996) J Org Chem 61: 3506
118. Adam W, Smerz AK (1995) Tetrahedron 51: 13039
119. Adam W, Dürr H, Haas W, Lohray BB (1986) Angew Chem Int Ed Engl 25: 101
120. a) Link JT, Danishefsky SJ, Schulte G (1994) Tetrahedron Lett 35: 9131 b) Adam W, Reinhardt D, Reissig H-U, Paulini K (1995) Tetrahedron 51: 12257 c) Bartels A, Jones PG, Liebscher J (1995) Tetrahedron Lett 36: 3673 d) Burgess LE, Gross EKM, Jurka J (1996) Tetrahedron Lett 37: 3255
121. Asensio G, González-Núñez ME, Bernardini CB, Mello R, Adam W (1993) J Am Chem Soc 115: 7250
122. Adam W, Prechtl F, Richter MJ, Smerz AK (1995) Tetrahedron Lett 36: 4991
123. Boehlow TR, Spilling CD (1996) Tetrahedron Lett 37: 2717
124. Adam W, Paredes R, Smerz AK, Veloza LA (1997) Liebigs Ann/Recueil 547
125. Nakajima M, Sasaki Y, Iwamoto H, Hashimoto S (1998) Tetrahedron Lett 39: 87
126. Adam W, Wirth T (1999) Acc Chem Res 32: 703
127. Sharpless KB, Michaelson RC (1973) J Am Chem Soc 95: 6136
128. Rudler H, Ribeiro Gregorio J, Denise B, Brégeault J-M, Deloffre A (1998) J Mol Cat, A: Chemical 133: 255
129. Lluch A-M, Sánchez-Baeza F, Messeguer A (1993) Tetrahedron 49: 6299
130. Bovicelli P, Lupattelli P, Mincione E (1994) J Org Chem 59: 4304
131. Adam W, Schuhmann RM (1996) J Org Chem 61: 874
132. a) Adam W, van Barneveld C, Golsch D (1996) Tetrahedron 52: 2377 b) Görth FC (1996) Diploma thesis, Universität Würzburg

133. a) Frampton R, Johnson CD, Katritzky AR (1971) Justus Liebigs Ann Chem 749: 12 b) Phillipson JD, Handa SS (1975) Phytochemistry 14: 2683 c) Seeman JI, Whidby JF (1976) J Org Chem 41: 3824
134. a) Paudler WW, Chen T (1971) J Org Chem 26: 787 b) Neunhoeffer H, Frühauf H-W (1972) Liebigs Ann Chem 758: 111 c) Neunhoeffer H (1984) 1,2,4-Triazines and their Benzo Derivatives. In: Boulton AJ, McKillop A (eds) Comprehensive Heterocyclic Chemistry, vol 3. Pergamon Press, Oxford, p 400
135. Voigt B, Porzel A, Golsch D, Adam W, Adam G (1996) Tetrahedron 52: 10653
136. Ballini R, Papa F, Bovicelli P (1996) Tetrahedron Lett 37: 3507
137. Herrmann WA, Correia JDG, Fischer RW, Adam W, Lin J, Saha-Möller CR, Shimizu M (1995) Hoechst AG EP-A 95100901.8
138. Pattenden G, Shuker AJ (1992) J Chem Soc Perkin Trans 1 1215
139. Vedejs E, Dent WH III, Kendall JT, Oliver PA (1996) J Am Chem Soc 118: 3556
140. a) Nicotra F, Panza L, Russo G (1991) Tetrahedron Lett 32: 4035 b) Timmers CM, van der Marel GA, van Boom JH (1993) Recueil Trav Chim Pays-Bas 112: 609 c) Seeberger PH, Bilodeau MT, Danishefsky SJ (1997) Aldrichim Acta 30: 75
141. Curci R, Detomaso A, Lattanzio ME, Carpenter GB (1996) J Am Chem Soc 118: 11089
142. a) Adam W, Müller M, Prechtl F (1994) J Org Chem 59: 2358 b) Adam W, Korb MN (1996) Tetrahedron 52: 5487
143. Schenk WA, Frisch J, Adam W, Prechtl F (1994) Angew Chem Int Ed Engl 33: 1609
144. Adam W, Prechtl F (1994) Chem Ber 127: 667
145. a) Herrmann WA, Kühn FE, Mattner MR, Artus GRJ, Geisberger MR, Correia JDG (1997) J Organomet Chem 538: 203 b) Tucker CE, Davenport KG (1997) Hoechst Celanese Corporation US-Patent 5618958 c) Gunaratne HQN, McKervey MA, Feutren S, Finlay J, Boyd J (1998) Tetrahedron Lett 39: 5655
146. a) Curci R, Fiorentino M, Serio MR (1984) J Chem Soc, Chem Commun 155 b) Curci R, D'Accolti L, Fiorentino M, Rosa A (1995) Tetrahedron Lett 36: 5831
147. Colonna S, Gaggero N (1989) Tetrahedron Lett 30: 6233
148. a) Adam W, Jekö J, Lévai A, Nemes C, Patonay T, Sebök P (1995) Tetrahedron Lett 36: 3669 b) Adam W, Jekö J, Lévai A, Majer Z, Nemes C, Patonay T, Párkányi L, Sebök P (1996) Tetrahedron: Asymmetry 7: 2437 c) Adam W, Fell RT, Lévai A, Patonay T, Peters K, Simon A, Tóth G (1998) Tetrahedron: Asymmetry 9: 1121 d) Lévai A, Adam W, Fell RT, Gessner R, Patonay T, Simon A, Tóth G (1998) Tetrahedron 54: 13105
149. Wang Z-X, Tu Y, Frohn M, Zhang J-R, Shi Y (1997) J Am Chem Soc 119: 11224
150. Armstrong A, Hayter BR (1998) J Chem Soc Chem Commun 621
151. a) Yang D, Yip Y-C, Tang M-W, Wong M-K, Zheng J-H, Cheung K-K (1996) J Am Chem Soc 118: 491 b) Yang D, Wang X-C, Wong M-K, Yip Y-C, Tang M-W (1996) J Am Chem Soc 118: 11311
152. Adam W, Zhao C-G (1997) Tetrahedron: Asymmetry 8: 3995
153. Song CE, Kim YH, Lee KC, Lee S, Jin BW (1997) Tetrahedron: Asymmetry 8: 2921
154. Adam W, Fell RT, Saha-Möller CR, Zhao C-G (1998) Tetrahedron: Asymmetry 9: 397

Nature of Activated Bleomycin

Richard M. Burger

Public Health Research Institute, 455 First Avenue, New York, NY 10016, USA
E-mail: burger@phri.nyu.edu

Activated bleomycin is the drug species seen to be kinetically competent to initiate DNA degradation in the reaction of bleomycin, Fe(II), and O_2. It also forms in reactions of bleomycin with Fe(III) and peroxide, or bleomycin with superoxide and either Fe(III) or Fe(II). Efforts to characterize this transient species proceeded by kinetic and spectroscopic strategies. Activated bleomycin now appears to be a drug-ferric-peroxide complex, but this may not be the proximate active drug species. Assuming that activated bleomycin peroxide cleavage yields a reactive product analogous to peroxidase compound I explains many characteristics of bleomycin-mediated DNA degradation reactions. DNA degradation is responsible for the cytotoxic and antitumor activities of this clinically useful antibiotic.

Keywords: Iron, Oxygen, Peroxide, Superoxide, DNA, Antitumor, Antibiotic

1 Introduction

The bleomycins are a clinically useful family of glycopeptide antibiotic congeners with antitumor activity [1–3]; secondary applications include wart removal [4] and genetic engineering [5]. Bleomycin is toxic to both prokaryotes and eukaryotes, although both can harbor genes conferring resistance. Cytotoxicity results from oxidative DNA degradation [6–14]. The chemistry of this degradation and the events leading to it have been studied extensively in vitro [15, 16].

Bleomycin and transition metal ions form complexes that react with dioxygen species and attack DNA [15, 17] and other less vital biomolecules [18, 19]. The first and most active bleomycin complex studied is that with Fe(II), which reacts with O_2 [20]. When the intermediates in this reaction pathway were distinguished by kinetic methodologies, one species proved kinetically competent to initiate the degradation of DNA [21, 22], and it was therefore named “activated bleomycin” [22]. Knowing the nature of activated bleomycin should help explain the complex DNA degradation process that it initiates as well as the action of the oxygenases it appears to model. However, this hope has not been completely realized, first because the oxygen structure of activated bleomycin resisted characterization for over a decade, and second because the proximate active species now seems to be an activated bleomycin product rather than activated bleomycin itself. It is now evident that the molecular species denoted “activated bleomycin” is a drug-ferric-peroxide complex [23, 24], with properties that may be compared to the shorter-lived metallo-enzyme peroxides.

The bleomycin O—O cleavage reaction is the rate-limiting step in DNA attack [24], but its role in bleomycin action remains to be clarified. While the O—O cleavage might be directly coupled to the attack on DNA, it more likely yields the proximate attacking agent(s), with reactive Fe—O structures thought to resemble those of peroxidase compounds I and II. Still to be understood are whether and how DNA affects activated bleomycin reactivity, what accounts for the unusual stability of this ferric peroxide, and what the immediate reaction products are.

Several lines of evidence indicate that activated bleomycin is a low-spin ferric peroxide drug complex [25]. It breaks down spontaneously to form ferric bleomycin [21, 24], releasing OH^- and, with DNA absent, O_2 [26]. The shorter-lived intermediates in this reaction have not been observed directly; thus how they interact with DNA to initiate its degradation has been inferred mainly from the products formed. Each of these points is considered below.

The structure of bleomycin is shown in Fig. 1. Bleomycin mixtures, which are products of *Streptomyces verticillis*, are obtained from the culture media as cupric complexes, but they are administered pharmaceutically in a metal-free form [3]. The bleomycins all share the chemical structure shown in Fig. 1, differing in the R group distal to the bithiazole moiety. Other bleomycin-family drugs, such as phleomycin [27] and tallysomycin [28], also differ in other aspects of structure. The structure and conformation of bleomycin, both alone and in complexes with DNA, are known in considerable detail, and they have recently been reviewed [29]. Bleomycin mainly occupies the minor groove of DNA, with the bithiazole group intercalating between the nucleic bases such that the R group emerges into the major groove. The metal ligands of bleomycin are arranged such that the metal-complexed oxygen is well approximated to the DNA 4′-hydrogen [30–32], which is the unique site of bleomycin attack.

The DNA degradation events that bleomycin initiates are understood in considerable detail (review completed 1997 [16]), but consideration here is limited to those that directly involve or reveal the activities of bleomycin.

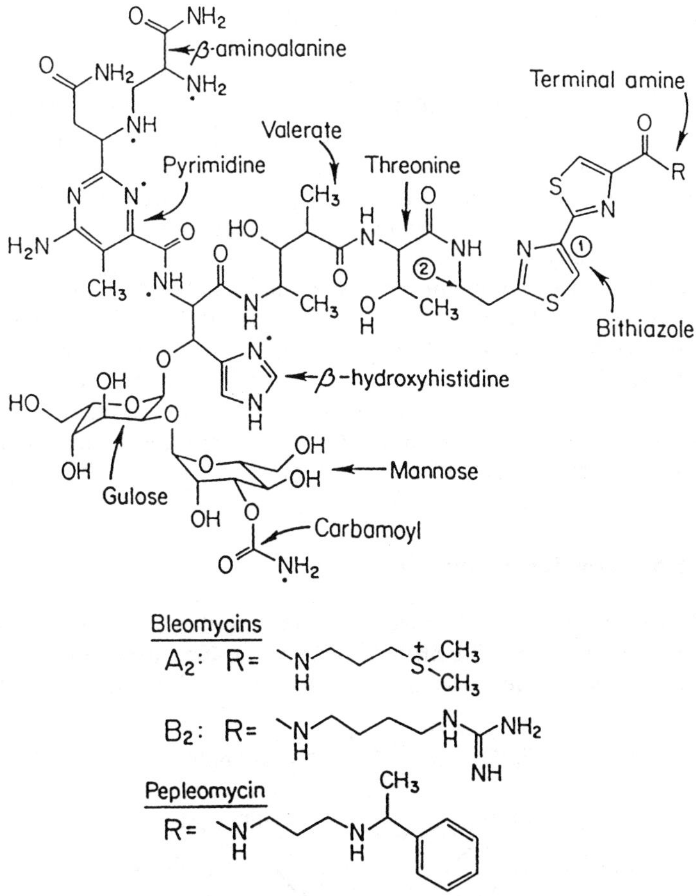

Fig. 1. Bleomycin structures. The structure shown is shared by all bleomycin-type drugs, which differ mainly by the R groups, shown for three variants. Phleomycins have the same structures except for lack of conjugation at thiazole (1). Tallysomycin differs by addition of a talose 1-glycoside at position (2) and substitution of a hydrogen for the valerate methyl group. All share the metal ligands indicated (●). Model compounds (not shown) have been prepared synthetically which contain only the metal-ligating moieties. Reprinted with kind permission from [13]. Copyright 1996 Elsevier Science

There are two modes of bleomycin-induced DNA degradation, and they have distinctive consequences [33–37] even though both are initiated by an attack on the deoxyribose H4′ [38, 39]. One mode leads to frank strand cleavage with the excision of base propenal [36], a C1′-C3′ fragment derivatized with nucleic base [35]. The gap left in DNA is bounded by 5′-phosphate [37] and 3′-phosphoglycolate [36] termini. The other mode leads to free base release [40] from a C4′-hydroxylated deoxyribose product [38, 41–43] that remains continuous but is alkali-labile [33]. The first mode requires O_2 in addition to the oxygen involved in forming activated bleomycin, while the second does not

[44]. In neither case is bleomycin-bound oxygen (or any other drug moiety) added to a DNA degradation product [45]. In addition to serving as a bleomycin target, DNA is also involved in two other bleomycin reactions: it inhibits the formation of activated bleomycin [22, 46], and it protects activated bleomycin from permanent self-destruction [22]. There is also spectral evidence that DNA binds to activated bleomycin, as it does to all observed bleomycin intermediates, but DNA has little, if any, influence on the rate of activated bleomycin consumption [22]. Thus DNA and activated bleomycin may not react directly with one another.

In summary, activated bleomycin is a moderately stable Fe(III)-peroxo complex that reacts as the attack on DNA is initiated. Its stability may mask a series of important shorter-lived drug products by limiting their steady-state concentrations. This, in turn, limits the accumulation of short-lived DNA degradation intermediates. Consequently, early steps in these reaction pathways are largely inferential. Activated bleomycin generation, form, fate, and effects are discussed below.

2 Formation of Activated Bleomycin

Activated bleomycin can be formed by several routes; at least one for each dioxygen species has been identified (Scheme 1). The first (and technically most direct) to be studied was the reaction of Fe(II)·bleomycin + O_2. This

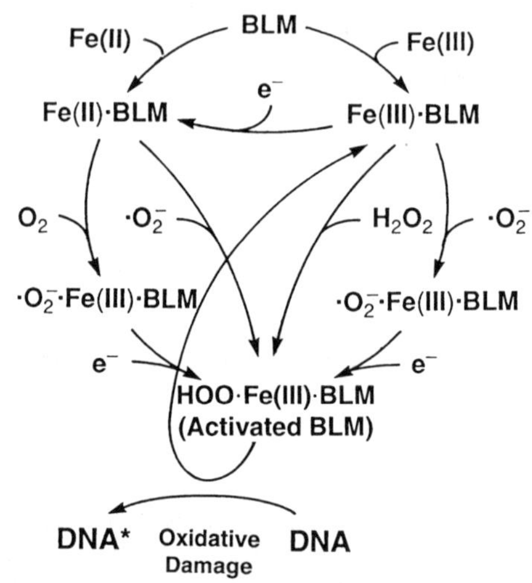

Scheme 1. Pathways of bleomycin activation. Activated bleomycin may be formed by the reactions of iron complexes with dioxygen species and, if necessary, a reductant. There is evidence for all the pathways shown here. Although activated bleomycin is shown initiating DNA damage, it is likely that an intermediate (not shown) in the conversion of activated bleomycin to Fe(III)·bleomycin is the proximate attacking species. A putative pathway of reactions initiating DNA degradation is given in Scheme 2. Reprinted with kind permission from [95]. Copyright 1996 Kluwer

pathway appeared at first to involve a single turnover to form Fe(III)·bleomycin, and the reactions of oxygenated intermediates were initially observed by means of optical static- [20, 47] and stopped-flow spectroscopy [48]. When the kinetics of bleomycin-induced DNA degradation were first characterized, a paradoxical discordance appeared between the rates of the optical transitions [48], which were relatively rapid, and the initiation of DNA degradation [49], which was relatively slow. The paradox was resolved by the discovery that the final optical species represented two chemical entities that were readily distinguishable by EPR spectroscopy [48]. One was the stable Fe(III)·bleomycin (g = 2.45, 2.18, 1.89) while the other was an unstable EPR species (g = 2.26, 2.17, 1.94) [21, 22, 50, 51] that converts to Fe(III)·bleomycin concurrently with the attack on DNA ($t_{1/2}$ about 8 s at room temperature and 2 min at 4 °C) [21, 22]. This observation focused interest on characterizing the transient EPR species, activated bleomycin.

The kinetics of the Fe(II)·bleomycin + O_2 pathway to activated bleomycin formation make it convenient to observe and manipulate. O_2 reacts avidly with Fe(II)·bleomycin (second order $k = 6.1 \times 10^3$ M^{-1} s^{-1} at 2 °C [48]), a high-spin compound [25] that reacts readily with O_2 whether DNA is present or absent, unlike oxygenases that require substrate binding for activity. The iron-drug complex and oxygen form a kinetic optical intermediate [48] that resembles complexes of Fe(II)·bleomycin with O_2 analogues such as CO, NO, and ethyl isocyanide [49]. The rapidity and avidity of this reaction make it possible to choose reaction conditions that render the mixture anaerobic upon drug-oxygen complexation [22, 44]. The first oxygenated species, oxyferrous bleomycin, is a low-spin [25], EPR-silent [48] precursor of activated bleomycin, analogous to peroxidase compound III. Mössbauer spectroscopy indicates that its dominant resonance form is superoxy-ferric-bleomycin [25], as shown in Scheme 1. This complex spontaneously dismutates to activated bleomycin [21] at less than one-tenth its rate of formation [48], so the former accumulates briefly.

$$\begin{aligned} 2O_2\cdot \mathrm{Fe(II)}\cdot \mathrm{bleomycin} + H^+ \rightarrow\ & HOO^-\cdot \mathrm{Fe(III)}\cdot \mathrm{bleomycin} \\ & + \mathrm{Fe(III)}\cdot \mathrm{bleomycin} + O_2 \end{aligned} \tag{1}$$

It can be trapped for characterization using stop-freeze techniques or by the presence of excess DNA. Under conditions that make its dismutation rate-limiting, the (bimolecular) formation of activated bleomycin and the consequent attack on DNA show second-order kinetics, as expected [52]. Without an excess of DNA present, activated bleomycin also is consumed more slowly than it forms, so it, too, accumulates and can be trapped and observed. Here the series of easily observable intermediates ends, for the stability of activated bleomycin exceeds that of any subsequent intermediates on the pathway to Fe(III)·bleomycin. The formation of activated bleomycin is consequently far better characterized than its consumption, which is addressed below under Dissolution and Products.

The second pathway, formation of activated bleomycin via reaction of Fe(III)·bleomycin with peroxide [21, 22], involves a more direct route than that of Fe(II)·bleomycin + O_2. Since it was long known that reductants can facilitate bleomycin-induced DNA degradation [15, 17] by reducing Fe(III)·bleomycin to Fe(II)·bleomycin [20, 47], it was assumed that H_2O_2 worked the same way, since peroxide can also function as a reductant. However, the redox potential of Fe(III)·bleomycin/Fe(II)·bleomycin was found to be incompatible with reduction by peroxide [53].

The activated bleomycin EPR spectrum was seen in reaction mixtures of bleomycin and peroxide, and the alternative to Fe(III)·bleomycin reduction was the formation of activated bleomycin by direct addition of H_2O_2 to Fe(III)·bleomycin. To demonstrate that the peroxide route to activated bleomycin formation involved an oxygen source other than O_2 [22], the reaction with peroxide and resultant DNA degradation were shown to proceed both in the absence of O_2 and in the presence of the O_2 analogue, ethyl isocyanide, which binds to Fe(II)·bleomycin, competing with O_2, but does not affect Fe(III)·bleomycin [44, 54]. To distinguish activated bleomycin from other possible compounds having similar metal ligation, its presence in the peroxide reaction mixtures was verified by confirming the expected kinetics of DNA attack and by demonstrating the expected reaction products: both Fe(III)·bleomycin and those of DNA breakdown.

Observing the kinetics of activated bleomycin breakdown in the presence of peroxide involved a technical complication, since in these reaction mixtures, unlike those with Fe(III)·bleomycin + O_2, activated bleomycin is regenerated from the Fe(III)·bleomycin product (Scheme 1). A relaxation experiment was therefore used, in which the reaction to accumulate activated bleomycin was run briefly and then squelched by addition of DNA, an inhibitor of the formation reaction. The conversion of activated bleomycin to Fe(III)·bleomycin was then seen by EPR to proceed with the expected kinetics, concurrent with DNA degradation [22].

Peroxides other than H_2O_2, such as ethyl hydroperoxide, also react with Fe(III)·bleomycin to produce the activated bleomycin EPR spectrum, breakdown kinetics, and DNA degradation, suggesting that a distal peroxide substituent need not interfere with bleomycin activity [22]. This may reflect the substituent's absence from a putative proximate active product derived from cleavage of the bleomycin peroxide.

The activation route with superoxide was the latest to be dissected, even though early evidence had indicated that superoxide enhanced bleomycin activity [55–60]. (In bacteria, endogenous superoxide paradoxically protects from bleomycin toxicity (unpublished results)). In vitro, superoxide relieved the inhibition of activated bleomycin formation imposed by high DNA:bleomycin ratios [61, 62]. This was attributed to superoxide providing an alternative route to activated bleomycin formation that did not require oxyferrous bleomycin disproportionation (Eq. 1). To do this, superoxide might serve either as a reductant of oxyferrous bleomycin or as a donor of reduced oxygen to iron-bleomycin complexes. There is evidence that both Fe(III)·bleomycin and Fe(II)·bleomycin can form superoxide adducts leading to activated bleomycin as shown in Scheme 1. The evidence is as follows. When a solution of superoxide (KO_2 in 8% 18-crown-6 in aprotic dimethylsulfoxide) is added to a solution of (anaerobic, aqueous) Fe(II)·bleomycin and DNA, the DNA is attacked with the production of base propenal. This attack could not result from a reaction of Fe(II)·bleomycin with the O_2 produced in the dismutation of superoxide in water, because the other dismutation product, H_2O_2, reacts even more avidly with Fe(II)·bleomycin and destroys it (unpublished result). Therefore the bleomycin activity observed is attributed to a reaction involving a superoxide adduct. When a similar experiment is done by adding the dry KO_2 solution to Fe(III)·bleomycin plus DNA, base propenal is again produced. This must result from addition of superoxide to Fe(III)·bleomycin, rather than reduction to Fe(II)·bleomycin, because this base propenal production is insensitive to additions of the low concentrations of H_2O_2 that destroy Fe(II)·bleomycin without significantly activating Fe(III)·-bleomycin.

The question of which route or routes to activated bleomycin are important in vivo is vexing. The high ratios of DNA:bleomycin expected in vivo inhibit the reaction with peroxide as they do the disproportionation of oxy-ferrous-bleomycin. Moreover, addition of marginally toxic H_2O_2 to bacteria (2–5 mmol/l) has no effect on their sensitivity to bleomycin (unpublished result), which is not surprising, considering that the apparent K_M of Fe(III)·bleomycin for peroxide is 20–30 mmol/l and that Fe(II)·bleomycin is destroyed by peroxide. The remaining routes involving superoxide are also suspect, since superoxide in bacteria protects against bleomycin. Thus none of the pathways in Scheme 1 is attractive without the intervention of some (unknown) cellular agent adept at reducing oxyferrous-bleomycin.

Attempts to make activated bleomycin with monooxygen donors have not been successful [63, 64]. This is not surprising, since activated bleomycin has now been shown to be a peroxide (see Sect. 3, Structure, below). However, it seems reasonable to expect that a monooxygen donor might succeed in producing an activated bleomycin O—O cleavage product capable of attacking DNA. A short-lived adduct thus formed might be difficult to detect but, if it formed rapidly, it might be more readily observable than the low concentrations expected via the relatively slow peroxide cleavage route. To establish the identity and importance of such a product, it would then be necessary to demonstrate the presence of a labile Fe—O linkage, as was done for activated bleomycin, as well as to substantiate the formation of typical DNA degradation products. Failure to produce or observe a monooxygenated bleomycin species is consistent with the idea that activated bleomycin may react directly with DNA, but it is not strong evidence against the existence of a short-lived intermediate.

3 Structure

Activated bleomycin was defined spectrally and kinetically [22] and, being unstable and paramagnetic, was difficult to characterize structurally. However, bleomycin also forms complexes with other transition metal ions that sometimes complex oxygen species that can then react with DNA [65]. The complex most nearly analogous to activated bleomycin is the Co(III)—OOH "green complex", that can lead to an attack on DNA. The green complex peroxide moiety is far more stable [32] than that in activated bleomycin, and it requires light to initiate the attack [66]. This stability and its diamagnetism have made the cobalt complex a useful analogue for activated bleomycin in NMR structural studies of bleomycin both alone [30, 67–70] and complexed to DNA [70–73]. Detailed NMR-derived structures are consistent with structural information obtained for Fe(II)·bleomycin by NMR [74] and activated bleomycin by ENDOR spectroscopy [75, 76]. The structures deduced for the green complex and its synthetic congeners are therefore assumed to be good models for activated bleomycin.

The iron-oxygen structure of activated bleomycin is, of course, its most interesting feature. The first properties discovered were deduced from iron-influenced spectral characteristics. The EPR spectrum of Fe(III)·bleomycin closely resembles those of heme-containing oxygenases and peroxidases, suggesting a similar ligation of the iron even though the conjugation of the bleomycin iron ligands is much less extensive than that in heme [49]. The parallel between bleomycin and heme-containing oxygenases was reinforced by resonance Raman studies of Fe(II)·bleomycin, CO·Fe(II)·bleomycin, and Fe(III)·bleomycin, which were interpreted as showing bond strengths and electronic distributions about the iron ligands analogous to those in the heme enzymes [77]. It was proposed that the conjugation between adjacent pyrimidine and amide ligand nitrogens (shown in Fig. 1) might permit the stabilization of a π-cation radical analogous to that on the peroxidase compound I heme [77]. There must be differences in the oxygen, however, because the relative stability of activated bleomycin and its putative O—O cleavage products are the inverse of those seen in its heme enzyme analogues, of which the longest lived and most easily observed are the cleavage products, compounds I and II.

Some of the earliest spectroscopic studies indicated that activated bleomycin contained an Fe—O bond, which was inferred to be at a redox level of V, like compound I [22]. The redox state of activated bleomycin was confirmed by direct titrations with the reductants I^-, NADH, and thio-NADH [78]. The ability of reductants to accelerate dramatically the decay of activated bleomycin contrasts with the inability of DNA to affect this reaction. Conversion of activated bleomycin to Fe(III)·bleomycin shows 2-electron stoichiometry, but conversion is also accomplished by 1-electron reductants. The ability of activated bleomycin to perform $1e^-$ oxidations is an important feature in the mechanisms proposed for its DNA degradation reactions.

The Fe(III) state of activated bleomycin iron was inferred from the ^{57}Fe-induced splitting of its g_{min} EPR feature [22]. Direct iron-oxygen ligation was inferred from broadening of Fe EPR features of activated bleomycin prepared with $^{17}O_2$ [22] and by ^{17}O ENDOR spectroscopy [75]. However, neither revealed a second, distal, oxygen. Still, the ^{17}O ENDOR characteristics of activated bleomycin differ distinctly from those of the monoxygenated horseradish peroxidase compound I [79], and they suggest quadrupolar couplings similar to those of ^{17}O-hydrogen peroxide [80]. The presence of both O_2 atoms (plus an extra H) in activated bleomycin was demonstrated by electrospray mass spectroscopy [23]. It has not yet been possible to observe bleomycin O—O vibrational spectra, but the presence of an O—O bond and the rate-limiting role of its breakage in activated bleomycin reactivity was established by the magnitude of an $^{18}O_2$ isotope effect on the rate of activated bleomycin decay [24]. These findings are consistent with X-ray absorption-edge and -fine structure analysis that showed no short Fe=O bond [81].

4 Dissolution and Products

Activated bleomycin can break down in the presence or absence of DNA; oxygen is not required. In the absence of DNA, the net oxygen consumption implies that 0.5 equivalents O_2 per activated bleomycin is released [21, 26]. This is as expected since without the participation of reductants other than Fe(II), only one O_2 can be reduced to water for every 4 equivalents of Fe(II)·bleomycin oxidized to Fe(III)·bleomycin.

$$2HOO^- \cdot Fe(III) \cdot bleomycin \rightarrow 2Fe(III) \cdot bleomycin + O_2 + 2OH^- \quad (2)$$

Since neither H_2O_2 nor superoxide are detected in significant amounts in activated bleomycin reaction mixtures, the reduced oxygen product must be OH^-. This was confirmed and quantitated by ^{17}O-NMR spectrometry, performed on reaction mixtures prepared with $^{17}O_2$ in ^{17}O-depleted water – the only labeled oxygen product was water [26]. In the absence of DNA, it accounted for half the oxygen in activated bleomycin. The remainder of the complexed oxygen must be regenerated as O_2. These findings would be consistent with a simple disproportionation of activated bleomycin, except that activated bleomycin consumption shows first-order, not the expected second-order, kinetics. In a separate experiment (unpublished result), the O_2 at the conclusion of the experiment was examined to determine its origin. When an Fe(II)·bleomycin + O_2 reaction was conducted with a mixture of ^{16}O-^{16}O and ^{18}O-^{18}O, no isotopic recombination was detected in the O_2 examined afterward by mass spectroscopy. Thus the bleomycin complex regenerating O_2 must have been dioxygenated, as would be the case for activated bleomycin itself. The bleomycin complex releasing OH^- should be a different entity to give the observed first-order reaction kinetics. Such a compound could be an O—O cleavage product of activated bleomycin. This cleavage is known to be the rate-limiting step in activated bleomycin decay. Thus, the yield, kinetics, and label distribution of O_2 regeneration is consistent with a catalase-type reaction in which an O—O cleavage product at the redox level of compound I reoxidizes the activated bleomycin peroxide.

$$\begin{aligned}[O = Fe \cdot bleomycin]^V &+ HOO^- \cdot Fe(III) \cdot bleomycin \\ &\rightarrow 2Fe(III) \cdot bleomycin + O_2 + OH^-\end{aligned} \quad (3)$$

It is difficult to show that the O—O cleavage is homolytic, as required for formation of a compound I homolog, but an indirect argument supports this assertion. Reoxidizing peroxide to O_2 is a $2e^-$ reaction, which is consistent with the bleomycin O—O cleavage product retaining both oxidizing equivalents of activated bleomycin. The alternative, heterolytic O—O cleavage, which would yield a $1e^-$ bleomycin oxidant plus •OH radical, is predicted to have a very different result: the hydroxyl radical is so reactive that it is unlikely to diffuse far enough to encounter and react with another activated bleomycin peroxide. Consequently, the regeneration of O_2 would be a grossly sub-stoichiometric reaction, contrary to experience. Direct experimental efforts to verify the mode of O—O cleavage have given ambiguous results for technical reasons [82, 83].

The presence or absence of excess O_2 in the reaction mixture determines which DNA degradation products will be formed, but their stoichiometry proved surprising: the yield of one alternative was not always at the expense of the other [44]. When DNA is present anaerobically, only the alkali-labilizing degradation occurs, and presumably no O_2 is released because the DNA is oxidized rather than activated bleomycin. The alkali-labilizing lesion is thought to develop from the reaction of water with a DNA-C4′-cation [84, 85], a $2e^-$ oxidized intermediate that could form with a stoichiometric conversion of activated bleomycin to Fe(III)·bleomycin. When DNA and O_2 are both present, the alkali-labilizing lesion is accompanied by the other, strand-cleaving excision. While the proportionality of the two events depends on $[O_2]$, when sufficient DNA is available, the amount of alkali-labilization is invariant [44]. Both events commence with an attack on H4′, but the partitioning events are somewhat mysterious. The bleomycin reactions proposed to account for them [86–89] involve oxygen-cleavage products of activated bleomycin [22] rather than activated bleomycin itself.

The vibrational spectra, often useful in characterizing metal—O or O—O bonds, were not available from activated bleomycin but, since these bleomycin bonds are scissile, an oxygen isotope effect was sought and used instead. Kinetic isotope effects can implicate the atoms involved in rate-determining reaction steps, and thereby have structural implications, too. The relationship between equilibrium and kinetic isotope effects can even provide information about the transition state intermediates [90, 91]. Where bond cleavage is not accompanied by new bond formation, kinetic effect measurements can exclude putative reactions with smaller equilibrium isotope effects, since without bond formation the kinetic effect cannot exceed the equilibrium effect [92]. A set of Fe- and O-equilibrium isotope effects calculated for some abundant isotopes is reproduced [24] in Table 1. Using these calculations and an observed kinetic ^{18}O-^{18}O isotope effect of 1.054 for activated bleomycin decay, it was demonstrated that a peroxide O—O cleavage was rate-limiting [24]. The relative stability of activated bleomycin made this kinetic isotope effect directly measurable without rapid-quench procedures. Unfortunately, this approach could not help distinguish between homolytic and heterolytic O—O cleavage in this case, as discussed above.

Another isotope effect is known to influence the stability of activated bleomycin, but it remains to be interpreted mechanistically. A solvent ^2H

Table 1. Equilibrium isotope effects on reactions of fully enriched oxygen and iron species. These estimates were calculated mainly from Fe—O and O—H stretching frequencies by G. Tian as described in [24] and others cited therein. Reprinted with permission from [24]. Copyright 1995 American Chemical Society

Reaction	$^{18}K_e$	$^{58/54}K_e$
$Fe^{III}{-}O_2^{\bullet -} \overset{H^+}{\longleftrightarrow} Fe^{III} + HO_2^{\bullet}$	0.9997	1.0039
$Fe^{III}{-}OOH \overset{H^+}{\longleftrightarrow} Fe^{III} + H_2O_2$	0.9951	1.0036
$Fe^{IV}{-}O \overset{e^- 2H^+}{\longleftrightarrow} Fe^{III} + H_2O$	0.9926	1.0091
$Fe^{III}{-}O_2^{\bullet -} \longleftrightarrow Fe^{III} + O_2^{\bullet -}$	1.0429	
$Fe^{III}{-}OOH \longleftrightarrow Fe^{III} + {}^{-}OOH$	1.0453	
$Fe^{IV}{-}O \overset{e^-}{\longleftrightarrow} Fe^{III} + O^{2-}$	1.0484	
$Fe^{III}{-}OOH \overset{e^- H^+}{\longleftrightarrow} Fe^{IV}{=}O + H_2O$	1.0322	0.9946
$Fe^{III}{-}O{-}OH \longleftrightarrow Fe^{III}{=}O^{\bullet} + {}^{\bullet}OH$	1.0831	
$HO_2^{\bullet} \overset{-H^+}{\longleftrightarrow} O_2^{\bullet -}$	1.0432	
$O_2 \overset{e^-}{\longleftrightarrow} O_2^{\bullet -}$	1.0673	
$O_2 \overset{e^- H^+}{\longleftrightarrow} HO_2^{\bullet}$	1.0231	
$H_2O_2 \overset{-H^+}{\longleftrightarrow} HO_2^-$	1.0504	
$O_2 \overset{2e^- H^+}{\longleftrightarrow} HO_2^-$	1.0692	
$O_2 \overset{2e^- 2H^+}{\longleftrightarrow} H_2O_2$	1.0179	
$Fe^{II} + O_2 \overset{2e^- 2H^+}{\longleftrightarrow} Fe^{IV}{=}O + H_2O$	1.0556	
$Fe^{II} + O_2 \overset{e^- H^+}{\longleftrightarrow} Fe^{III}{-}OOH$	1.0227	
$Fe^{II} + O_2 \longleftrightarrow Fe^{III}{-}O_2^{\bullet -}$	1.0108	

kinetic isotope effect of 2.4–2.8 is seen (B. Birdsall and R. Burger, unpublished data) that may be due either to exchangeable bleomycin or to solvent hydrogens. If the former is the case, the hydrogen exchange must occur in seconds or less, because a change in activated bleomycin decay rate, monitored at 1 min intervals, occurs without a detectable lag when the solvent isotopic composition is rapidly changed. Exchangeable hydrogens have been found in the vicinity of the metal-peroxide complex, both by ENDOR [76] and by NMR [32] spectroscopy, but they are probably different atoms. The ENDOR proton signals are tentatively assigned to an axial metal ligand, probably β-aminoalanine, while the NMR protons are assigned to hydrogen bonds linking the peroxide with the threonine and methylvalerate moieties. The three exchangeable hydrogens seen by NMR as hydrogen-bonding to the peroxide may well be jointly involved in its reactivity, for the ^{2}H kinetic isotope effect seen is proportional to a higher power than the square of ^{2}H enrichment (unpublished result).

5 Consequences to DNA

So much about bleomycin reactions has been inferred from the DNA degradation chemistry that a brief comment on their relationship seems in order (Scheme 2) [16]. DNA degradation proceeds in steps, and some of these occur slowly enough to permit resolution of a succession of chemical events that have helped to verify a series of mechanistic hypotheses. The DNA degradation reactions are considered only for the light they shed on the early events that directly involve bleomycin; the reactions responsible for resolving, rather than inflicting, DNA lesions are discussed elsewhere [16].

DNA degradation by bleomycin is complex and paradoxical, as illustrated by a series of difficult and seemingly unrelated questions. For example, how does a unique initial lesion resolve in two ways that involve putative intermediates at different oxidation levels? Or how does O_2 enhance one outcome without diminishing the other, and how does omitting O_2 leave no trace of an unresolved DNA lesion? Why does activated bleomycin decay display first-order kinetics, and why does the presence or absence of DNA have no effect on this rate? How is it that a certain fraction of lesions are found as closely-spaced pairs in opposing DNA strands and also in the same strand, and how is it that both single and double strand cleavage develops with the same kinetics? The current mechanistic model appears to resolve many of these paradoxes in a fairly simple way.

Stated simply, the current model posits that soon after the proximate active drug product reacts with DNA, the products of both drug and DNA react again in one of two ways depending on whether the resulting DNA radical has first encountered O_2. In Scheme 2 this is shown beginning with the abstraction of H-4′ from DNA (c), leaving a 4′-radical and a 1e$^-$-reduced oxy-bleomycin comparable to peroxidase compound II. If the latter intermediate reacts directly to further oxidize the DNA radical (reaction e), it is reduced to Fe(III)·bleomycin, and a DNA 4′-carbocation is formed that reacts with water to yield the base-releasing lesion. If, however, the DNA radical first reacts with O_2 to form a peroxyl-DNA radical adduct (reaction f), this adduct reoxidizes the bleomycin (reaction d) to the proximate active drug product, which can therefore repeat the DNA attack (reaction c). The reduced DNA product (of d) is a 4′-alkoxy cation [93] (shown) or, according to another hypothesis, a 4′-peroxide [87, 94]. Either may be a precursor of base propenal and consequently of strand cleavage. The reactions with O_2 + •DNA leading to DNA cleavage constitute a chain in which the proximate active bleomycin species is regenerated. Chain termination occurs when the initial bleomycin and DNA products react again without O_2 intervening [95].

This reaction scheme reconciles many paradoxical features of bleomycin-initiated DNA degradation. First, the spontaneous breakage of the activated

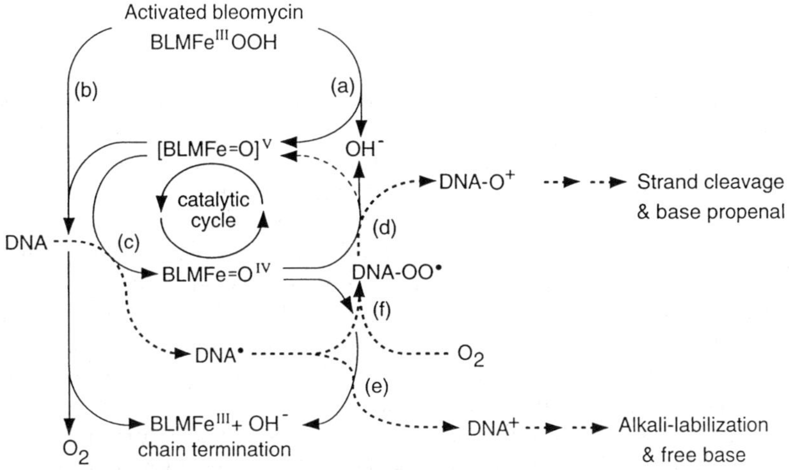

Scheme 2. Catalytic DNA cleavage by bleomycin as an O_2-driven chain reaction. Except for the DNA-independent consumption of activated bleomycin (reaction b), the nucleic base release should be stoichiometric with activated bleomycin consumption, and the DNA cleavage stoichiometric with the O_2 consumed subsequent to activated bleomycin formation. The individual reactions (a–f) are described in the text. Reprinted with permission from [16]. Copyright 1998 American Chemical Society

bleomycin peroxide bond (a), to form a short-lived monooxygenated proximate active species that can react with either DNA or activated bleomycin, is consistent with this reaction being rate-limiting, first order, and independent of DNA. Second, by attacking DNA in stepwise fashion (c,d,e), bleomycin provides a unique primary DNA lesion that can react either with O_2 or with the reduced bleomycin intermediate, giving DNA intermediates at different oxidation states subsequent to the unique initial event. Third, the re-oxidation of this same bleomycin intermediate by a DNA-peroxyl radical produces both a DNA intermediate committed to the cleavage pathway and a reactive bleomycin ready to inflict another lesion nearby, with these consequences jointly dependent on O_2. Both DNA lesions of a closely-spaced pair are initiated at nearly the same time, since the regenerated proximate active intermediate is thought to be short-lived, consistent with the observation of nearly identical rates of single- and double-strand cleavage [96, 97]. Finally, the O_2-insensitivity of free nucleic base yield [44] is consistent with base liberation representing a chain-terminating event: every simple chain, however long, comes to an end.

Notably missing from the supporting data is direct evidence for the monooxygenated bleomycin intermediates analogous to peroxidase compounds I and II. To produce these intermediates in detectable quantities, it may be necessary to induce synchronous cleavage of the activated bleomycin peroxide bond or to present a suitable monooxygen donor and, perhaps, to trap the product.

6 Conclusion

Kinetic methodologies have revealed many details of the reaction pathways leading to drug action and DNA degradation. We now know what activated bleomycin is and how it forms. A plausible model for how it reacts has also evolved. The model is general, consistent, and simple. However, it invokes some fleeting intermediates that will require ingenuity to verify.

The structural details that determine how bleomycin reactivities parallel or diverge from those of its enzymatic counterparts, the oxygenases and peroxidases, are just beginning to be addressed [98]. Powerful new methods for preparing modified bleomycins are in use, making incremental structural adjustments available for evaluation [99]. The conformational details of diamagnetic bleomycin complexes in solution have become accessible by NMR. ENDOR is giving complementary information on structures involved in iron ligation, and it is applicable to species stable only at low temperatures. The nature of the bonds about the iron-ligands is inferred from resonance Raman studies, and the iron-oxygen bonding consequences of drug structural modifications should be evident in Mössbauer spectra. Provided these artificial models are recognized as approximations, they may be very helpful in revealing the relationship of bleomycin structures to the reactivity of its complexed oxygen.

Acknowledgements. The author was partially supported as Principal Investigator of NIH grants EY11724 and AI43798, and by GM35103 to Prof. Charles P. Scholes. Helpful discussions with Drs. Scholes and Karl Drlica are gratefully acknowledged. The author holds concurrent appointments as Chief Scientist of the Redox Pharmaceutical Corporation, of Southampton, New York, Visiting Research Professor of Biology at New York University, Volunteer Professor of Chemistry at the City College of the City University of New York, and Adjunct Professor of Chemistry at the State University of New York at Albany.

7 References

1. Umezawa H, Maeda K, Takeuchi T, Okami Y (1966) J Antibiot (Tokyo) Ser A 19: 200
2. Umezawa H, Suhara Y, Takita T, Maeda K (1966) J Antibiot (Tokyo) Ser A 19: 210
3. Lazo JS, Chabner BA (1996) Bleomycin. In: Chabner BA, Longo DL (eds) Cancer chemotherapy and biotherapy: principals and practice. Lippincott-Raven, Philadelphia, PA, p 379
4. Munn SE, Higgins E, Marshall M, Clement M (1996) Br J Dermatol 135: 969
5. Drocourt D, Calmels T, Reynes J-P, Baron M, Tiraby G (1990) Nucl Acids Res 18: 4009
6. Suzuki H, Nagai K, Yamaki H, Tanaka N, Umezawa H (1968) J Antibiot (Tokyo) 21: 379
7. Onishi T, Shimada K, Takagi Y (1973) Biochem Biophys Acta 312: 248
8. Gudas LJ, Pardee AB (1976) J Mol Biol 101: 459
9. Moore CW (1978) Mutation Res 58: 41
10. Murray V, Martin RF (1985) J Biol Chem 260: 10,389
11. Povirk LF, Goldberg IH (1987) Biochimie 69: 815

12. Umezawa K, Haresaku M, Muramatsu M, Matsushima T (1987) Biomed Pharmacother 41: 214
13. Povirk (1996) Mutation Res 355: 71
14. Levin JD, Demple B (1996) Nucleic Acids Res 24: 885
15. Nagai N, Yamaki H, Suzuki H, Tanaka N, Umezawa H (1969) Biochim Biophys Acta 179: 165
16. Burger RM (1998) Chem Rev 98: 1153
17. Nagai K, Suzuki H, Tanaka N, Umezawa H (1969) J Antibiot (Tokyo) 22: 624
18. Carter BJ, de Vroom E, van der Marel GA, van Boom JH, Hecht SM (1990) Proc Natl Acad Sci USA 87: 9373
19. Lim ST, Jue CK, Moore CW, Lipke PN (1995) J Bacteriol 177: 3534
20. Sausville EA, Peisach J, Horwitz SB (1976) Biochem Biophys Res Commun 73: 814
21. Kuramochi H, Takahashi K, Takita T, Umezawa H (1981) J Antibiot (Tokyo) 34: 578
22. Burger RM, Peisach J, Horwitz SB (1981) J Biol Chem 256: 11,636
23. Sam JW, Tang X-J, Peisach J (1994) J Am Chem Soc 116: 5250
24. Burger RM, Tian G, Drlica K (1995) J Am Chem Soc 117: 1167
25. Burger RM, Kent TA, Horwitz SB, Münck E, Peisach J (1983) J Biol Chem 258: 1559
26. Barr JR, Van Atta RB, Natrajan A, Hecht SM, van der Marel GA, van Boom JH (1990) J Am Chem Soc 112: 4058
27. Maeda K, Kosaka H, Yagishita K, Umezawa H (1956) J Antibiot (Ser A) 9: 82
28. Kawaguchi H, Tsukuira H, Tomita K, Konishi M, Saito K, Kobaru S, Numata K, Fujisawa K, Miyaski T, Hatori M, Koshiyama H (1977) J Antibiot (Tokyo) 30: 779
29. Claussen CA, Long EC (1999) Chem Rev 99: 2797
30. Xu RX, Nettesheim D, Otvos JD, Petering DH (1994) Biochemistry 33: 907
31. Caceres-Cortes J, Sugiyama H, Ikudome K, Saito I, Wang AH-J (1997) Eur J Biochem 244: 818
32. Wu W, Vanderwall DE, Turner CJ, Kozarich JW, Stubbe J (1996) J Am Chem Soc 118: 1281
33. Povirk LF, Wübker W, Köhnlein W, Hutchinson F (1977) Nucleic Acids Res 4: 3573
34. Povirk LF, Köhnlein W, Hutchinson F (1978) Biochim Biophys Acta 521: 126
35. Burger RM, Berkowitz AR, Peisach J, Horwitz SB (1980) J Biol Chem 255: 11,832
36. Giloni L, Takeshita M, Johnson F, Iden C, Grollman AP (1981) J Biol Chem 256: 8608
37. Sugiyama H, Kilkuskie RE, Hecht SM, van der Marel GA, van Boom JH (1985) J Am Chem Soc 107: 7765
38. Wu JC, Kozarich JW, Stubbe JA (1983) J Biol Chem 258: 4694
39. Absalon MJ, Kozarich JW, Stubbe J (1992) Nucleic Acids Res 20: 4179
40. Haidle CW, Weiss KK, Kuo MT (1972) Mol Pharmacol 8: 531
41. Wu JC, Stubbe JA, Kozarich JW (1985) Biochemistry 24: 7569
42. Sugiyama H, Xu C, Murugesan N, Hecht SM (1985) J Am Chem Soc 107: 4104
43. Sugiyama H, Xu C, Murugesan N, Hecht SM, van der Marel GA, van Boom JH (1988) Biochemistry 27: 58
44. Burger RM, Peisach J, Horwitz SB (1982) J Biol Chem 257: 3372
45. McGall GH, Rabow LE, Stubbe J, Kozarich JW (1987) J Am Chem Soc 109: 2836
46. Albertini JP, Garnier-Suillerot A, Tosi L (1982) Biochem Biophys Res Commun 104: 557
47. Sausville EA, Peisach J, Horwitz SB (1978) Biochemistry 17: 2740
48. Burger RM, Horwitz SB, Peisach J, Wittenberg JB (1979) J Biol Chem 254: 12,299
49. Burger RM, Peisach J, Blumberg WE, Horwitz SB (1979) J Biol Chem 254: 10,906
50. Sugiura Y, Kikuchi T (1978) J Antibiot (Tokyo) 31: 1310
51. Dabrowiak JC, Greenaway FT, Santillo FS, Crooke ST (1979) Biochem Biophys Res Commun 91: 721
52. Fulmer P, Petering DH (1994) Biochemistry 33: 5319
53. Melnyk D, Horwitz SB, Peisach J (1981) Biochemistry 20: 5327
54. Burger RM, Peisach J, Horwitz SB (1978) J Biol Chem 253: 4830
55. Ishida R, Takahashi T (1975) Biochem Biophys Res Commun 66: 1432

56. Bickers DR, Dixit R, Mukhtar H (1984) Biochim Biophys Acta 781: 265
57. Sausville EA, Stein RW, Peisach J, Horwitz SB (1978) Biochemistry 17: 2746
58. Lown JW (1979) Contribution of the superoxide anion-hydroxyl radical pathway to the cleavage of DNA by bleomycin. In: Hecht SM (ed) Bleomycin: chemical, biochemical, and biological aspects. Springer, Berlin Heidelberg New York, p 184
59. Yamanaka N, Kato T, Nishida K, Ota K (1978) Cancer Res 38: 3900
60. Trush MA (1983) Chem-Biol Interactions 45: 65
61. Ciriolo MR, Magliozzo RS, Peisach J (1987) J Biol Chem 262: 6290
62. Ciriolo MR, Peisach J, Magliozzo RS (1989) J Biol Chem 264: 1443
63. Sam JW, Tang X-J, Magliozzo RS, Peisach J (1995) J Am Chem Soc 117: 1012
64. Lippai I, Magliozzo RS, Peisach J (1999) J Am Chem Soc 121: 780
65. Petering DH, Byrnes RW, Antholione WE (1990) Chem Biol Interact 73: 133
66. Chang C-H, Meares CF (1982) Biochemistry 21: 6332
67. Haasnoot CAG, Pandit UK, Kruk C, Hilbers CW (1984) J Biomolec Struct Dyn 2: 449
68. Akkerman MAJ, Haasnoot CAG, Hilbers CW (1988) Eur J Biochem 173: 211
69. Akkerman MAJ, Neijman EJWF, Wijmenga SS, Hilbers CW, Bermel W (1990) J Am Chem Soc 112: 7462
70. Wu W, Vanderwall DE, Lui SM, Tang X-J, Turner CJ, Kozarich JW, Stubbe J (1996) J Am Chem Soc 118: 1268
71. Manderville RA, Ellena JF, Hecht SM (1994) J Am Chem Soc 116: 10,851
72. Manderville RA, Ellena JF, Hecht SM (1995) J Am Chem Soc 117: 7891
73. Caceres-Cortes J, Sugiyama H, Ikudome K, Saito I, Wang AH-J (1997) Biochemistry 36: 9995
74. Lehmann TE, Ming L-J, Rosen ME, Que L Jr (1997) Biochemsitry 36: 2807
75. Veselov A, Sun H, Sienkiewicz A, Taylor H, Burger RM, Scholes CP (1995) J Am Chem Soc 117: 7508
76. Veselov A, Burger RM, Scholes CP (1998) J Am Chem Soc 120: 1030
77. Takahashi S, Sam JW, Peisach J, Rousseau DL (1994) J Am Chem Soc 116: 4408
78. Burger RM, Blanchard JS, Horwitz SB, Peisach J (1985) J Biol Chem 260: 15,406
79. Roberts JE, Hoffman BM, Rutter R, Hager LP (1981) J Am Chem Soc 103: 7654
80. Lumpkin O, Dixon WT (1971) J Chem Phys 71: 3550
81. Westre TE, Loeb KE, Zaleski JM, Hedman B, Hodgson KO, Solomon EI (1995) J Am Chem Soc 117: 1309
82. Padbury G, Sligar S (1987) Fed Proc 46: 2265
83. Natrajan A, Hecht SM, Van der Marel GA, Van Boom JH (1990) J Am Chem Soc 112: 4532
84. Rabow LE, Stubbe J, Kozarich JW (1990) J Am Chem Soc 112: 3196
85. McGall GH, Rabow LE, Ashley RG, Wu SH, Kozarich JW, Stubbe J (1992) J Am Chem Soc 114: 4958
86. Rabow LE, McGall GH, Stubbe J, Kozarich JW (1990) J Am Chem Soc 112: 3203
87. Absalon MJ, Wu W, Kozarich JW, Stubbe J (1995) Biochemistry 34: 2076
88. Sugiyama H, Sera T, Dannoue Y, Marumoto R, Saito I (1991) J Am Chem Soc 113: 2290
89. Sugiyama H, Ohmori K, Saito I (1994) J Am Chem Soc 116: 10,326
90. Bigeleisen J (1949) J Chem Phys 17: 675
91. Fry A (1970) Heavy atom isotope effects in organic reaction mechanism studies. In: Collins CJ, Bowman NS (eds) Isotope effects in chemical reactions. Van Nostrand Reinhold, New York, p 364
92. Klinman JP (1978) Adv Enzymol 46: 415
93. Rashid R, Langfinger D, Wagner R, Schuchmann H-P, von Sonntag C (1999) Int J Radiat 75: 101
94. Takeshita M, Grollman AP (1979) A molecular basis for the interaction of bleomycin with DNA. In: Hecht SM (ed) Bleomycin: chemical, biochemical, and biological aspects. Springer, Berlin Heidelberg New York, p 207

95. Burger RM, Drlica K (1996) Bleomycin reaction pathways: kinetic approaches. In: Meunier B (ed) DNA and RNA cleavers and chemotherapy of cancer and viral disease. Kluwer, Dordrecht, Netherlands, p 91
96. Burger RM, Projan SJ, Horwitz SB, Peisach J (1986) J Biol Chem 261: 15,955
97. Burger RM, Drlica K, Birdsall B (1994) J Biol Chem 269: 25,978
98. Solomon EI, Brunold TC, Davis MI, Kemsley JN, Lee S-K, Lehnert N, Neese F, Skulan AJ, Yang Y-S, Zhou J (2000) Chem Rev 100: 235
99. Boger DL, Ramsey TM, Cai H, Hoehn ST, Kozarich JW, Stubbe J (1998) J Am Chem Soc 120: 53

Author Index Volumes 1–97

Printing (Computer to Film): Saladruck, Berlin
Binding: Lüderitz & Bauer, Berlin